Moscow Lectures

Volume 7

More information about this series at http://www.springer.com/series/15875

Boris M. Makarov • Anatolii N. Podkorytov

Smooth Functions and Maps

Skoltech
Skolkovo Institute of Science and Technology

Springer

Boris M. Makarov
St. Petersburg State University
St. Petersburg, Russia

Anatolii N. Podkorytov
St. Petersburg State University
St. Petersburg, Russia

1. Translated from the Russian by Natalia Tsilevich. Originally published as Гладкие функции и отображения by МССМЕ 2020
2. под ред. О.Л.Виноградова

1st edition: © МССМЕ Moscow, Москва Издательство МЦНМО 2020

ISSN 2522-0314   ISSN 2522-0322 (electronic)
Moscow Lectures
ISBN 978-3-030-79437-8   ISBN 978-3-030-79438-5 (eBook)
https://doi.org/10.1007/978-3-030-79438-5

Mathematics Subject Classification (2020): 54C30, 57H35, 57B50

This Springer imprint is published by the registered company Springer Nature Switzerland AG
The registered company address is: Gewerbestrasse 11, 6330 Cham, Switzerland

In loving memory of our friend and colleague Victor Petrovich HAVIN.

Preface to the English Translation

This is an English translation of the book *Smooth Functions and Maps* published by MCCME in June 2020. It reflects the authors' long experience in teaching at the Department of Mathematics and Mechanics of the St. Petersburg State University. The book gives a systematic and rather comprehensive presentation of the theory of smooth functions and maps in the part dealing with differential calculus (problems related to integration are not addressed here). It contains numerous examples illustrating the results, as well as more than 150 exercises.

Having mastered the material of the book, the reader will be able to read consciously any literature based on the differential calculus of functions and maps of several variables.

We are grateful to the Springer publishing house for publishing the translation of our book, which hopefully will significantly expand its readership.

When preparing the English translation of the book, we benefited from the assistance of many colleagues, most notably Andrei Lodkin and Fedor Nazarov. We are grateful to the translator Natalia Tsilevich not only for her hard work, but also for her helpful comments on our text.

Our special thanks go to Aya Makarova for her patient and highly competent help with our correspondence with the publisher.

St. Petersburg

Boris M. Makarov
Anatolii N. Podkorytov

Preface

This book is addressed primarily to students who wish to become acquainted with the theory of smooth functions and maps to an extent which roughly corresponds to a university course for mathematics majors, as well as to professors teaching mathematical analysis.

It is based on the lecture courses taught by the authors for many years at the Department of Mathematics and Mechanics of the St. Petersburg State University. The material of the first three chapters corresponds to a slightly extended syllabus of the first half of the third semester. The prerequisite here, apart from a familiarity with the basics of differential calculus, is a knowledge of the principal results of the theory of continuous maps in Euclidean spaces; to make the text complete, they are presented (with some proofs) in the introduction. We also assume that the reader is familiar with the basics of linear algebra. For convenience, the related simple results are stated in the concluding sections of the introduction.

The first chapter (except the last section) is devoted to the traditional subjects of the theory of smooth functions: differentiability, Taylor's formula, extrema, implicit functions.

In presenting the theory of smooth maps (mostly in Chapter II), we tried to emphasize the role of the open mapping theorem, choosing to use it whenever several alternatives are available. Although this could be done much earlier, we deliberately postponed the discussion of relative extrema until the end of the third chapter, in order to use the notion of manifold and provide the reader with an opportunity to get a geometric understanding and appreciate the role of the assumption that the functional matrix corresponding to the constraint equations must have maximal rank.

In an effort to fill in some gaps in the existing textbooks, we included in the book some nontraditional subjects. Among those are, most notably, problems related to Whitney's extension theorem, including Stein's theorem on domains with minimally smooth boundary (Section I.11). These and some other sections (the global invertibility theorem going back to Hadamard (Section II.5), etc.) are marked by a star (∗) and can be omitted in a first reading.

Chapters I–III contain more than 150 examples and exercises, which constitute an important supplement to the main text.

The fourth chapter is very different in character from the previous ones. Here we consider an important, yet more specific, problem concerning the massiveness of the set of critical values of a smooth map. The material of this chapter (Morse's, Sard's, Federer's theorems) is, of course, beyond the scope of a compulsory course, but can form the basis of elective courses and seminars. It requires more prerequisites, in particular, the reader is assumed to be familiar with the basics of measure theory (in this chapter we use the notion of Hausdorff measure and Hausdorff dimension; the necessary background material is presented in Sections 3–5 of Chapter V).

The last, fifth, chapter is an addendum. It contains some necessary results (mostly with proofs) whose subjects go beyond the scope of the book. Sections 1–2 are used in Section I.11, and the remaining ones, in Chapter IV.

The whole time while working on this book we enjoyed continued support from our colleagues at the Chair of Mathematical Analysis, to whom we are sincerely grateful. The meticulous work of our scientific editor O. L. Vinogradov was highly useful and allowed us to correct numerous faults. A. B. Alexandrov read one of the first versions of the manuscript, and his enthusiastic reaction was very encouraging. In addition to diverse extra-professional help, which had a beneficial effect on the fate of the book, K. P. Kokhas prepared electronic versions of all figures, realizing our (not always well articulated) wishes; due to his suggestions, some illustrations turned out to be more expressive than we could hope. The material of the book, especially that of the last two chapters, has been repeatedly discussed with A. A. Florinsky. We highly appreciate the contribution of all these colleagues. Special thanks go to our friend and colleague F. L. Nazarov for his advice and comments which allowed us to make some significant improvements. In particular, he suggested the proof of Lemma 2 in Section V.5.3, which is reproduced here with his consent. Last but not least, thanks are due to N. V. Tsilevich whose encouragement and advice have been very helpful.

The chapters of the book (except the introduction) are numbered with Roman numerals. They are divided into sections, which, in turn, consist of subsections numbered with two Arabic numerals. The first of them indicates the number of the section, and the second one, the number of the subsection within it. All statements in a given subsection are referred to in the same way as the subsection itself. References from within the same chapter include only the number of the subsection (for example, "by Theorem 2.1" refers to Theorem 2.1 in the given chapter). References to statements from other chapters are preceded by the number of the chapter (for example, "see Lemma III.2.3" refers to the lemma in Section 2.3 of Chapter III). Equations are numbered continuously throughout each section. The end of a proof is marked by the symbol □.

St. Petersburg   *Boris M. Makarov*
April 2019   *Anatolii N. Podkorytov*

Contents

Basic notation

Sets

$\varnothing$	the empty set;
$A \times B$	the Cartesian product of sets A, B;
$\bar{A}$	the closure of a set A;
$\mathrm{diam}(A)$	the diameter of a set A;
$\dim_H(A)$	the Hausdorff dimension of a set A;
$\mathrm{dist}(x, A)$	the distance from a point x to a set A;
∂A	the boundary of a set A;
$\mathrm{Int}(A)$	the interior of a set A;
$\mathbb{N}$	the set of positive integers;
$\mathbb{Z}$	the set of integers;
$\mathbb{R}$	the set of real numbers;
$\mathbb{R}^m$	the m-dimensional arithmetic space;
$\mathbb{R}^m_+$	the subset in $\mathbb{R}^m$ consisting of points with positive coordinates;
$\mathbb{Z}^m$	the subset in $\mathbb{R}^m$ consisting of points with integer coordinates;
$\mathbb{Z}^m_+$	the subset in $\mathbb{R}^m$ consisting of points with positive integer coordinates (multiindices);
(a, b)	an open interval;
$[a, b)$	a semi-open interval;
$[a, b]$	a closed interval;
$B(a, r)$	the open ball of radius $r > 0$ centered at a;
$\overline{B}(a, r)$	the closed ball of radius $r > 0$ centered at a;
$B(r), B^m(r)$	the ball $B(0, r)$ in $\mathbb{R}^m$;
B^m	the ball $B^m(1)$;
S^{m-1}	the unit sphere (the boundary of the ball B^m) in $\mathbb{R}^m$;
C_a	the Cantor-like set with parameter a.

Maps and Functions

$T: X \to Y$	a map from X to Y;
$T(A)$	the image of a set A under a map T;
$T^{-1}(B)$	the full preimage of a set B under a map T;
$T \circ S$	the composition of maps T and S;
$T|_A$	the restriction of a map T to a set A;
$x \mapsto T(x)$	a map T sends a point x to $T(x)$;
Γ_f	the graph of a function f;
$\|x\|$	the norm of a vector x;
$\langle x, y \rangle$	the scalar product of vectors x, y.

Smooth Functions
(below, $\alpha = (\alpha_1, \alpha_2, \ldots, \alpha_m) \in \mathbb{Z}_+^m$ is a multiindex)

$|\alpha| = \alpha_1 + \ldots + \alpha_m$;

$\alpha! = \alpha_1! \alpha_2! \ldots \alpha_m!$;

$x^\alpha = x_1^{\alpha_1} \ldots x_m^{\alpha_m}$ $\quad (x = (x_1, x_2, \ldots, x_m) \in \mathbb{R}^m)$;

$\frac{\partial f}{\partial x_k}(x)$	the partial derivative of a function f with respect to the kth coordinate at a point x;
$\frac{\partial f}{\partial \ell}(x)$	the derivative of a function f along a vector ℓ at a point x;
$\frac{\partial^{|\alpha|} f}{\partial x^\alpha}(x), f^{(\alpha)}(x)$	the partial derivative of a function f at a point x corresponding to a multiindex α;
grad f	the gradient of a function f;
$d_a f(h)$	the value at a vector h of the differential of a function f at a point a;
$d_a^n f(h)$	the value at a vector h of the nth differential of a function f at a point a;
N_F, N	the critical set of a map F;
$\mathcal{K}_F$	the set of critical values of a map F;
$T_n(f, a), T_n(f, a; x)$	the Taylor polynomial of order n of a function f at a point a;
$\Phi'(x)$	the Jacobian matrix of a map Φ at a point x.

Sets of Functions

$C(X), C^0(X)$	the set of continuous functions on a topological space X;
$C^r(O)$ $(C^r(O; \mathbb{R}^n))$	the set of r times ($r = 1, 2, \ldots, +\infty$) continuously differentiable functions ($\mathbb{R}^n$-valued maps) defined on an open subset O of $\mathbb{R}^m$;
$C^t(O)$	the set of t-smooth functions in O;
$LC^r(O)$	the set of functions of class $C^r(O)$ whose all derivatives of order r satisfy the local Lipschitz condition in O.

Measures

α_m	the volume (Lebesgue measure) of the unit ball in $\mathbb{R}^m$;
λ_m	the m-dimensional Lebesgue measure;
μ_p	the Hausdorff measure of dimension p.

Introduction. Preliminaries

0.1 The Space $\mathbb{R}^m$

As we have already mentioned in the preface, the reader is assumed to be familiar with basic notions and results of the theory of continuous maps, which can be found, e.g., in Chapter VII of [19]. However, for convenience we recall basic notation and results, without motivations but with some selected proofs.

Although many of the results we discuss are purely topological and hold in more general situations, we restrict ourselves to the framework of Euclidean[1] spaces, which is sufficient for our purposes.

By definition, the set $\mathbb{R}^m$ is the Cartesian[2] product of m copies of the real line:

$$\mathbb{R}^m = \mathbb{R} \times \ldots \times \mathbb{R} \quad (m \text{ factors}).$$

Thus, a point from $\mathbb{R}^m$ is an ordered m-tuple of real numbers, called its coordinates. Usually, the coordinates of a point $x \in \mathbb{R}^m$ are denoted by the same letter with subscripts:

$$x = (x_1, \ldots, x_m).$$

Here we depart (for purely technical reasons) from the tradition established in linear algebra to arrange the coordinates of a vector as a column

$$x = \begin{pmatrix} x_1 \\ \ldots \\ x_m \end{pmatrix}$$

and use this notation only sporadically.

In some cases, we identify the space $\mathbb{R}^m$ in a standard way with the product of spaces of smaller dimension: $\mathbb{R}^m = \mathbb{R}^k \times \mathbb{R}^{m-k}$ for $1 \le k < m$.

---

[1] **Euclid** (fl. 300 BC) was an ancient Greek mathematician.

[2] René **Descartes** (Latinized: Renatus Cartesius; 1596–1650) was a French philosopher, mathematician, and scientist.

1

Given $x, y \in \mathbb{R}^m$, $x = (x_1, \ldots, x_m)$, $y = (y_1, \ldots, y_m)$, $a \in \mathbb{R}$, we define

$$x + y := (x_1 + y_1, \ldots, x_m + y_m), \quad ax := (ax_1, \ldots, ax_m).$$

With this definition of algebraic operations, the set $\mathbb{R}^m$ becomes a vector space. Its points will also be called vectors. The zero vector in $\mathbb{R}^m$ is the vector with zero coordinates (the origin), which is denoted, as in the scalar (one-dimensional) case, by 0. The vectors with one coordinate equal to 1 and the other coordinates equal to 0 form the *canonical basis* of the space $\mathbb{R}^m$.

The vector of this basis whose kth coordinate is equal to 1 is denoted[3] by e_k ($k = 1, 2 \ldots, m$). The expansion of an arbitrary vector $x = (x_1, \ldots, x_m) \in \mathbb{R}^m$ with respect to this basis has the form

$$x = x_1 e_1 + \ldots + x_m e_m.$$

The *scalar product* of vectors $x = (x_1, \ldots, x_m)$ and $y = (y_1, \ldots, y_m)$ from $\mathbb{R}^m$ is the sum

$$\langle x, y \rangle := x_1 y_1 + \ldots + x_m y_m.$$

In particular, $x_k = \langle x, e_k \rangle$ for $k = 1, \ldots, m$.

The scalar product is linear in each of its arguments and invariant under interchanging them. It generates the *Euclidean norm*

$$\|x\| := \sqrt{\langle x, x \rangle} = \sqrt{x_1^2 + \ldots + x_m^2}.$$

The set $\mathbb{R}^m$ equipped with this norm is called a *Euclidean space*.

The scalar product and the Euclidean norm are related by the important *Cauchy's inequality*[4]

$$|\langle x, y \rangle| \le \|x\| \, \|y\| \text{ for any } x, y \in \mathbb{R}^m.$$

Indeed, set

$$P(t) = \|x + ty\|^2 = \langle x + ty, x + ty \rangle$$

for every real t. Then

$$0 \le P(t) = \|x\|^2 + 2t\langle x, y \rangle + t^2 \|y\|^2.$$

Assuming that $y \ne 0$, we see that P is a nonnegative quadratic polynomial. Therefore, its discriminant is nonpositive, which is equivalent to Cauchy's inequality.

Note also that the linear independence of vectors x and y means that the polynomial P is strictly positive, which is equivalent to the fact that its discriminant is negative, i.e., to the strict Cauchy inequality. Hence, Cauchy's inequality becomes an equality only if the vectors x and y are linearly dependent.

The norm satisfies the triangle inequality:

---

[3] One should disambiguate this notation when dealing simultaneously with canonical bases of spaces of different dimensions.

[4] Augustin-Louis **Cauchy** (1789–1857) was a French mathematician.

$$\|x + y\| \leq \|x\| + \|y\|.$$

Indeed, since

$$\|x + y\|^2 = \langle x + y, x + y \rangle = \|x\|^2 + 2\langle x, y \rangle + \|y\|^2,$$

Cauchy's inequality yields

$$\|x + y\|^2 \leq \|x\|^2 + 2\|x\|\,\|y\| + \|y\|^2 = (\|x\| + \|y\|)^2,$$

which is equivalent to the desired inequality. The above calculation also shows that the triangle inequality becomes an equality only if the vectors x and y are positive scalar multiples of one another.

By definition, the *distance between points* $x, y \in \mathbb{R}^m$ is equal to

$$\|x - y\| = \left(\sum_{k=1}^{m} (x_k - y_k)^2 \right)^{\frac{1}{2}}.$$

In what follows, we always assume that the space $\mathbb{R}^m$ is equipped with this (Euclidean) metric and regard it as a metric space.

If $\langle x, y \rangle = 0$, then the vectors x and y are said to be *orthogonal*. Such vectors satisfy the following equation (Pythagorean[5] theorem):

$$\|x + y\|^2 = \|x\|^2 + 2\langle x, y \rangle + \|y\|^2 = \|x\|^2 + \|y\|^2.$$

One can easily see that a similar equation holds for the sum of several pairwise orthogonal vectors.

0.2 Open and Closed Sets

The *open and closed balls* $B(a, r)$ and $\overline{B}(a, r)$ of radius $r > 0$ centered at a point $a \in \mathbb{R}^m$ are defined as

$$B(a, r) = \{x \in \mathbb{R}^m \mid \|x - a\| < r\}, \quad \overline{B}(a, r) = \{x \in \mathbb{R}^m \mid \|x - a\| \leq r\}.$$

A subset in $\mathbb{R}^m$ is bounded if it is contained in a ball. The triangle inequality implies that one may always assume that the center of this ball is at the origin. Note also that, by the simple double inequality

$$\max\{|x_1|, \ldots, |x_m|\} \leq \|x\| \leq |x_1| + \ldots + |x_m|,$$

---

[5] **Pythagoras** (c. 570–500) was an ancient Greek philosopher and mathematician.

the boundedness of a set $E \subset \mathbb{R}^m$ is equivalent to its coordinatewise boundedness: for every $k = 1, \ldots, m$, the set $\{\langle x, e_k \rangle \mid x \in E\}$ formed by the kth coordinates of all points of E is a bounded subset of the real line.

In many cases, to describe the size of a set A in a Euclidean space, it is convenient to use the value $\mathrm{diam}(A)$ defined as

$$\mathrm{diam}(A) = \sup_{x, y \in A} \|x - y\|.$$

It is called the *diameter* of A.

A point x_0 is an *interior point* of a set $E \subset \mathbb{R}^m$ if $B(x_0, r) \subset E$ for some $r > 0$. The set of all interior points of E (the *interior* of E) is denoted by $\mathrm{Int}(E)$. A set E is said to be *open* if all its points are interior points, i.e., $E = \mathrm{Int}(E)$ (the empty set is open by definition). If $x_0 \in E$ and E is open, then E is also called a *neighborhood* of x_0.

The triangle inequality immediately implies that for every point $x_0 \in B(a, r)$, a sufficiently small ball $B(x_0, \delta)$ is contained in $B(a, r)$ (one must take δ not exceeding the difference $r - \|x_0 - a\|$). Hence, every open ball $B(a, r)$ is an open set. Since every point x_0 is an interior point of any neighborhood of x_0, every such neighborhood contains a ball $B(x_0, \delta)$, which is called the δ-*neighborhood* of x_0.

Obviously, the intersection of finitely many open sets and the union of arbitrarily many open sets are open.

A subset in $\mathbb{R}^m$ is said to be *closed* if its complement is open. Using the well-known identities (duality formulas)

$$X \setminus \bigcup_{\alpha \in A} E_\alpha = \bigcap_{\alpha \in A} (X \setminus E_\alpha) \quad \text{and} \quad X \setminus \bigcap_{\alpha \in A} E_\alpha = \bigcup_{\alpha \in A} (X \setminus E_\alpha),$$

one can easily check that both the union of finitely many closed sets and the intersection of arbitrarily many closed sets are closed.

Note that a closed ball is a closed set. Indeed, the complement of such a ball is open, since if a point x_0 does not belong to $\bar{B}(a, r)$, i.e., if $\|x_0 - a\| > r$, then for $\delta \le \|x_0 - a\| - r$ the complement of $\bar{B}(a, r)$ contains the ball $B(x_0, \delta)$, which follows directly from the triangle inequality.

Definition The *closure* of a set $E \subset \mathbb{R}^m$ is the set

$$\bar{E} = \bigcap_{\substack{F \text{ is closed,} \\ E \subset F}} F.$$

The family of sets to be intersected is, of course, nonempty: it contains the entire space $\mathbb{R}^m$. Being the intersection of closed sets, $\bar{E}$ is closed. Therefore, $\bar{E}$ is the smallest closed set containing E. It follows that a set E is closed if and only if it coincides with $\bar{E}$.

The points making up the closure of a set E have a clear geometric description. We say that x is an *adherent point* of E if

$$E \cap B(x, r) \neq \emptyset \quad \text{for every } r > 0.$$

Note that every point from E is an adherent point of this set, and all the other adherent points of E (lying outside E) can be approximated with arbitrary accuracy by points of E.

Lemma *The closure $\bar{E}$ of every set $E \subset \mathbb{R}^m$ coincides with the set of its adherent points.*

Proof First, we verify that every point from $\bar{E}$ is an adherent point. Indeed, otherwise in $\bar{E}$ there is a point x such that $E \cap B(x, r) = \emptyset$ for some $r > 0$. But then E is contained in the closed set $F_0 = \mathbb{R}^m \setminus B(x, r)$, which contains $\bar{E}$ and does not contain x, a contradiction. Therefore, all points of $\bar{E}$ are adherent points.

Now, we verify the converse. If an adherent point x does not belong to $\bar{E}$, then $x \notin F$ for some closed set F containing E. Therefore, x belongs to the open set $\mathbb{R}^m \setminus F$. But in this case $B(x, r) \subset \mathbb{R}^m \setminus F$ for some $r > 0$, and hence $B(x, r) \cap E = \emptyset$. Thus, x is not an adherent point of E. □

An adherent point of a set E that is not an interior point of E is called a boundary point. The set of all such points, i.e., the difference $\bar{E} \setminus \text{Int}(E)$, is called the *boundary* of E and denoted by ∂E. For example, one can easily check that the balls $B(a, r)$ and $\bar{B}(a, r)$ share the common boundary, which is the sphere $S(a, r)$ of radius r centered at a, i.e., the set $\{x \in \mathbb{R}^m : \|x - a\| = r\}$.

Along with the notion of a neighborhood of a point, it is useful to introduce the more general notion of a neighborhood of a set. A *neighborhood of a subset E* in a Euclidean space is any open set containing E. Correspondingly, the *δ-neighborhood* of E is the union of the δ-neighborhoods of all its points. Since any union of open sets is open, any δ-neighborhood is an open set.

Note that, in contrast to neighborhoods of points, a neighborhood of a set does not necessarily contain some δ-neighborhood of this set (see Exercises 3, 4).

Generalizing the notion of the distance between points, we introduce the distance between a point and a set.

Definition Let E be a nonempty subset in $\mathbb{R}^m$ and $x \in \mathbb{R}^m$. The value

$$\text{dist}(x, E) = \inf_{z \in E} \|z - x\|$$

is called the *distance between x and E*.

The distance between a point and a set not containing this point is not always positive. We leave it to the reader to verify that $\text{dist}(x, E) = 0$ if and only if x an adherent point of E. Thus, the set of points at zero distance from a set E coincides with its closure. In a similar way, it is not difficult to verify that the δ-neighborhood of E coincides with the set of points at distance less than δ from E, i.e., with the set $\{x \in \mathbb{R}^m : \text{dist}(x, E) < \delta\}$.

0.3 The Limit of a Sequence in $\mathbb{R}^m$

A sequence of points of a set X is an arbitrary map from $\mathbb{N}$ to X. It is usually denoted by $\{x_j\}_{j\in\mathbb{N}}$, where x_j is the value of the map at the point j. We also use the shorter notation $\{x_j\}_j$.

A sequence is said to be convergent if $\|x_j - x_0\| \underset{j\to\infty}{\longrightarrow} 0$ for some $x_0 \in \mathbb{R}^m$. The vector x_0 is called the *limit of the sequence* $\{x_n\}$. In this case, one writes $\lim x_j = x_0$ or $x_j \to x_0$. The double inequality for the norm mentioned in Section 0.2 implies that the convergence of a sequence is equivalent to its coordinatewise convergence, i.e., the convergence of m numerical sequences of coordinates of the vectors x_j. In particular, all terms of a convergent sequence lie in a sufficiently large ball (since the coordinate sequences are bounded).

Using the coordinatewise convergence, it is easy to check that

$$\text{if } x_j \to x_0 \text{ and } y_j \to y_0, \text{ then } \langle x_j, y_j \rangle \to \langle x_0, y_0 \rangle.$$

In particular,

$$\|x_j\| = \sqrt{\langle x_j, x_j \rangle} \to \|x_0\| \text{ and } \langle x_j, a \rangle \to \langle x_0, a \rangle \text{ for any } a \in \mathbb{R}^m.$$

By analogy with the one-dimensional case, the convergence of a sequence in a Euclidean space can be described by introducing the notion of a Cauchy sequence. A sequence $\{x_j\}$ of points of a Euclidean space is called a *Cauchy sequence* if its terms become arbitrarily close to each other for sufficiently large indices, i.e., if

$$\forall \varepsilon > 0 \quad \exists N \in \mathbb{N} \quad \forall j, k > N : \|x_j - x_k\| < \varepsilon$$

(another notation: $\|x_j - x_k\| \underset{j,k\to\infty}{\longrightarrow} 0$).

Obviously, every convergent sequence is a Cauchy sequence; on the other hand, since the convergence in $\mathbb{R}^m$ is equivalent to the coordinate convergence, the Cauchy criterion for convergence of real numbers implies the converse.

It is easy to describe adherent points in the "language of sequences." Indeed, considering a sequence of balls that converge to an adherent point x_0 of a set E and picking a point of E in each of them, we construct in E a sequence converging to x_0 (if $x_0 \in E$, then we may take a constant sequence). Of course, the converse is also true: the limit of a convergent sequence of points of E is an adherent point of E. Thus, we obtain a description of adherent points in the "language of sequences": these are the limits of convergent sequences whose all elements lie in E.

This characterization of adherent points implies that a set is closed if and only if it is "stable" under limits. More exactly, a set $F \subset \mathbb{R}^m$ is closed if and only if it has the following property: the limit of every convergent sequence of points of F belongs to F. The reader is urged to verify this.

0.4 Lindelöf's Lemma

We need a geometric result based on a simple observation.

Lemma *Every open subset G in $\mathbb{R}^m$ is the union of the balls of rational radii centered at points with rational coordinates contained in G.*

Proof Since every point x of G is the center of a ball $B(x, r)$ contained in G, it suffices to observe that for $0 < \rho < \frac{r}{2}$ and $\|x - y\| < \rho$, we have $x \in B(y, \rho) \subset B(x, r)$. Obviously, ρ and the coordinates of y can be chosen to be rational. $\qquad\square$

Theorem (Lindelöf's[6] lemma) *For every family $\{G_\alpha\}_{\alpha \in A}$ of open sets in $\mathbb{R}^m$ there exists a subfamily $\{G_\alpha\}_{\alpha \in A_0}$, where $A_0 \subset A$ and A_0 is at most countable, with the same union:*

$$\bigcup_{\alpha \in A} G_\alpha = \bigcup_{\alpha \in A_0} G_\alpha.$$

Proof Consider all open balls of rational radii centered at points with rational coordinates contained in at least one of the sets G_α. The set of such balls is countable. Let $\{B_j\}_j$ be an arbitrary numbering of this set. By the choice of these balls, for every j there exists an index $\alpha_j \in A$ such that $B_j \subset G_{\alpha_j}$. Besides, by the previous lemma, every set G_α is covered by these balls:

$$G_\alpha \subset \bigcup_{j \in \mathbb{N}} B_j \quad \text{for every index } \alpha \in A.$$

Therefore,

$$\bigcup_{\alpha \in A} G_\alpha \subset \bigcup_{j \in \mathbb{N}} B_j \subset \bigcup_{j \in \mathbb{N}} G_{\alpha_j}.$$

Since, obviously, $\bigcup_{j \in \mathbb{N}} G_{\alpha_j} \subset \bigcup_{\alpha \in A} G_\alpha$, for A_0 we can take the set of all α_j. $\qquad\square$

0.5 Compact Sets

We have already observed that every convergent sequence is bounded. However, the converse is definitely false. Nevertheless, a partial converse is available. This important result will be called, as in the one-dimensional case, the Bolzano[7]–Weierstrass[8] theorem. It says that a bounded sequence of vectors in $\mathbb{R}^m$ contains a convergent subsequence.

Indeed, since the boundedness of a sequence $\{x_j\}_j$ implies its coordinatewise boundedness, by the classical Bolzano–Weierstrass theorem, the bounded (numerical) sequence formed by the first coordinates contains a convergent subsequence.

---

[6] Ernst Leonard **Lindelöf** (1870–1946) was a Finnish mathematician.

[7] Bernard **Bolzano** (1781–1848) was a Czech mathematician.

[8] Karl Theodor Wilhelm **Weierstrass** (1815–1897) was a German mathematician.

Consider the bounded subsequence of the corresponding second coordinates. It also contains an (even thinner) convergent subsequence. Repeating this argument, after m steps we obtain a subsequence $\{j_k\}_k$ of indices for which every coordinate has a finite limit. Since convergence in $\mathbb{R}^m$ is equivalent to coordinatewise convergence, this is a required subsequence.

Definition A set $K \subset \mathbb{R}^m$ is said to be *compact* if every family $\{G_\alpha\}_{\alpha \in A}$ of open sets that is a cover of K (i.e., such that $K \subset \bigcup_{\alpha \in A} G_\alpha$) contains a finite subcover:

$$K \subset G_{\alpha_1} \cup \ldots \cup G_{\alpha_N} \quad \text{for some } \alpha_1, \ldots, \alpha_N \in A.$$

A cover of a set E by a family of open sets is called an open cover of E.

We note two properties of compact sets.

1. A closed subset F of a compact set K is compact. Indeed, taking an arbitrary open cover of F and adding to it the open set $G = \mathbb{R}^m \setminus F$, we obtain an open cover of K. It has a finite subcover. Removing from it the set G (if it belongs to this subcover), which is disjoint with F, we obtain an open subcover of F.

2. If K is a nonempty compact subset of the real line, then $\min K$ and $\max K$ exist. Indeed, let $L = \sup K$. If $L \notin K$, then the open half-lines $(-\infty, \ell)$ with $\ell < L$ form a cover of K which, obviously, contains no finite subcover. The existence of $\min K$ can be proved in a similar way.

Compact sets play an important role in various problems. That is why, it is useful to have their equivalent descriptions.

Theorem *Let $K \subset \mathbb{R}^m$. The following three statements are equivalent.*

I. *The set K is compact.*

II. *The set K is bounded and closed.*

III. *The following selection principle holds: every sequence of points of K contains a convergent subsequence whose limit also belongs to K.*

Proof The proof follows the scheme $I \Rightarrow II \Rightarrow III \Rightarrow I$.

$I \Rightarrow II$. Since

$$K \subset \mathbb{R}^m = \bigcup_{j=1}^{\infty} B(0, j),$$

the balls $B(0, j)$ form a cover of K. By definition, it contains a finite subcover, hence K is contained in a ball (the largest ball in the subcover). Thus, K is bounded.

Now we verify that the complement $\mathbb{R}^m \setminus K$ is open, i.e., every point x not in K is an interior point of this complement. Indeed, the complements

$$G_j = \mathbb{R}^m \setminus \bar{B}\left(x, \frac{1}{j}\right), \quad j \in \mathbb{N},$$

of closed balls are open and, as one can easily see, form a cover of K. Hence, there exists a finite subcover. Since the sets G_j become larger as j increases, this means that

$$K \subset G_N = \mathbb{R}^m \setminus \bar{B}\left(x, \frac{1}{N}\right)$$

for sufficiently large N. Hence, the ball $\bar{B}\left(x, \frac{1}{N}\right)$, which contains x, lies in $\mathbb{R}^m \setminus K$. Therefore, x is an interior point of this complement. Since x is an arbitrary point from $\mathbb{R}^m \setminus K$, we see that this set is open.

II $\Rightarrow$ III. By condition II, the set K is bounded. Hence, every sequence of points from K is also bounded. By the Bolzano–Weierstrass theorem, it contains a convergent subsequence, and the limit of this subsequence belongs to K, because this set is closed by condition II.

III $\Rightarrow$ I. Consider a family of open sets that form a cover of K. Lindelöf's lemma allows us to assume, without loss of generality, that this family is a sequence:

$$K \subset \bigcup_{j=1}^{\infty} G_j,$$

where every set G_j is open. We must prove that

$$K \subset \bigcup_{j=1}^{N} G_j \quad \text{for some index } N.$$

Assume the contrary. Then for every j the set K contains a point x_j outside $G_1 \cup \ldots \cup G_j$. Using condition III, we extract from $\{x_j\}_j$ a convergent subsequence $\{x_{j_i}\}_i$ whose limit x_0 belongs to K. Since $x_0 \in K \subset \bigcup_{j=1}^{\infty} G_j$, we have $x_0 \in G_N$ for some index N. By the openness of G_N and the convergence $x_{j_k} \to x_0$, we see that $x_{j_k} \in G_N$ for all sufficiently large k, which is impossible, since, by construction, the set G_N does not contain the points x_j with $j \geq N$. Hence, our assumption is false, i.e., $K \subset \bigcup_{j=1}^{N} G_j$ for some index N. $\qquad\square$

0.6 The Limit of a Map

A point $a \in \mathbb{R}^m$ is called an *accumulation point* (or a limit point) of a set $E \subset \mathbb{R}^m$ if within an arbitrarily small distance of a there is a point from E different from a, i.e., if the intersection $E \cap (B(a, r) \setminus \{a\})$ is not empty for every $r > 0$. Using the notion of a neighborhood of a point, we can also say that a is an accumulation point of E if for every neighborhood $U(a)$ the punctured neighborhood $\mathring{U}(a) = U(a) \setminus \{a\}$ has a nonempty intersection with E.

Every accumulation point of a set E is an adherent point of this set. The converse is false, since E may contain points in some neighborhoods of which there are no other points from E. Such points are called *isolated points* of E.

In the one-dimensional case ($m = 1$), it is convenient to consider the extended real line $\overline{\mathbb{R}}$, i.e., the set obtained from $\mathbb{R}$ by adjoining two elements $+\infty$ and $-\infty$ (points at infinity). The punctured neighborhoods of these points are the half-lines $(C, +\infty)$

and $(-\infty, C)$, respectively. Extending the definition of accumulation points to $\overline{\mathbb{R}}$, we may say that $+\infty$ (respectively, $-\infty$) is an accumulation point of a set E of numbers if and only if E is not bounded from above (respectively, from below).

In many cases, the following description of accumulation points in the language of sequences, similar to that for adherent points, is useful:

a is an accumulation point of a set $E \subset \mathbb{R}^m$ if and only if in $E \setminus \{a\}$ there is a sequence $\{x_j\}_j$ such that $x_j \to a$ (if $m = 1$, the case $a = \pm\infty$ is not excluded).

With every map $F : E \to \mathbb{R}^n$ we can associate functions $F_1, \ldots, F_n$ defined by the formulas

$$F_k(x) = \langle F(x), e_k \rangle,$$

where e_k $(k = 1, \ldots, n)$ are the canonical basis vectors in $\mathbb{R}^n$. Thus,

$$F(x) = (F_1(x), \ldots, F_n(x)),$$

i.e., $F_k(x)$ is the kth coordinate of the vector $F(x)$. These functions are called the *coordinate functions of F*.

A map $F : E \to \mathbb{R}^n$ is said to be bounded if the image of the set E is bounded in $\mathbb{R}^n$. As already mentioned, this condition is equivalent to the coordinatewise boundedness of the image. Thus, a map is bounded if and only if all its coordinate functions are bounded.

Definition Let a be an accumulation point of a set $E \subset \mathbb{R}^m$, and let F be a map from E to $\mathbb{R}^n$. One says that a point $L \in \mathbb{R}^n$ is the *limit of F* at a if for every neighborhood V of L there exists a neighborhood U of a such that $F(E \cap \dot{U}) \subset V$ (in other words, the value $F(x)$ is arbitrarily close to L as long as the argument $x \neq a$ is sufficiently close to a). In this case, we write $L = \lim\limits_{x \to a} F(x)$, $F(x) \xrightarrow[x \to a]{} L$, or $F(x) \to L$ as $x \to a$.

If $\lim\limits_{x \to a} F(x) = 0$, then the map F is said to be *infinitesimal* as $x \to a$. The relations $F(x) \xrightarrow[x \to a]{} L$ and $F(x) - L \xrightarrow[x \to a]{} 0$ are equivalent. Thus,

$$F(x) \xrightarrow[x \to a]{} L \in \mathbb{R}^m \quad \text{if and only if} \quad F(x) = L + \alpha(x)$$

where α is infinitesimal as $x \to a$.

The definition of limit can be extended in an obvious way to the cases $L = \pm\infty$ (for $n = 1$) and $a = \pm\infty$ (for $m = 1$).

The double inequality for the norm in $\mathbb{R}^n$ shows that for a map F with coordinate functions $F_1, \ldots, F_n$, the relation

$$F(x) \xrightarrow[x \to a]{} L, \quad \text{where } L = (L_1, \ldots, L_n),$$

is equivalent to n scalar relations

$$F_1(x) \xrightarrow[x \to a]{} L_1, \ldots, F_n(x) \xrightarrow[x \to a]{} L_n.$$

We will not discuss the basic properties of limits, uniqueness and locality, as well as algebraic operations on limits. Note only that the existence of a finite limit at a point a implies that the map is locally bounded (i.e., its restriction to a sufficiently small neighborhood of a is bounded).

If a is an accumulation point of a set E, then $F(x) \xrightarrow[x \to a]{} L$ if and only if $F(x_j) \to L$ for every sequence of points $x_j \in E \setminus \{a\}$ converging to a. The proof is the same as in the one-dimensional case, and we leave it to the reader as an easy exercise.

0.7 Continuity

In what follows, a fundamental role is played by the notion of a continuous map. As in the one-dimensional case, the idea behind the definition of continuity is that sufficiently small changes in the argument result in arbitrarily small changes in the value.

Definition Let F be a map defined on a set $E \subset \mathbb{R}^m$ with values in $\mathbb{R}^n$. One says that F is *continuous at a point $x_0 \in E$* if

> for every $\varepsilon > 0$ there exists $\delta > 0$ such that
>
> $\|F(x) - F(x_0)\| < \varepsilon$ as long as $x \in E$ and $\|x - x_0\| < \delta$.

Clearly, continuity is a local property.

If x_0 is an accumulation point of E, then, obviously, the continuity of F at x_0 is equivalent to the limit relation

$$\lim_{x \to x_0} F(x) = F(x_0).$$

If x_0 is an isolated point of E, then every map defined on E is continuous at x_0 (since for sufficiently small positive δ, the condition $\|x - x_0\| < \delta$ in E is satisfied only for the point $x = x_0$, for which $\|F(x) - F(x_0)\| = \|F(x_0) - F(x_0)\| = 0 < \varepsilon$). Thus, the continuity of a map at a given point either can be described in terms of limits, or holds automatically. The characterization of limit in terms of sequences mentioned at the end of the previous subsection leads to a similar characterization of continuity: a map F is continuous at a point x_0 if and only if $F(x_j) \to F(x_0)$ for every sequence of points x_j from E converging to x_0.

The continuity of F at a point x_0 means that the F-image of the intersection of E with a sufficiently small ball in $\mathbb{R}^m$ centered at x_0 is contained in an arbitrarily small ball in $\mathbb{R}^n$ centered at $F(x_0)$. Since every neighborhood of x_0 contains some ball $B(x_0, r)$, the continuity of F at x_0 means that for any given neighborhood of $F(x_0)$, the image of the intersection of E with a sufficiently small neighborhood of x_0 is contained in this neighborhood. More exactly, the continuity of a map

$F: E \to \mathbb{R}^n$ at a point $x_0 \in E$ means that for every neighborhood V of $F(x_0)$ there exists a (sufficiently small) neighborhood U of x_0 such that $F(E \cap U) \subset V$.

The continuity of a map is equivalent to the continuity of all its coordinate functions. Obviously, a linear combination of continuous functions is continuous. Observe also that the composition of continuous maps is continuous: if $F: E \to \mathbb{R}^n$ with $E \subset \mathbb{R}^m$ and $G: \widetilde{E} \to \mathbb{R}^d$ with $\widetilde{E} \subset \mathbb{R}^n$, and if the composition $G \circ F$ makes sense (i.e., $F(E) \subset \widetilde{E}$), then it is continuous at a point $x_0 \in E$ as long as the maps F and G are continuous (at the points x_0 and $F(x_0)$, respectively).

By $C(E; \mathbb{R}^n)$ we denote the class of all *continuous maps* from E to $\mathbb{R}^n$ (i.e., the class of maps continuous at all points of E). Obviously, $C(E; \mathbb{R}^n)$ is a vector space. For the class of continuous functions on E (i.e., in the case $n = 1$), we use the simpler notation $C(E)$. This space contains, along with any two elements, not only their linear combinations, but also their product and quotient (provided that the denominator does not vanish).

The next theorem gives a description of maps that are continuous at all points of a given set. Here and in what follows, it is convenient to use "relative" notions, that of relatively open and relatively closed sets and of a relative neighborhood.

Let E be an arbitrary subset in $\mathbb{R}^m$. A subset of E is said to be *relatively open* (*in E*), or *open in E*, if it can be represented as the intersection of E with an open subset of $\mathbb{R}^m$. In a similar way one defines a *relatively closed* set (or a set *closed in E*). A relative neighborhood of a point $a \in E$ is any relatively open set containing a, i.e., the intersection of a neighborhood of a (in $\mathbb{R}^m$) with E.

Theorem *Let $E \subset \mathbb{R}^m$ and $F: E \to \mathbb{R}^n$. The following four statements are equivalent.*

I. *The map F belongs to $C(E; \mathbb{R}^n)$.*

II. *For every point $a \in E$ and every neighborhood V of $F(a)$, the preimage of V is a relative neighborhood of a.*

III. *For every set V open in $\mathbb{R}^n$, its F-preimage is relatively open in E.*

IV. *For every set V closed in $\mathbb{R}^n$, its F-preimage is relatively closed in E.*

Proof The equivalence of I and II directly follows from the definition of continuity, as the reader can easily check.

Conditions III and IV are equivalent, since a subset of E is closed in E if and only if its complement in E is open in E. Hence, we must check only the equivalence of I and III.

I $\Rightarrow$ III. If V is an open set such that $F^{-1}(V) = \emptyset$, there is nothing to prove. In what follows, we assume that $H := F^{-1}(V) \neq \emptyset$. For every x from H, the value $F(x)$ belongs to the open set V. Hence, there is a neighborhood U_x of x such that the image of $U_x \cap E$ is contained in V. Thus, $U_x \cap E \subset F^{-1}(V) = H$. Therefore,

$$\bigcup_{x \in H} U_x \cap E \subset H.$$

Set $U = \bigcup_{x \in H} U_x$. Clearly, U is open, and

$$U \cap E = \bigcup_{x \in H} U_x \cap E \subset H.$$

On the other hand, $H \subset U$; thus, we conclude that $U \cap E = H$, as required.

III $\Rightarrow$ II. Let a be an arbitrary point from E and V be an arbitrary neighborhood of $F(a)$. By assumption, $F^{-1}(V) = U \cap E$ where the set U is open. Hence, $F^{-1}(V)$ is a relative neighborhood of a, as required.   □

Given a function $f : E \to \mathbb{R}$, consider the sets of the form $E(f < t)$, $E(f > t)$, $E(f \le t)$, $E(f \ge t)$ with $t \in \mathbb{R}$, where

$$E(f < t) = f^{-1}((-\infty, t)) = \{x \in E : f(x) < t\}$$

and the other sets are defined in a similar way. The above theorem immediately implies that if f is continuous on E, then the sets $E(f < t)$ and $E(f > t)$ are open in E for every t, while the sets $E(f \le t)$ and $E(f \ge t)$ are closed in E. The set $E(f = t) = f^{-1}(\{t\})$ is also closed in E.

0.8 Continuity on a Compact Set

The following theorems are of fundamental importance in analysis. In what follows, by a "continuous image" of a set we mean its image under a continuous map.

Theorem *A continuous image of a compact set is compact.*

Proof Let K be a compact subset in $\mathbb{R}^m$ and $F \in C(K; \mathbb{R}^n)$. We must prove that the set $F(K)$ is compact. By the third assertion of Theorem 0.5, it suffices to check that every sequence $\{y_j\}_j$ of points from $F(K)$ contains a subsequence converging in $F(K)$. Since $y_j \in F(K)$ ($j \in \mathbb{N}$), we have $y_j = F(x_j)$ for some x_j from K. The points x_j form a sequence in the compact set K. By Theorem 0.5, it contains a subsequence $\{x_{j_k}\}_k$ converging to some point x from K. Since the map F is continuous at this point, we have

$$y_{j_i} = F(x_{j_i}) \xrightarrow[i \to \infty]{} F(x) \in F(K).$$

Thus, $\{y_{j_i}\}$ is a required subsequence.   □

Here is a very important special case of this theorem.

Theorem (Weierstrass extreme value theorem) *A function continuous on a compact set K attains a maximum and a minimum on K. In particular, it is bounded.*

This is a direct corollary of the above theorem and the property of a compact subset of the real line (the existence of a maximum and a minimum) mentioned in Section 0.5.

Theorem (Heine[9]–Cantor[10]) *Let K be a compact set in $\mathbb{R}^m$ and $F \in C(K; \mathbb{R}^n)$. Then for every $\varepsilon > 0$ there exists $\delta > 0$ such that*

$$\|F(x'') - F(x')\| < \varepsilon \quad \text{as long as} \quad x', x'' \in K \quad \text{and} \quad \|x'' - x\| < \delta.$$

Proof Assume the contrary. Then there exists a positive number ε such that for every $\delta > 0$ in K there are points x', x'' such that $\|x'' - x'\| < \delta$ and $\|F(x'') - F(x')\| \geq \varepsilon$. Successively taking δ equal to $1, \frac{1}{2}, \ldots, \frac{1}{j}, \ldots$, we construct sequences $\{x'_j\}_j$ and $\{x''_j\}_j$ in K such that

$$\|F(x''_j) - F(x'_j)\| \geq \varepsilon \quad \text{although} \quad \|x''_j - x'_j\| \to 0.$$

Here we may assume that the sequence $\{x'_j\}_j$ converges to a point x_0 from K (otherwise we should use the third assertion of Theorem 0.5). Since $\|x'_j - x_0\| \to 0$ and $\|x''_j - x'_j\| \to 0$, it is clear that $x''_j \to x_0$. Hence,

$$F(x''_j) \to F(x_0) \quad \text{and} \quad F(x'_j) \to F(x_0)$$

by the continuity of F at x_0, whence $F(x''_j) - F(x'_j) \to 0$. But this is impossible, since $\|F(x''_j) - F(x'_j)\| \geq \varepsilon$ for all j. Thus, our assumption leads to a contradiction. $\square$

The Heine–Cantor theorem can be interpreted geometrically using the notion of the diameter of a set (see Section 0.2). The theorem says that the image of a set of a sufficiently small diameter is contained in a ball of a uniformly small radius. In other words, for every $\varepsilon > 0$ there exists $\delta > 0$ such that the image $F(e)$ of a set $e \subset K$ is contained in a ball of radius ε as long as $\text{diam}(e) < \delta$.

The following statement, which is sometimes called Lebesgue's[11] number lemma, generalizes the Heine–Cantor theorem.

Theorem (Lebesgue) *Let F be a continuous map from a compact set $K \subset \mathbb{R}^m$ to $\mathbb{R}^n$. For every open cover $\{G_\alpha\}_\alpha$ of the set $F(K)$ there is $\delta > 0$ such that if $e \subset K$ and $\text{diam}(e) < \delta$, then $F(e)$ is contained in one of the sets G_α.*

We note two special cases of this theorem.

Considering the cover of a set by all balls of radius $\varepsilon/2$ and applying Lebesgue's lemma to the two-point sets $\{x, y\}$, we obtain the Heine–Cantor theorem.

In the case where F is the identity map, Lebesgue's lemma implies that for every open cover $\{G_\alpha\}_\alpha$ of a compact set K in a Euclidean space there exists $\delta > 0$ such that every subset of K with diameter less than δ is contained in one of the sets G_α.

Proof Assume the contrary. Then for every j there is a subset e_j in K such that $\text{diam}(e_j) < 1/j$ but $F(e_j)$ is not contained in any G_α. Fix a point x_j in each set e_j. Passing if necessary to a subsequence, we may assume that the sequence $\{x_j\}_j$

---

[10] Heinrich Eduard **Heine** (1821–1881) was a German mathematician.

[10] Georg Ferdinand Ludwig Philipp **Cantor** (1845–1918) was a German mathematician.

[11] Henri Léon **Lebesgue** (1875–1941) was a French mathematician.

converges to a point x_0 from K. Let G_{α_0} be a set from the open cover containing the point $y_0 = F(x_0)$. By the continuity of F at x_0, there exists $\delta > 0$ such that

$$F(K \cap B(x_0, \delta)) \subset G_{\alpha_0}.$$

If $\|x_j - x_0\| < \frac{\delta}{2}$ and $\operatorname{diam}(e_j) < \frac{\delta}{2}$, then $e_j \subset B(x_0, \delta)$. Hence, $F(e_j) \subset G_{\alpha_0}$ for sufficiently large j, which contradicts the choice of e_j. □

0.9 Uniform Continuity

The Heine–Cantor theorem establishes an important property valid not only for continuous maps on compact sets. Given the importance of this property in many problems, it is reasonable to introduce the following definition.

Definition A map $F \colon E \to \mathbb{R}^n$, where $E \subset \mathbb{R}^m$, is said to be *uniformly continuous* on E if for every $\varepsilon > 0$ there is $\delta > 0$ such that

$$\|F(x'') - F(x')\| < \varepsilon \quad \text{as long as } x', x'' \in E \text{ and } \|x'' - x'\| < \delta.$$

Clearly, a map uniformly continuous on a set E is continuous on E. The converse is definitely false (for instance, the function $x \mapsto x^2$ is continuous on $\mathbb{R}$, but not uniformly continuous). However, the Heine–Cantor theorem shows that continuity on a compact set implies uniform continuity. Also note that the uniform continuity of a map is equivalent to the uniform continuity of all its coordinate functions.

Remark A map uniformly continuous on E transforms every Cauchy sequence in E into a sequence convergent in $\mathbb{R}^n$. Hence, such a map can be extended by continuity to the closure of E, and this extension is unique.

It is convenient to use a quantitative measure for estimating the increment of a map in terms of an increment of the argument. For this, with a map $F \colon E \to \mathbb{R}^n$, where $E \subset \mathbb{R}^m$, we associate its *modulus of continuity*, that is, the function

$$\omega_F \colon [0, +\infty) \to [0, +\infty]$$

defined by the formula

$$\omega_F(t) = \sup_{\substack{x, y \in E \\ \|x - y\| \le t}} \|F(x) - F(y)\|.$$

One can easily see that F is uniformly continuous if and only if ω_F is continuous at 0, i.e.,

$$\omega_F(t) \xrightarrow[t \to 0]{} \omega_F(0) = 0.$$

We say that a map $F: E \rightarrow \mathbb{R}^n$ satisfies the *Lipschitz*[12] *condition of order* α $(0 < \alpha \leq 1)$ if $\omega_F(t) \leq Lt^\alpha$ for $t \geq 0$ (the constant L depends on F), i.e., if

$$\|F(x) - F(y)\| \leq L\|x - y\|^\alpha \quad \text{for all } x, y \in E.$$

If $\alpha = 1$, we say "Lipschitz condition" without specifying the order. For example, the triangle inequality implies that the funciton $x \mapsto \|x\|$ satisfies the Lipschitz condition (with constant $L = 1$). This statement can be generalized by replacing the distance from an arbitrary point to the origin with the distance to some set $E \subset \mathbb{R}^m$.

Indeed, for any points $x, y \in \mathbb{R}^m$, we have

$$\text{dist}(y, E) = \inf_{z \in E} \|z - y\| \leq \inf_{z \in E} (\|z - x\| + \|x - y\|) = \text{dist}(x, E) + \|x - y\|.$$

Thus, $\text{dist}(y, E) - \text{dist}(x, E) \leq \|x - y\|$. Since the points x and y in this inequality are interchangeable, it follows that

$$|\text{dist}(y, E) - \text{dist}(x, E)| \leq \|x - y\|,$$

i.e., the function $x \mapsto \text{dist}(x, E)$ satisfies the Lipschitz condition in $\mathbb{R}^m$ (with constant $L = 1$).

If the restriction of F to a subset $E_0 \subset E$ satisfies the Lipschitz condition of order α, we say that F satisfies this condition on E_0. In particular, this allows us to define a local version of the condition: a map $F: E \rightarrow \mathbb{R}^n$ *satisfies the local Lipschitz condition of order* α if it satisfies this condition in a sufficiently small (relative) neighborhood of every point of E. We leave it to the reader to verify that a map defined on a compact set and satisfying the local Lipschitz condition of order α satisfies this condition on the entire set (see Exercise 14).

0.10 The Continuity of the Inverse Map

In the one-dimensional case, a condition for the existence and continuity of the inverse function is quite easy to state: if a function is continuous on an interval, then it is necessary and sufficient for the existence and continuity of the inverse function that the original function is strictly monotone. In the multidimensional case, we are unable to use monotonicity considerations. However, for invertible maps defined on "good" sets, the result is quite similar to the one-dimensional one: the continuity of the original map implies the continuity of the inverse map.

Theorem *Let K be a compact subset in $\mathbb{R}^m$ and $F \in C(K; \mathbb{R}^n)$ be a one-to-one map. Then the inverse map is continuous on $F(K)$.*

Proof Let E be an arbitrary closed subset in K. As observed in Section 0.5, it is compact. The continuous image $F(E)$ of E is compact, and therefore closed. But

---

[12] Rudolf Otto Sigismund **Lipschitz** (1832–1903) was a German mathematician.

$F(E)$ is the preimage of E for the map F^{-1}. Thus, for F^{-1} the preimages of all closed sets are closed, which, by Theorem 0.7, is equivalent to the continuity of F^{-1}.   □

Recall that an invertible map F is called a *homeomorhism* if both F and F^{-1} are continuous. Obviously, the composition of homeomorphisms is a homeomorphism. Two sets are said to be homeomorphic if one is the image of the other under a homeomorphism. By the above theorem, a map defined on a compact set is a homeomorphism if it is continuous and invertible. Checking the continuity usually presents no problems, while checking the invertibility is often quite a difficult task (see, e.g., Section II.2).

An important special case of a continuous map defined on a compact set is a *path*, that is, a continuous map from a closed interval to a Euclidean space.

The set $\Gamma = \gamma([a, b])$ of all values of a path $\gamma \colon [a, b] \to \mathbb{R}^n$ is called the support of γ. The points $\gamma(a)$ and $\gamma(b)$ are called the beginning and the end of γ, respectively; one says that γ connects these points. Also, one says that γ lies in a set if this set contains the support of γ.

By definition, a path is simple if

$$\gamma(t') \neq \gamma(t'') \quad \text{for } t' \neq t'' \ (t', t'' \in [a, b]),$$

i.e., if the map γ is one-to-one. By the above theorem, a simple path is a homeomorphism, so its support, also called a simple curve, or a Jordan curve, is homeomorphic to an interval of the real line. An example of a simple curve is the graph of a continuous function f defined on a closed interval: it is the support of the path $x \mapsto (x, f(x))$.

0.11 Connectedness

A subset E in a Euclidean space is said to be *connected* if it cannot be represented as a disjoint union of two nonempty sets open in E (an equivalent definition can be obtained in terms of a disjoint union of nonempty sets closed in E). A characteristic property of a connected set E is that on E one cannot define a continuous function taking only two (different) values, since the preimages of the corresponding one-point sets are closed in E. It easily follows that the only connected sets in $\mathbb{R}$ are intervals. In spaces of higher dimension, the class of connected subsets is much wider. For example, a torus and an annulus (a set bounded by two concentric circles) are connected.

If E is an open set, then its relatively open subsets are open. Thus, an open set is connected if it cannot be represented as a union of nonempty open sets without common points. A connected open set is called a *domain*.

It is easy to check that for an open set, connectedness is equivalent to the intuitive geometric property of *path connectedness*, which means that any two points of the set can be connected by a path in this set (see Exercise 5).

Theorem *A continuous image of a connected set is connected.*

Proof The proof is almost obvious. Indeed, let E be a connected subset in $\mathbb{R}^m$ and $F \in C(E; \mathbb{R}^n)$. We must check that the set $F(E)$ is connected. Assume to the contrary that it can be divided into two nonempty subsets open in $F(E)$. Then their preimages are nonempty, form a partition of the set E, and are open in E by Theorem 0.7. This means that, contrary to the assumption, E is not connected. Thus, we have arrived at a contradiction.   □

In particular ($n = 1$), the set of all values of a function continuous on a connected set is an interval. In other words, such a function satisfies the classical intermediate value theorem.

We also mention another result related to connected sets. A map defined on a set $E \subset \mathbb{R}^m$ is said to be *locally constant* on E if it is constant near every point of E (there exists a neighborhood U of this point such that the map is constant on the intersection $U \cap E$).

Proposition *A map locally constant on a connected set E is constant on E.*

Proof Let F be a locally constant map on a connected set $E \subset \mathbb{R}^m$ and $a \in E$. Divide E into the disjoint subsets

$$A = \{x \in E : F(x) = F(a)\} \quad \text{and} \quad B = \{x \in E : F(x) \neq F(a)\}.$$

They are open in E, because F is locally constant. Since E is connected and $A \neq \varnothing$, we have $B = \varnothing$, which proves the theorem.   □

0.12 Convex Sets in $\mathbb{R}^m$

A *line* through a point x_0 with *direction vector* $\ell \neq 0$ is the set $\{x_0 + t\ell \,|\, t \in \mathbb{R}\}$. To obtain the line passing through two (different) points a and b, we can take, for example, $x_0 = a$, $\ell = b - a$. Its part lying "between" a and b, i.e., the set

$$[a, b] = \{a + t(b - a) : t \in [0, 1]\} = \{(1 - t)a + t b : t \in [0, 1]\},$$

is called the segment with endpoints a and b. In the case where $a = b$, we set $[a, a] = \{a\}$.

A subset E in $\mathbb{R}^m$ is said to be *convex* if it contains with any two points the segment between them, i.e., $[a, b] \subset E$ for any points $a, b \in E$ (the empty set is convex by definition). Obviously, a convex set is connected.

It is easy to check that the intersection of an arbitrary collection of convex sets is convex. Using the triangle inequality, one can readily show that a ball (open or closed) is convex. Now we consider other examples of convex sets which will repeatedly appear throughout the book.

Let $x_0, v \in \mathbb{R}^m$, $v \neq 0$. The set

$$H = \{x \in \mathbb{R}^m : \langle x - x_0, v \rangle = 0\}$$

is called a *plane* through x_0; the vector v is a *normal* to H (in the two-dimensional case, i.e., if $m = 2$, a "plane" is a line).

Since the scalar product is continuous, every plane is a closed set. If $\langle x_0, v \rangle = 0$, then H contains the origin and, as one can easily check, is a vector space (of dimension $m - 1$) consisting of the vectors orthogonal to the normal.

Obviously, a plane contains with every pair of different points the line through them. In particular, every plane is convex. Clearly, the difference of any two vectors in a plane is orthogonal to the normal to this plane. It easily follows that two planes sharing the same normal either coincide or are disjoint. In the second case, they are said to be parallel. Parallel planes can be obtained from each other by a translation. In particular, a plane H such that $0 \notin H$ can be obtained by translating the parallel plane passing through the origin. In more detail: the plane $\langle x - x_0, v \rangle = 0$ can be obtained from the subspace $\langle x, v \rangle = 0$ consisting of the vectors orthogonal to v by the translation by the vector x_0.

Let us find the distance between a point and a plane (recall that the distance between a point x and a set E is defined as $\mathrm{dist}(x, E) = \inf\{\|x - y\| : y \in E\}$ (see Section 0.2).

Lemma *Let H be a plane defined by an equation $\langle x - x_0, v \rangle = 0$. Then for every point $w \in \mathbb{R}^m$,*

$$\mathrm{dist}(w, H) = \frac{|\langle w - x_0, v \rangle|}{\|v\|}.$$

Proof We may assume without loss of generality that $\|v\| = 1$ (otherwise we replace v with $\frac{v}{\|v\|}$).

First, we find the orthogonal projection $\bar{w}$ of a point w to the plane H. Clearly, $\bar{w} = w + c\,v$ where $c \in \mathbb{R}$. To find the coefficient c, multiply both sides of the equality $\bar{w} - x_0 = w - x_0 + c\,v$ by v. Then $\langle \bar{w} - x_0, v \rangle = 0$, because $\bar{w} \in H$, and hence

$$0 = \langle w - x_0, v \rangle + c\|v\|^2 = \langle w - x_0, v \rangle + c.$$

Therefore, $c = -\langle w - x_0, v \rangle$.

Since $\bar{w} \in H$, the difference $\bar{w} - x$ is orthogonal to the normal v for every $x \in H$. Hence, the Pythagorean theorem yields

$$\|w - x\| = \|(\bar{w} - x) - c\,v\| = \sqrt{\|\bar{w} - x\|^2 + c^2\|v\|^2} = \sqrt{\|\bar{w} - x\|^2 + c^2}.$$

Thus,

$$\mathrm{dist}(w, H) = \min_{x \in H} \|w - x\| = \min_{x \in H} \sqrt{\|\bar{w} - x\|^2 + c^2} = |c| = |\langle w - x_0, v \rangle|. \qquad \square$$

One can see from the proof that a point closest to w in H is unique (the reader is encouraged to check that this point is $\bar{w}$, i.e., the orthogonal projection of w to H). This is true also in a more general setting, see Exercise 9.

An open half-space, i.e., a set defined by a strict inequality $\langle x - x_0, v \rangle > 0$, is open, since it is a set of the form $E(f > t)$ (see the end of Section 0.7) for the continuous function

$$x \mapsto \langle x - x_0, v \rangle.$$

In a similar way, a closed half-space, defined by a non-strict inequality $\langle x - x_0, v \rangle \geq 0$, is a closed set. It follows from the linearity of the scalar product in the first argument that both open and closed half-spaces are convex.

One can show that a convex open (respectively, closed) set is the intersection of a family of open (respectively, closed) half-spaces.

0.13 Linear Maps Between Euclidean Spaces

Here we briefly review the simplest properties of linear maps used in what follows.

A map $A \colon \mathbb{R}^m \to \mathbb{R}^n$ is said to be linear if it sends a linear combination of two vectors to the analogous linear combination of the images:

$$A(cx + c'x') = cA(x) + c'A(x') \quad \text{for all } x, x' \in \mathbb{R}^m \text{ and } c, c' \in \mathbb{R}.$$

Fixing bases in the spaces $\mathbb{R}^m$ and $\mathbb{R}^n$, with every linear map $A \colon \mathbb{R}^m \to \mathbb{R}^n$ we can associate a rectangular $n \times m$ matrix, the matrix representation of A. Now we describe the construction of this matrix, assuming that the bases under consideration are the canonical bases (see Section 0.1).

Writing a vector $x = (x_1, \ldots, x_m)$ in the form $x_1 e_1 + x_2 e_2 + \ldots + x_m e_m$, where $e_1 = (1, 0, \ldots, 0), \ldots, e_m = (0, \ldots, 0, 1)$ are the canonical basis vectors, we arrive at the equality

$$y = A(x) = \sum_{k=1}^{m} x_k g_k, \quad \text{where } g_k = A(e_k). \tag{1}$$

Let $a_{1k}, a_{2k}, \ldots, a_{nk}$ $(k = 1, \ldots, m)$ be the coordinates of g_k and $y_1, \ldots, y_n$ be the coordinates of y. Writing (1) in coordinates, we see that

$$y_j = \sum_{k=1}^{m} a_{jk} x_k \quad (j = 1, \ldots, n).$$

Identifying a vector with the column of its coordinates, we can rewrite these equalities in matrix form:

$$A(x) = \begin{pmatrix} y_1 \\ \ldots \\ y_n \end{pmatrix} = \begin{pmatrix} a_{11} & \ldots & a_{1m} \\ \ldots & \ldots & \ldots \\ a_{n1} & \ldots & a_{nm} \end{pmatrix} \begin{pmatrix} x_1 \\ \ldots \\ x_m \end{pmatrix}. \tag{2}$$

The $n \times m$ matrix

$$\widetilde{A} = \begin{pmatrix} a_{11} & \ldots & a_{1m} \\ \ldots & \ldots & \ldots \\ a_{n1} & \ldots & a_{nm} \end{pmatrix}$$

appearing in this formula is called the *matrix representation* of the map A (in the canonical bases). Matrix representations in other bases will not be considered here.

Of course, given an arbitrary real $n \times m$ matrix, one can use (2) to construct a linear map from $\mathbb{R}^m$ to $\mathbb{R}^n$. Thus, there is a one-to-one correspondence between the linear maps from $\mathbb{R}^m$ to $\mathbb{R}^n$ and the real $n \times m$ matrices, which allows one to use the same symbol for a linear map and the corresponding matrix representation. As one can easily see, the sum or the composition of maps corresponds to the sum or the product of their matrix representations, respectively, and the inverse map (if it exists) corresponds to the inverse matrix. The identity map in $\mathbb{R}^m$ corresponds to the identity $m \times m$ matrix.

The columns of the matrix representation of a linear map A consist of the coordinates of the vectors $g_k = A(e_k)$, the images of the canonical basis vectors in $\mathbb{R}^m$. Their coordinates a_{jk} can be written as scalar products: $a_{jk} = \langle A(e_k), \widetilde{e}_j \rangle$, where $\widetilde{e}_j$ ($j = 1, \ldots, n$) are the canonical basis vectors in $\mathbb{R}^n$.

Note that since algebraic operations are continuous, it follows from (1) that every linear map is continuous. Besides, from (1) and Cauchy's inequality we deduce that

$$\|A(x)\| \le \sum_{k=1}^{m} |x_k| \, \|g_k\| \le C \, \|x\|$$

for every $x = (x_1, \ldots, x_m)$, where $C = \left(\sum_{k=1}^{m} \|g_k\|^2 \right)^{1/2} = \left(\sum_{k=1}^{m} \sum_{j=1}^{n} a_{jk}^2 \right)^{1/2}$. This estimate is rather crude (say, for the identity map, which preserves the norm of a vector, $C = \sqrt{m}$), and it can be refined by introducing a value denoted by $\|A\|$ and called the *norm of the map* A. By definition,

$$\|A\| = \sup_{\|x\|=1} \|A(x)\|.$$

Using the positive homogeneity of the norm ($\|cx\| = c\|x\|$ for $c > 0$), one can easily see that

$$\|A(x)\| \le \|A\| \, \|x\| \quad \text{for all } x \tag{3}$$

and

$$\|A\| = \sup_{x \neq 0} \frac{\|A(x)\|}{\|x\|}.$$

Thus, the norm of a linear map A is the largest factor by which A can increase the norm of a (nonzero) vector. Clearly, if $\|A(x)\| \le K\|x\|$ for all x, then $\|A\| \le K$, i.e., (3) is the best possible estimate of this kind. If A is invertible (so $m = n$), then there is also a similar lower estimate. Indeed,

$$\|A^{-1}(y)\| \le \|A^{-1}\| \, \|y\| \quad \text{for every } y \in \mathbb{R}^m.$$

It follows that for $y = A(x)$ we have $\|x\| \le \|A^{-1}\| \, \|Ax\|$, i.e.,

$$\|A(x)\| \ge \frac{1}{\|A^{-1}\|} \, \|x\| \quad \text{for all } x \in \mathbb{R}^m. \tag{4}$$

We note several other properties of the norm of a map, leaving the proof to the reader:

1) if $\|A\| = 0$, then A is the zero map;
2) $\|cA\| = |c|\,\|A\|$ $(c \in \mathbb{R})$;
3) $\|A + B\| \le \|A\| + \|B\|$;
4) $\|A \circ B\| \le \|A\|\,\|B\|$.

A remarkable class of linear maps from $\mathbb{R}^m$ to $\mathbb{R}^m$ consists of the *orthogonal maps*. These are linear maps A preserving the lengths of vectors:

$$\|A(x)\| = \|x\| \quad \text{for all } x \in \mathbb{R}^m.$$

An orthogonal map is invertible, since its kernel is trivial. Moreover, it is clear that the inverse map also preserves the norms of vectors, i.e., is orthogonal.

An orthogonal map has an important property: it preserves not only the norm, but also the scalar product of vectors:

$$\langle A(x), A(y) \rangle = \langle x, y \rangle \quad \text{for all } x, y \in \mathbb{R}^m$$

(since $4\langle x, y \rangle = \|x + y\|^2 - \|x - y\|^2$). This equality has a clear geometric interpretation: an orthogonal map preserves angles between vectors and, in particular, sends orthogonal vectors to orthogonal vectors. Thus, one may say that the action of an orthogonal map amounts to a rotation (in a wide sense, i.e., reflections are also considered to be rotations).

Since the columns of the matrix representation of A consist of the coordinates of the vectors $A(e_1), \ldots, A(e_m)$, the matrix representation $\widetilde{A}$ of an orthogonal map A is a square matrix in which all columns have unit length and different columns are orthogonal. It is easy to see that the product of such a matrix with its transpose coincides with the identity matrix and, consequently, the matrix representation of A^{-1} coincides with $(\widetilde{A})^T$. It follows, in particular, that the rows of the matrix representation of an orthogonal map have the same property as the columns: they are pairwise orthogonal and have unit length.

Another very simple (but not linear) distance-preserving map is the map defined by the formula $x \mapsto a + x$. It is called a *translation* (by a). The image of a set $E \subset \mathbb{R}^m$ under this map, i.e., the set $\{a + x : x \in E\}$, is denoted by $a + E$ (the translate of E by a). A translation is a bijection in $\mathbb{R}^m$, the inverse map is the translation by $-a$.

The composition of a translation and an orthogonal map preserves distances between vectors. Such a map is called a *rigid motion*. Two subsets in $\mathbb{R}^m$ are congruent if one of them coincides with the image of the other under a rigid motion.

An (m-dimensional) rectangular parallelepiped P is the Cartesian product of nondegenerate intervals $\Delta_1, \ldots, \Delta_m$ or a set congruent to such a product. The lengths of the intervals are called the edge lengths of P. If all these intervals are open (closed), then the parallelepiped is said to be open (closed). The reader can readily check that open (closed) parallelepipeds are open (closed) sets. A parallelepiped is an arbitrary linear one-to-one image of a rectangular parallelepiped. In what follows, we mainly consider rectangular parallelepipeds, and thus omit the adjective "rectangular." Note that every parallelepiped is a convex set, being a linear image of a product of convex sets.

Exercises

1. Show that $\partial E = \partial(\mathbb{R}^m \setminus E)$ and $\bar{E} = E \cup \partial E$ for every subset E in $\mathbb{R}^m$.
2. Show that the intersection of all neighborhoods of a set E coincides with E, while the intersection of all its δ-neighborhoods coincides with $\bar{E}$.
3. Let E be the x-axis and $U = \{(x, y) \in \mathbb{R}^2 : |y| < e^x\}$. Clearly, U is a neighborhood of E. Verify that U contains no δ-neighborhood of E.
4. Show that every neighborhood of a compact set contains a δ-neighborhood of this set.
5. Show that for an open set E to be connected it is necessary and sufficient that any two points of E can be joined by a polygonal line in E.
6. Check that a nonempty open set is a union of at most countably many domains.
7. Show that the union of an arbitrary family of connected sets having a common point is connected.
8. Check that the closure and the interior of a convex set are convex.
9. Let $x \in \mathbb{R}^m$ and E be a nonempty closed subset in $\mathbb{R}^m$. Show that in E there exists a best approximation to x, i.e., a point z_x such that

$$z_x \in E \quad \text{and} \quad \text{dist}(x, E) = \|x - z_x\|.$$

   Check that if E is convex, then a best approximation is unique.
10. Let $x \in \mathbb{R}^m$, $E \subset \mathbb{R}^m$ be a convex set, and $\{x_n\}_{n \geq 1}$ be a sequence of points of E such that $\|x_n - x\| \to \text{dist}(x, E)$ (a minimizing sequence). Show that this sequence converges and its limit is the best approximation to x in $\bar{E}$.
11. Let A and B be nonempty closed subsets in $\mathbb{R}^m$ at least one of which is compact. Show that there exist points $a \in A$, $b \in B$ such that

$$\|a - b\| \leq \|x - y\| \quad \text{for all } x \in A, \ y \in B.$$

    Is this assertion true if we drop the compactness condition?
    HINT. To obtain a counterexample, consider a hyperbola and its asymptote.
12. Let F be a map defined on a convex set. Show that the modulus of continuity ω_F is subadditive: $\omega_F(t+t') \leq \omega_F(t)+\omega_F(t')$ for any nonnegative numbers t and t'. Can we drop the convexity condition?
13. Let $f(x) = \sqrt{x}$ for $x \geq 0$ and $g(x) = \sqrt[3]{x}$ for $x \in \mathbb{R}$. Check that $\omega_f(t) = \sqrt{t}$ and $\omega_g(t) = \sqrt[3]{4t}$ for all $t \geq 0$.
14. Show that a map defined on a compact set and satisfying the local Lipschitz condition of order α satisfies this condition on the entire set.

15. a) Assume that a function continuous on an interval $\Delta = \Delta_1 \cup \Delta_2$ is uniformly continuous on each of the two subintervals Δ_1, Δ_2. Show that it is uniformly continuous on Δ.

b) Is the analogous result true for functions of several variables if we replace intervals with arbitrary connected closed sets?

HINT. To obtain a counterexample, consider the function $f(x, y) = \operatorname{sign} x$ on the set $\{(x, y): x^2 y \geq 1\}$. The domain of definition can be made connected, e.g., by additionally setting $f(x, 1) = x$ for $x \in (-1, 1)$.

16. a) Let f be a function of several variables satisfying the Lipschitz condition of order α on convex compact sets K_1 and K_2 having a common interior point. Show that f satisfies this condition (in general, with a larger constant) on the union $K_1 \cup K_2$.

b) Is this assertion true if the common point is a boundary point?

HINT. Consider the sets $K_1 = \{(x, 0): 0 \leq x \leq 1\}$ and $K_2 = \{(x, y): x^2 \leq y \leq x\}$. Check that the function

$$f(x, y) = \begin{cases} 0 & \text{if } y = 0, \\ x & \text{if } y \neq 0 \end{cases}$$

satisfies the Lipschitz condition of order 1 on each of the sets K_1, K_2, while on the union $K_1 \cup K_2$ it satisfies the Lipschitz condition of order $\frac{1}{2}$, but does not satisfy the Lipschitz condition of a larger order.

c) In Exercise 16 (b), replace K_2 with a convex set K_2' that is so close to the x-axis at the origin that on $K_1 \cup K_2'$ the function f is continuous, by does not satisfy the Lipschitz condition of order α for any $\alpha > 0$.

17. Show that a function defined on a subset of a Euclidean space and satisfying the Lipschitz condition can be extended to a function satisfying this condition on the entire space.

HINT. Show that a required extension of such a function $f : E \to \mathbb{R}$ can be defined by the formula

$$y \mapsto \inf_{x \in E} (f(x) + L\|x - y\|),$$

where L is the constant from the definition of the Lipschitz condition.

18. Show that a map from a Euclidean space to itself that preserves distances between points is a rigid motion.

HINT. Assuming that the origin is mapped to itself, check that such a map preserves not only the norm, but also the scalar product, so the image of an orthonormal basis is also an orthonormal basis.

Chapter I. Differentiable Functions

1 Partial Derivatives and Increments

1.1 Partial Derivatives

Given the fundamental importance of the notion of derivative in the study of functions of one variable, it is natural to want to extend it in some way to functions of several variables. One of the most obvious steps in this direction is to fix all coordinates except one and consider the derivative of the resulting function of one variable. Now we describe this procedure in more detail.

Let f be a function defined on a subset X of $\mathbb{R}^m$ and $a = (a_1, \ldots, a_m)$ be an interior point of X. Then the points $a + h$ lie in X for all h with sufficiently small norm. Set $h = te_k$, where $t \in \mathbb{R}$ and e_k $(k = 1, \ldots, m)$ is a canonical basis vector in $\mathbb{R}^m$. Fixing all coordinates except the kth, consider the corresponding increment of f:

$$f(a + te_k) - f(a)$$
$$= f(a_1, \ldots, a_{k-1}, a_k + t, a_{k+1}, \ldots, a_m) - f(a_1, \ldots, a_{k-1}, a_k, a_{k+1}, \ldots, a_m).$$

Definition The *partial derivative* of the function f at the point a with respect to the kth coordinate is the limit

$$\lim_{t \to 0} \frac{f(a + te_k) - f(a)}{t}$$

(if it exists).

As can be seen from the definition, this derivative measures the rate of change of a function of several variables in the kth coordinate. In what follows, speaking of the existence of partial derivatives, we always assume that they are finite, unless otherwise stated.

© The Author(s), under exclusive license to Springer Nature Switzerland AG 2021
B. M. Makarov, A. N. Podkorytov, *Smooth Functions and Maps*, Moscow Lectures 7,
https://doi.org/10.1007/978-3-030-79438-5_1

The partial derivative with respect to x_k is denoted by f'_{x_k} or $\frac{\partial f}{\partial x_k}$. If the coordinates of vectors are denoted by other symbols, then the corresponding notation carries over to partial derivatives. In some cases, to make the notation for partial derivatives independent of that for the coordinates of vectors, the derivative with respect to the kth coordinate is denoted by $\partial_k f$.

According to the definition, the partial derivative of a function f at a point a with respect to the kth coordinate is nothing else than the ordinary derivative at the point a_k of the function

$$u \mapsto f(a_1, \ldots, a_{k-1}, u, a_{k+1}, \ldots, a_m),$$

which is defined, obviously, for all u sufficiently close to a_k. Hence, in the computation of partial derivatives we can use the ordinary differentiation rules for sums, products, and quotients. For example (we assume that $x > 0, y > 0$),

$$\frac{\partial}{\partial x}\left(\frac{x}{y}\right)^z = \frac{z}{y}\left(\frac{x}{y}\right)^{z-1}, \quad \frac{\partial}{\partial y}\left(\frac{x}{y}\right)^z = -\frac{z}{y}\left(\frac{x}{y}\right)^z, \quad \frac{\partial}{\partial z}\left(\frac{x}{y}\right)^z = \left(\frac{x}{y}\right)^z \log\frac{x}{y}.$$

A simple but important property of partial derivatives is that they (like ordinary derivatives) commute with translations. This means that

$$f'_{x_k}(x+c) = (f(x+c))'_{x_k}.$$

In more detail: if f has a partial derivative $f'_{x_k}(a)$ at a point a, then the "translated" function

$$x \mapsto g(x) = f(x+c)$$

has a partial derivative with respect to x_k at the point $a - c$ and these derivatives coincide. For example, $f'_{x_k}(c) = g'_{x_k}(0)$. This property, which immediately follows from the definition of a partial derivative, will be repeatedly used throughout the book.

1.2 Lagrange's Mean Value Theorem

Using partial derivatives, one can obtain a generalization of an important formula of differential calculus, Lagrange's[1] mean value theorem, to the case of functions of several variables. As in the one-dimensional case, this formula is most commonly used to estimate increments of functions from above. Later, in Section 5.4, we will establish Lagrange's inequality, which is more convenient for this purpose. For brevity, we say that a vector $\widetilde{h} = (\widetilde{h}_1, \ldots, \widetilde{h}_m)$ lies between 0 and a vector $h = (h_1, \ldots, h_m)$ if every coordinate of $\widetilde{h}$ lies between 0 and the corresponding coordinate of h, i.e., $\widetilde{h}_1 = \theta_1 h_1, \ldots, \widetilde{h}_m = \theta_m h_m$ where $\theta_1, \ldots, \theta_m \in [0, 1]$.

---

[1] Joseph-Louis **Lagrange** (1736–1813) was a French mathematician.

Theorem *Let f be a function defined on an open parallelepiped $P \subset \mathbb{R}^m$ that has finite partial derivatives with respect to all variables at all points of P. Then for any points a and $a + h$ (with $h = (h_1, \ldots, h_m)$) from P,*

$$f(a + h) - f(a) = \sum_{k=1}^{m} f'_{x_k}(a + h^{(k)})h_k \tag{1}$$

where $h^{(1)}, \ldots, h^{(m)}$ are vectors lying between 0 and h.

Formula (1) will be called the *mean value theorem*.

Proof We represent the increment of f as the sum of increments caused by varying only one coordinate of the argument:

$$f(a + h) - f(a) = \Delta_1 + \ldots + \Delta_m,$$

where (below, $e_1, \ldots, e_m$ are the canonical basis vectors)

$$\Delta_1 = f(a + h_1 e_1) - f(a)$$

and

$$\Delta_k = f(a + h_1 e_1 + \ldots + h_k e_k) - f(a + h_1 e_1 + \ldots + h_{k-1} e_{k-1}) \quad \text{for } k > 1.$$

Replacing h_k in Δ_k by a variable t, we obtain the function

$$t \mapsto f(a + h_1 e_1 + \ldots + h_{k-1} e_{k-1} + t e_k) - f(a + h_1 e_1 + \ldots + h_{k-1} e_{k-1}),$$

whose increment over the interval with endpoints 0 and h_k is equal to Δ_k. By the classical Lagrange's mean value theorem,

$$\Delta_k = f'_{x_k}(a + h_1 e_1 + \ldots + h_{k-1} e_{k-1} + \theta_k h_k e_k)h_k$$

for some $\theta_k \in (0, 1)$. Thus,

$$f(a + h) - f(a) = \sum_{k=1}^{m} \Delta_k = \sum_{k=1}^{m} f'_{x_k}(a + h^{(k)})h_k,$$

where

$$h^{(1)} = \theta_1 h_1 e_1 \quad \text{and} \quad h^{(k)} = h_1 e_1 + \ldots + h_{k-1} e_{k-1} + \theta_k h_k e_k \quad (k > 1)$$

are vectors lying between 0 and h. $\qquad\square$

Note the special case of (1) for functions of two variables:

$$f(x + u, y + v) - f(x, y) = f'_x(x + \theta_1 u, y)u + f'_y(x + u, y + \theta_2 v)v \tag{1'}$$

for some $\theta_1, \theta_2 \in [0, 1]$.

1.3 The Criterion of Constancy

Here is an important corollary of the above theorem.

Corollary *Let f be a function defined in a domain O that has finite partial derivatives with respect to all coordinates at all points of O. Then it is constant in O if and only if all its partial derivatives vanish in O:*

$$\frac{\partial f}{\partial x_1}(x) = \ldots = \frac{\partial f}{\partial x_m}(x) = 0 \quad \text{for all } x \in O.$$

Proof The "only if" part is obvious. Let us prove the "if" part.

Since all partial derivatives of f vanish in O, the mean value theorem implies that f is constant near every point of O. But O is a connected set, so it suffices to apply Proposition 0.11 about locally constant functions.   □

Remark Later, we will need a slightly more general result, when only some partial derivatives vanish. In this case, it is natural to expect that the function does not depend on the corresponding variables. However, this is not true for an arbitrary domain O (see Exercise 2). Still, the conclusion is true if the domain is convex. Indeed, assume without loss of generality that the partial derivatives with respect to the variables $x_1, \ldots, x_n$, $1 \leq n < m$, vanish in O. Fixing the remaining variables, we obtain a function defined in a connected (and even convex) domain (in an n-dimensional cross section of O) whose all partial derivatives vanish. By the above corollary, it is constant. Hence, the values of f are determined only by the variables $x_{n+1}, \ldots, x_m$.

1.4 The Lipschitz Condition

Here we show that a function with bounded partial derivatives satisfies (as in the one-dimensional case) the Lipschitz condition (see Section 0.9).

Theorem *Let f be a function defined in an open parallelepiped P that has partial derivatives with respect to all coordinates at all points of P. If these partial derivatives are bounded in P, then f satisfies the Lipschitz condition of order 1 in P.*

Proof By assumption, for some $C_1, \ldots, C_m$ we have

$$|f'_{x_1}(x)| \leq C_1, \ |f'_{x_2}(x)| \leq C_2, \ldots, \ |f'_{x_m}(x)| \leq C_m \quad \text{for all } x \in P.$$

Assuming that $x, \widetilde{x} \in P$, $\widetilde{x} - x = h = (h_1, \ldots, h_m)$, we use the mean value theorem (see (1)) to obtain

$$|f(\widetilde{x}) - f(x)| \leq \sum_{k=1}^{m} |f'_{x_k}(x + h^{(k)})| \, |h_k| \leq \sum_{k=1}^{m} C_k \|h\| = \left(\sum_{k=1}^{m} C_k \right) \|\widetilde{x} - x\|.$$

Thus, f satisfies the Lipschitz condition with constant $L = \sum\limits_{k=1}^{m} C_k$.   □

As the reader can verify, the constant L can be easily reduced to $\left(\sum\limits_{k=1}^{m} C_k^2 \right)^{1/2}$, but, in general, cannot be reduced further.

Exercises

1. Show that the function

$$f(x, y) = \begin{cases} \dfrac{xy^2}{x^2 + y^2} & \text{if } x^2 + y^2 > 0, \\ 0 & \text{if } (x, y) = (0, 0) \end{cases}$$

   has partial derivatives and satisfies the Lipschitz condition of order 1 in the entire plane.
2. Give an example of a function of two variables showing that the convexity of O in the remark in Section 1.3 is an indispensable condition.
3. Refine Theorem 1.2 by proving that one can take

$$h^{(k)} = h_1 e_1 + \ldots + h_{k-1} e_{k-1} + \theta h_k e_k$$

   with $\theta \in (0, 1)$ not depending on k.

2 The Definition of a Differentiable Function

Throughout this section, we assume that the functions under consideration are defined on a subset in $\mathbb{R}^m$. Recall that every linear function defined in $\mathbb{R}^m$ can be written as a scalar product.

2.1 Differentiability

The definition of differentiability of a function $f(x)$ of several variables at a given point relies (as in the case of one variable) on the study of increments of f for small increments of x. However, the "almost proportionality" of the increments of f to small increments of x, which underlies the definition of differentiability in the case of one variable, must be replaced by a more general condition, since it is no longer meaningful in the case of many variables. We proceed from the assumption that, as before, the principal part of the increment of f corresponding to an increment

$h = h' + h''$ (or $h = t\widetilde{h}$) of x is made up of two principal parts corresponding to the increments of f caused by h' and h'' (respectively, is proportional to the increment caused by $\widetilde{h}$). These two assumptions, that the principal parts of increments of f are additive and that they are proportional for proportional increments of the variable, suggest that the principal part of the increment of the function must depend linearly on an increment of the argument. The definition of differentiability is based on an analysis of the difference between the increment of f and various linear functions $h \mapsto \langle A, h \rangle$. The differentiability of f means the existence of a linear function most accurately approximating the increment of f. Thus, we arrive at the following definition.

Definition Let $X \subset \mathbb{R}^m$, $a \in \text{Int}(X)$. A function f defined on X is said to be *differentiable* at the point a if there exists a vector A such that the difference

$$\alpha(h) = (f(a + h) - f(a)) - \langle A, h \rangle \qquad (1\,a)$$

(defined for all sufficiently small h) satisfies the condition

$$\alpha(h) \underset{h \to 0}{=} o(h), \qquad (1\,b)$$

that is, $\alpha(h)$ is an infinitesimal of higher order than h, i.e., $\frac{\alpha(h)}{\|h\|} \xrightarrow[h \to 0]{} 0$.

Writing the scalar product in terms of the coordinates of the vectors $A = (A_1, \ldots, A_m)$ and $h = (h_1, \ldots, h_m)$, we can restate the definition of differentiability as follows:

a function f defined on a set X is said to be *differentiable* at an (interior) point a of X if there exist numbers $A_1, \ldots, A_m$ such that the difference

$$\alpha(h) = (f(a + h) - f(a)) - (A_1 h_1 + \ldots + A_m h_m) \qquad (1')$$

(defined for all sufficiently small h) satisfies the condition $\alpha(h) \underset{h \to 0}{=} o(h)$.

According to this definition, we speak of the differentiability of f only at interior points of the set on which f is defined. However, if $m = 1$ and X is an interval, we keep it in the case where a is a boundary point belonging to X.

The condition $\alpha(h) \underset{h \to 0}{=} o(h)$ can also be rewritten as follows:

$$\alpha(h) = \|h\| \, \omega(h) \quad \text{where} \quad \omega(h) \xrightarrow[h \to 0]{} 0.$$

At the point $h = 0$, the function ω can be defined arbitrarily, but we always assume that $\omega(0) = 0$, to ensure the continuity of ω at 0.

Differentiability is a local property: it depends on the "behavior" of the function in an arbitrarily small neighborhood of the point under consideration; changing the function outside this neighborhood does not affect the differentiability at the point.

Importantly, a vector A satisfying the differentiability condition is determined uniquely. Indeed, assume that along with $(1\,a)$ we also have

$$f(a + h) - f(a) - \langle B, h \rangle = \beta(h) \quad \text{where } \beta(h) \underset{h \to 0}{=} o(h).$$

Then, in view of (1 b), for $h = t(A - B)$ we conclude (since $\|h\| = |t| \, \|A - B\|$) that $\beta(h) - \alpha(h) = t \, \|A - B\|^2 \underset{t \to 0}{=} o(t)$. This, clearly, implies that $B = A$.

This uniqueness justifies the following definition.

Definition If a function f is differentiable at a point a, then the principal linear part of its increment, i.e., the linear function

$$h \mapsto \langle A, h \rangle \quad (h \in \mathbb{R}^m)$$

satisfying conditions (1 a), (1 b), is called the *differential of* f at a and denoted by $d_a f$.

Clearly, using the coordinates of the vectors A and h, one can write $d_a f(h)$ as the sum $A_1 h_1 + \ldots + A_m h_m$. Thus, (1′) can be rewritten in the form

$$\alpha(h) = (f(a + h) - f(a)) - d_a f(h).$$

It follows immediately from the definition of differentiability that a linear combination of differentiable functions is differentiable and its differential is equal to the corresponding linear combination of the differentials of these functions, so differentiation is a linear operation. A translation operator, which replaces f with the function

$$x \mapsto f_c(x) = f(x + c) \quad \text{where } c \in \mathbb{R}^m,$$

preserves differentiability: if f is differentiable at a point a, then f_c is differentiable at the point $a - c$ and $d_a f = d_{a-c} f_c$. Besides, differentiability is preserved under linear changes of variables U: the function $g = f \circ U$ is differentiable at the point $b = U^{-1}(a)$. We leave it to the reader to verify this, using the definition and the inequality $\|U(h)\| \le \|U\| \, \|h\|$. In Section 5.3 below, we will establish a much more general result on the differentiation of composite functions.

Here is a simple but useful example of computing the differential. Let C be a symmetric $m \times m$ matrix and f be the corresponding quadratic form, i.e., the function $x \mapsto f(x) = \langle C x, x \rangle$ $(x \in \mathbb{R}^m)$. Let us find the differential of f at a point $a \in \mathbb{R}^m$. Since $\langle C h, a \rangle = \langle C a, h \rangle$ by the symmetry of C, we have

$$f(a + h) - f(a) = \langle C(a + h), a + h \rangle - \langle Ca, a \rangle = \langle Ca, h \rangle + \langle C h, a \rangle + \langle C h, h \rangle$$
$$\underset{h \to 0}{=} 2\langle Ca, h \rangle + o(h).$$

Hence, by definition, $d_a f(h) = 2\langle Ca, h \rangle$.

2.2 Differentiability: a Necessary Condition

One can easily deduce a necessary condition for differentiability directly from the definition.

Theorem *If a function f is differentiable at a point a, then at this point it is continuous and has finite partial derivatives with respect to all coordinates.*

Proof The continuity of f at a follows directly by definition, since, clearly, the increment of f caused by an infinitesimal increment of the argument is infinitesimal.

To prove the existence of partial derivatives, fix an arbitrary index $j = 1, \ldots, m$ and set $h = t\,e_j$ in (1 a), assuming that t is sufficiently small in absolute value, $t \neq 0$. Dividing by t, we obtain

$$\frac{f(a + te_j) - f(a)}{t} = A_j + \frac{\alpha(te_j)}{t}.$$

Since $\alpha(te_j) = o(t)$, the right-hand side tends to A_j as $t \to 0$, which proves that the partial derivative $\frac{\partial f}{\partial x_j}(a)$ exists and is finite.   □

As a by-product, we have established that $\frac{\partial f}{\partial x_j}(a) = A_j$ for $j = 1, \ldots, m$. This reaffirms the uniqueness of coefficients $A_1, \ldots, A_m$ satisfying conditions (1 a), (1 b).

2.3 Differentiability: a Sufficient Condition

For a function of several variables, in contrast to the case of one variable, the existence of finite partial derivatives is not sufficient for differentiability. Moreover, it does not even guarantee a much weaker property, that of continuity. For example, the function

$$f(x, y) = \begin{cases} \dfrac{2xy}{x^2 + y^2} & \text{if } x^2 + y^2 > 0, \\ 0 & \text{if } (x, y) = (0, 0) \end{cases}$$

has finite partial derivatives at every point (in particular, $f_x'(0,0) = f_y'(0,0) = 0$), but it is discontinuous at $(0,0)$, since $f(x,x) = 1$ for $x \neq 0$. The necessary condition for differentiability established above (the existence of partial derivatives together with the continuity of the function) is not sufficient either. To see this, consider the function $f(x,y) = \sqrt[3]{xy}$. It is continuous in the entire plane and has zero partial derivatives at the point $(0,0)$, but is not differentiable at this point. Indeed, otherwise we would have

$$\sqrt[3]{xy} = f(x,y) - f(0,0) \underset{x,y \to 0}{=} A_1 x + A_2 y + o\left(\sqrt{x^2 + y^2}\right).$$

Besides, $A_1 = f'_x(0,0) = 0$ and $A_2 = f'_y(0,0) = 0$, as mentioned after the proof of Theorem 2.2. Therefore,

$$\sqrt[3]{xy} \underset{x,y \to 0}{=} o\left(\sqrt{x^2 + y^2}\right).$$

But this relation fails as $x = y \to 0$.

To obtain differentiability, one must assume that the partial derivatives not only exist, but are continuous.

Theorem *Let f be a function that has finite partial derivatives with respect to all variables in a neighborhood of a point a. If these partial derivatives are continuous at a, then f is differentiable at this point.*

Note that the continuity of partial derivatives is not necessary for the differentiability of f (see Exercise 3).

Proof We may assume without loss of generality that the neighborhood of a in which the partial derivatives exist is an open parallelepiped P. Let $a + h \in P$, $h = (h_1, \ldots, h_m)$. By the mean value theorem (see Theorem 1.2), we have

$$f(a+h) - f(a) = \sum_{k=1}^{m} f'_{x_k}(a + h^{(k)}) h_k \qquad (2)$$

where $a + h^{(k)} \in P$ and $\|h^{(k)}\| \le \|h\|$ for all $k = 1, \ldots, m$. Set

$$f'_{x_k}(a + h^{(k)}) = f'_{x_k}(a) + \omega_k(h),$$

where $\omega_k(h) \xrightarrow[h \to 0]{} 0$, because the derivative f'_{x_k} is continuous at a. Substituting these equalities into (2), we obtain

$$f(a+h) - f(a) = \sum_{k=1}^{m} (f'_{x_k}(a) + \omega_k(h)) h_k = \sum_{k=1}^{m} f'_{x_k}(a) h_k + \alpha(h), \qquad (3)$$

where $\alpha(h) = \sum_{k=1}^{m} \omega_k(h) h_k$. Clearly,

$$\frac{\alpha(h)}{\|h\|} = \sum_{k=1}^{m} \omega_k(h) \frac{h_k}{\|h\|} \xrightarrow[h \to 0]{} 0,$$

since the functions $\omega_k(h)$ are infinitesimal and the quotients $h_k/\|h\|$ are bounded. Thus, the first term in the right-hand side of (3) is linear in h and the second term is an infinitesimal of higher order than $\|h\|$, which proves that f is differentiable at a. □

2.4 Gradient

Now we introduce an important vector associated with a differentiable function, which, as we will see later, largely determines its properties.

Definition Let f be a function that has finite partial derivatives with respect to all coordinates at a point a. The vector $\left(f'_{x_1}(a), \ldots, f'_{x_m}(a)\right)$ is called the *gradient* of f (at a) and denoted by grad $f(a)$.

   The notion of gradient appears in mathematical formulations of many physical laws. Suffice it to say that the gravitational field of a point mass at a coincides, up to a constant factor, with the gradient of the function $x \mapsto \frac{1}{\|x-a\|}$.

   Using the notion of gradient, the differentiability condition for a function f at a point a can be rewritten in the form

$$f(a+h) - f(a) \underset{h \to 0}{=} \langle \operatorname{grad} f(a), h \rangle + o(h). \tag{4}$$

Hence, the differential satisfies the equation

$$d_a f(h) = \langle \operatorname{grad} f(a), h \rangle.$$

Example (the gradient of a quadratic form) Returning to the example considered at the end of Section 2.1, we see that the gradient of a quadratic form with symmetric matrix C at a point a is the vector $2Ca$.

Exercises

1. Let f, g be functions differentiable at a point a. Show that

$$d_a(fg) = g(a)d_a f + f(a)d_a g$$

   and, therefore,

$$\operatorname{grad}(fg)(a) = g(a) \operatorname{grad} f(a) + f(a) \operatorname{grad} g(a).$$

2. Show that the function

$$f(x, y) = \begin{cases} \dfrac{xy^2}{x^2 + y^2} & \text{if } x^2 + y^2 > 0, \\ 0 & \text{if } (x, y) = (0, 0) \end{cases}$$

   is not differentiable at $(0, 0)$ (although it has partial derivatives at all points).
3. Check that the function

$$f(x) = \begin{cases} \|x\|^2 \sin \dfrac{1}{\|x\|^2} & \text{if } x \neq 0, \\ 0 & \text{if } x = 0 \end{cases}$$

is differentiable at the origin, although its partial derivatives are not bounded in any neighborhood of the origin.

4. For what positive values of p is the function $x \mapsto \|x\|^p$ differentiable at the origin? Find its differential at the origin and at a point $x_0 \neq 0$.

5. Is the function $x \mapsto e^{\|x\|} - \|x\|$ differentiable at the origin?

6. Identifying the set of square $m \times m$ matrices with the space $\mathbb{R}^{m^2}$, consider the function $X \mapsto f(X) = \det(X)$ in this space.

   a) Show that the value of the differential $d_I f$ (where I is the identity matrix) at a point H is equal to the trace of the matrix H (the sum of its diagonal elements).

   b) Check that $d_A f(H)$ coincides with the trace of the product $\widetilde{A} H$ where $\widetilde{A}$ is the transpose of the matrix consisting of the cofactors of the corresponding elements of A (if this matrix is invertible, then $\widetilde{A} = \det(A) A^{-1}$).

3 Directional Derivatives

Throughout this section, we assume that the functions under consideration are defined on some subset in $\mathbb{R}^m$.

3.1 The Derivative Along a Vector

The partial derivatives of a function f at a given point measure the rate of change of f in the directions parallel to the coordinate axes. It is useful to generalize this notion by considering increments of the argument along arbitrary directions.

Let f be a function defined on a subset $X \subset \mathbb{R}^m$, a be an interior point of X, and ℓ be an arbitrary nonzero vector. Consider the line $L = \{a + t\ell : t \in \mathbb{R}\}$ through a with direction vector ℓ. Clearly, $a + t\ell \in X$ for t sufficiently small in absolute value, say for $|t| < \delta$. For such t, we can consider the increment $f(a + t\ell) - f(a)$ of the function f along the line L. If the limit

$$\lim_{t \to 0} \frac{f(a + t\ell) - f(a)}{t}$$

exists, then it is called the *directional derivative* of f *along the vector* ℓ at the point a and denoted by $\frac{\partial f}{\partial \ell}(a)$. If $\|\ell\| = 1$, then it is also called the *derivative in the direction of* ℓ. Obviously, the partial derivative $f'_{x_k}(a)$ is the directional derivative along $\ell = e_k$ (as usual, e_k is a canonical basis vector): $f'_{x_k}(x) = \frac{\partial f}{\partial e_k}(a)$.

The directional derivative along ℓ measures the rate of change of f as its argument moves along the line L. More formally, it is the derivative at 0 of the function $t \mapsto \varphi(t) = f(a + t\ell)$, i.e.,

$$\frac{\partial f}{\partial \ell}(a) = \varphi'(0).$$

This formula implies that in the computation of the directional derivative along a fixed vector ℓ, one can use the ordinary differentiation rules for sums, products, and quotients.

3.2 Computing Directional Derivatives

Now we obtain a formula for directional derivatives of a differentiable function.

Theorem *Let $\ell \neq 0$ be an arbitrary vector. If a function f is differentiable at a point a, then*

$$\frac{\partial f}{\partial \ell}(a) = \langle \operatorname{grad} f(a), \ell \rangle. \tag{1}$$

We see that the derivative of a differentiable function along a vector depends linearly on this vector.

Proof By formula (4) from Section 2.4,

$$f(a + h) - f(a) \underset{h \to 0}{=} \langle \operatorname{grad} f(a), h \rangle + o(h).$$

Taking $h = t\ell$, we obtain

$$f(a + t\ell) - f(a) \underset{t \to 0}{=} t \langle \operatorname{grad} f(a), \ell \rangle + o(t),$$

which implies the desired equality after dividing by t and taking the limit.   □

Making an appropriate linear change of variables, we can reduce a directional derivative to one partial derivative (see Exercise 4). This allows one, in particular, to solve in quadratures a differential equation of the form

$$c_1 f'_{x_1}(x) + \ldots + c_m f'_{x_m}(x) = \varphi(x)$$

with a known function φ.

Note that the existence of directional derivatives of a function f along all vectors does not imply not only the differentiability of f, but even its continuity. The first fact can be verified by looking at the function f from Exercise 2 of the previous section:

$$f(x, y) = \begin{cases} \dfrac{xy^2}{x^2 + y^2} & \text{if } x^2 + y^2 > 0, \\ 0 & \text{if } (x, y) = (0, 0). \end{cases}$$

It is not differentiable, but has derivatives in all directions.

Surprisingly, a function that at a point a has derivatives in all directions is not necessarily continuous at a (though, of course, its restriction to every line passing through a is continuous at this point). The reason is that for a function to be continuous, its increments must be small for all sufficiently small increments of the argument in all directions simultaneously, which does not follow from the smallness of increments along all lines. To see this, look at the following function g, which vanishes at all points of $\mathbb{R}^2$ except a parabola with the vertex removed:

$$g(x, x^2) = 1 \quad \text{if } x \neq 0, \quad g(0, 0) = g(x, y) = 0 \quad \text{if } y \neq x^2.$$

It is clear that g is discontinuous at the origin and that every line passing through $(0, 0)$ contains at most one point at which g does not vanish. It follows that $\frac{\partial g}{\partial \ell}(0, 0) = 0$ for every $\ell \neq 0$.

3.3 A Coordinate-Free Description of the Gradient

Using directional derivatives, we can establish a coordinate-free characteristic property of the gradient. In the theorem below, we exclude the trivial case of zero gradient, when (by Theorem 3.2) the partial derivatives in all directions vanish.

Theorem *Let f be a function defined on a set X and differentiable at a point $a \in \text{Int}(X)$, $\text{grad } f(a) \neq 0$, and $\ell_0 = \frac{\text{grad } f(a)}{\|\text{grad } f(a)\|}$. Then*

$$\|\text{grad } f(a)\| = \frac{\partial f}{\partial \ell_0}(a) > \left|\frac{\partial f}{\partial \ell}(a)\right| \quad \text{for all } \ell \neq \pm\ell_0, \|\ell\| = 1. \tag{2}$$

Thus, the theorem gives a characterization of the gradient that does not depend on the coordinate system: the nonzero gradient indicates the direction of steepest ascent of the function, and its norm is equal to the maximum growth rate. This property underlies, in particular, a number of computational methods ("gradient methods").

Proof Below we assume, as in the statement of the theorem, that ℓ is a normalized vector: $\|\ell\| = 1$. By Theorem 3.2 and Cauchy's inequality,

$$\left|\frac{\partial f}{\partial \ell}(a)\right| = \left|\langle \text{grad } f(a), \ell \rangle\right| \leq \|\text{grad } f(a)\| \cdot \|\ell\| = \|\text{grad } f(a)\|,$$

and the inequality is strict if the vectors ℓ and $\text{grad } f(a)$ are linearly independent (see Section 0.1). The last condition is equivalent to $\ell \neq \pm\ell_0$. Hence, for such ℓ we have

$$\left|\frac{\partial f}{\partial \ell}(a)\right| < \|\text{grad } f(a)\|.$$

Besides, obviously,

$$\frac{\partial f}{\partial \ell_0}(a) = \langle \operatorname{grad} f(a), \ell_0 \rangle = \left\langle \operatorname{grad} f(a), \frac{\operatorname{grad} f(a)}{\|\operatorname{grad} f(a)\|} \right\rangle = \|\operatorname{grad} f(a)\|,$$

which, together with the previous inequality, completes the proof.   □

Exercises

1. Let $x, y \in \mathbb{R}^m$, $\ell = y - x \neq 0$. Prove the following generalization of the mean value theorem: if a function f is differentiable at all points of the interval

$$[x, y] = \{(1 - t)x + ty : 0 \le t \le 1\},$$

then there exists a point $c \in [x, y]$ such that

$$f(y) - f(x) = \frac{\partial f}{\partial \ell}(c) = \langle \operatorname{grad} f(c), y - x \rangle.$$

2. Show that if $\ell_1 = c\ell \neq 0$ ($\ell \in \mathbb{R}^m, c \in \mathbb{R}$), then $\frac{\partial f}{\partial \ell_1}(a) = c \frac{\partial f}{\partial \ell}(a)$ (provided that the derivative along ℓ exists).

3. Let f be a function differentiable at a point $a \in \mathbb{R}^2$, and let $\frac{\partial f}{\partial \ell_1}(a) = A$, $\frac{\partial f}{\partial \ell_2}(a) = B$ where $\ell_1 = (1, 2)$, $\ell_2 = (2, 3)$. Find $\frac{\partial f}{\partial \ell}(a)$ for $\ell = (1, 3)$.

4. Let f be a function differentiable at a point $a \in \mathbb{R}^m$ and $\ell = (c_1, \ldots, c_m)$ be a unit vector. Find the image of the sum $c_1 f'_{x_1}(a) + \ldots + c_m f'_{x_m}(a)$ under the change of variables $x = U(y)$ where U is an orthogonal matrix with first column ℓ.

5. Show that the continuous function $f(x, y, z) = (xyz)^{\frac{1}{3}}$ at the origin has finite derivatives in all directions, but is not differentiable.

6. Let

$$f(x, y) = \frac{xy^2}{x^2 + y^2}, \quad g(x, y) = \frac{x^3}{x^2 + y^2} \quad \text{if } x^2 + y^2 > 0,$$

$$f(0, 0) = g(0, 0) = 0.$$

Show that these functions are continuous and at the origin have finite derivatives in all directions, but are not differentiable.

7. Show that the function

$$\varphi(x, y) = \begin{cases} \dfrac{xy^3}{x^2 + y^4} & \text{if } x^2 + y^2 > 0, \\ 0 & \text{if } (x, y) = (0, 0) \end{cases}$$

is continuous everywhere and at the origin has finite derivatives in all directions, but is not differentiable. Verify that, in contrast to the functions from the previous exercise, the derivatives of φ along a vector depend linearly on this vector (they are identically zero).

4 The Tangent Plane to a Level Surface

4.1 Level Sets

We often encounter sets of constancy of various functions. For example, listening to a weather forecast, we hear about isotherms and isobars, which are lines connecting points with the same temperature and the same pressure, respectively. Looking at a geographical map, we see isobaths and isohypses, which are lines of equal depths and equal heights. The term "line" here is rather loose: for example, plateau regions lying at the same height must belong to one such "line." Avoiding for the moment the terms "line" and "surface," we introduce the following general definition.

Definition Let F be a function defined on a set $X \subset \mathbb{R}^m$ and C be an arbitrary number. The set
$$X_C = \{x \in X \colon F(x) = C\}$$
is called a *level set* of F.

In other words, a level set is the preimage $X_C = F^{-1}(\{C\})$ of a one-point set $\{C\}$, for which we also use the shorter notation $F^{-1}(C)$.

In the two-dimensional case, one also uses the term "level curve", and in the case of higher dimensions, "level surface." As we have already mentioned, one should be aware that these terms are loose: if F is constant on a subset in X, then this subset is entirely contained in the corresponding level "curve" (or "surface"), even if it has, for example, interior points. Level curves and surfaces are frequent in descriptions of geometric objects. For example, straight lines (in the plane) are level curves of functions of the form $F(x, y) = \alpha x + \beta y$, circles are level curves of functions of the form $F(x, y) = (x - \alpha)^2 + (y - \beta)^2$, ellipsoids are level surfaces of positive definite quadratic forms, etc.

Clearly, every point $a \in X$ belongs to some level set (with $C = F(a)$), and different level sets are disjoint. Thus, the domain of definition of F becomes sort of "stratified" into level sets (see Fig. 1).

Level sets give us considerable information about the function in a visualized form. For example, parts of the domain of definition where level sets are close together correspond to regions of rapid growth of F, while parts where level sets are sparse correspond to regions where F changes more slowly.

Let us consider some examples.

Example 1 Let $X = \mathbb{R}^2$, $F(x, y) = x^2 + y^2$. For $C > 0$, the level set X_C is the circle of radius $\sqrt{C}$ (centered at the origin). For $C < 0$, the level set is empty, and for $C = 0$, it contains only the origin.

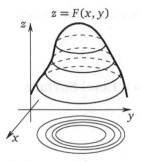

$$z = F(x, y)$$

Fig. 1

Example 2 Let $X = \mathbb{R}^2$, $F(x, y) = (x^2 + y^2)(x^2 + y^2 - 1)$. The level set with $C = 0$ consists of the unit circle and its center. Thus, level "lines" may contain not only "good" curves, but also isolated points. Later, we will find a condition that guarantees the absence of isolated points in a level set.

Example 3 Let φ be the function defined as follows: $\varphi(t) = e^{-1/t}$ if $t > 0$, $\varphi(t) = 0$ if $t \leq 0$. It is well known that φ is infinitely differentiable on $\mathbb{R}$. Set $F(x, y) = \varphi(x) \cdot \varphi(y)$. Then the zero level set of F coincides with the complement of the set $\{(x, y) \in \mathbb{R}^2 : x > 0, y > 0\}$.

Example 4 Let us describe the level sets $F(x, y) = C$ of the function

$$F(x, y) = (x^2 + y^2 + a^2)^2 - 4a^2 x^2 \quad (a > 0).$$

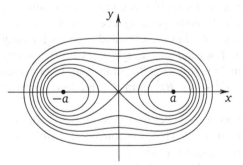

Fig. 2

A simple calculation shows that $F(x, y)$ is the squared product of the distances from a point (x, y) to the points $(a, 0)$ and $(-a, 0)$. For $C = 0$, the level set consists of the two points $(a, 0)$ and $(-a, 0)$. For $C > 0$, the level sets are called Cassini[1] ovals, and

---

[1] Giovanni Domenico **Cassini** (1625–1712) was a French astronomer.

the points $(\pm a, 0)$ are called their foci. The curve obtained for $C = a^4$ is called the lemniscate of Bernoulli[2].

For $0 < C < a^4$, the Cassini oval is disconnected and consists of two parts lying in the left and right half-planes. For $C \geq C_0$, it is a convex curve, while for $a^4 \leq C < C_0$, it is a nonconvex curve (which has a "waist," see Fig. 2). We leave it to the reader to find C_0.

4.2 The Graph as a Level Set

An important special case of a level set is the graph of a function. Now we discuss this case in more detail. Let f be a function defined on a set $X \subset \mathbb{R}^m$. By definition, its *graph* Γ_f is the set

$$\Gamma_f = \{(x, y) \in \mathbb{R}^{m+1} : x \in X, \; y = f(x)\}.$$

Let

$$F(x, y) = y - f(x) \quad \text{for } x \in X, y \in \mathbb{R}.$$

Thus, the function F is defined on the set $X \times \mathbb{R}$ and

$$(x, y) \in \Gamma_f \iff y = f(x) \iff F(x, y) = 0,$$

so it is clear that Γ_f coincides with the zero level set of F.

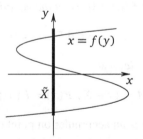

Fig. 3

Consider a slightly more general situation. Fix a number k $(1 \leq k \leq m)$ and assume that a function $\widetilde{f}$ is defined on some subset $\widetilde{X}$ of the coordinate plane $x_k = 0$ in $\mathbb{R}^{m+1}$. Writing $x = (x_1, \ldots, x_{k-1}, x_{k+1}, \ldots, x_m, x_{m+1})$, let

$$\widetilde{\Gamma}_{\widetilde{f}} = \{(x_1, \ldots, x_{k-1}, \widetilde{f}(x), x_{k+1}, \ldots, x_m, x_{m+1}) : x \in \widetilde{X}\}$$

---

[2] Jacob **Bernoulli** (1654–1705) was a Swiss mathematician.

The set $\widetilde{\Gamma}_{\tilde{f}}$ (see Fig. 3 for the two-dimensional case) will be called a *wide-sense graph*.

Obviously, a wide-sense graph is also a level set of a function defined in the cylinder with base $\widetilde{X}$ and generator parallel to the x_k-axis:

$$\widetilde{\Gamma}_{\tilde{f}} = \left\{ (x_1, \ldots, x_k, \ldots, x_{m+1}) \in \mathbb{R}^{m+1} : x_k - \widetilde{f}(x) = 0, \ x \in \widetilde{X}, x_k \in \mathbb{R} \right\}.$$

4.3 Tangent Plane

Let M be an arbitrary subset in $\mathbb{R}^m$, $m \geq 2$, containing an accumulation point a. The definition of a tangent plane relies on the idea of a plane that "closely approaches" M in a neighborhood of a.

Definition A plane H passing through a is called a *tangent plane* to M at this point if

$$\text{dist}(x, H) = o(\|x - a\|) \quad \text{as } x \to a, x \in M,$$

i.e., $\frac{\text{dist}(x,H)}{\|x-a\|} \to 0$ as $x \to a$, $x \in M$ (in the two-dimensional case, we speak of a tangent line, or simply a *tangent* to M).

Note the role of the last condition: as we define a tangent plane to M, the distances from x to H and a must be compared, of course, only for $x \in M$.

Now we show that under natural assumptions a level set has a tangent plane.

Theorem *Let F be a function continuous in a neighborhood of a point $a \in \mathbb{R}^m$ and differentiable at a, with $\text{grad}\, F(a) \neq 0$. Then the equation*

$$\langle x - a, \text{grad}\, F(a) \rangle = 0$$

defines a tangent plane to the level set

$$M = \{ x \in X : F(x) = F(a) \}.$$

Proof First, we verify that a is an accumulation point of M. Let F be continuous in a ball $B(a, r)$ and $v = \text{grad}\, F(a)$. Then, by Theorem 3.2,

$$0 < \| \text{grad}\, F(a) \|^2 = \frac{\partial F}{\partial v}(a) = \lim_{t \to 0} \frac{F(a + tv) - F(a)}{t}.$$

Hence, for sufficiently small positive t we have $a \pm tv \in B(a, r)$ and

$$F(a - tv) < F(a) < F(a + tv).$$

Connect the points $a \pm tv$ with a curve that does not pass through a (for instance, the semicircle of radius $t\|v\| < r$ centered at a). By the intermediate value theorem, it contains a point x_t such that $F(x_t) = F(a)$, i.e., $x_t \in M$. Thus, arbitrarily close to a there are points from M different from a, so a is an accumulation point of M.

Now, we claim that the plane H defined by the equation $\langle x - a, v \rangle = 0$ is tangent to M at the point a. Indeed, since F is differentiable, we have

$$F(x) - F(a) = \langle x - a, v \rangle + \alpha(x) \quad \text{where} \quad \frac{\alpha(x)}{\|x - a\|} \xrightarrow[x \to a]{} 0.$$

For $x \in M$, this gives $0 = F(x) - F(a) = \langle x - a, v \rangle + \alpha(x)$. Thus,

$$|\langle x - a, v \rangle| = |\alpha(x)| = o(\|x - a\|) \quad \text{as } x \to a, x \in M.$$

However, by Lemma 0.12, the left-hand side is proportional to the distance from x to H, hence $\text{dist}(x, H) = o(\|x - a\|)$ as $x \to a$, $x \in M$. □

As we have mentioned in Section 4.2, the graph of a function is a special case of a level set. Keeping the same notation, we see that if a function f defined on a subset of $\mathbb{R}^m$ is differentiable at a point a, then the function $F(x, y) = f(x) - y$ is differentiable at the point $c = (a, f(a))$ and the tangent plane to the zero level set of F (i.e., to the graph of f) is given by the equation

$$\langle \text{grad } F(c), z - c \rangle = 0$$

(hereafter, $z = (x, y)$ where $x \in \mathbb{R}^m$, $y \in \mathbb{R}$). We now use the special nature of the function F to transform this equation. Since $\text{grad } F = (\text{grad } f, -1)$, we can rewrite it, letting $x = (x_1, \ldots, x_m)$ and $a = (a_1, \ldots, a_m)$, in the form

$$\sum_{k=1}^{m} \frac{\partial f}{\partial x_k}(a) (x_k - a_k) + (-1)(y - f(a)) = 0,$$

or

$$y - f(a) = \sum_{k=1}^{m} \frac{\partial f}{\partial x_k}(a) (x_k - a_k) = \langle \text{grad } f(a), x - a \rangle. \tag{1}$$

Note that the plane defined by (1) is not parallel to the last coordinate axis. Such planes will be called *oblique*. Thus, we have the following corollary.

Corollary (tangent plane to a graph) *If f is a function differentiable at a point $a = (a_1, \ldots, a_m)$, then the graph of f at the point $c = (a, f(a))$ has an oblique tangent plane, which is defined by equation (1).*

Remark So far, we have not addressed the issue of whether a tangent plane is unique. As one can easily check, in general the answer is negative. For example, consider the surface obtained by rotating the graph of the function $y = \sqrt{x}$ ($x \geq 0$) about the y-axis; then every plane containing this axis is tangent to this surface at the origin.

Without going into detail, we note, however, that the situation changes dramatically if we speak of an oblique tangent plane to the graph of a function f defined at least in a neighborhood of a point a. In this case, a tangent plane to the graph at the point $c = (a, f(a))$ is unique (if it exists). We leave the proof to the reader. The above corollary on the existence of an oblique tangent plane to the graph of a differentiable function can be reversed: the existence of an oblique tangent plane to the graph at the point c implies that the function is differentiable at the point a and the vector $(\text{grad } f(a), -1)$ is orthogonal to the tangent plane, which simultaneously proves the uniqueness of an oblique tangent plane to the graph.

Exercises

1. Find all level sets of the function F if
   a) $F(x, y) = \frac{xy}{x+y}$ for $(x, y) \in \mathbb{R}^2$, $x + y \neq 0$;
   b) $F(x, y) = \frac{x^2+y^2}{1+y^2}$ for $(x, y) \in \mathbb{R}^2$.

2. Show that the function

$$F(x, y) = \begin{cases} 2y - x^2 - y^2 & \text{if } x^2 + y^2 \leq 2y, \\ 2y & \text{if } y \leq 0, \\ 0 & \text{if } 0 \leq 2y \leq x^2 + y^2 \end{cases}$$

   is differentiable at $(0, 0)$ and $\text{grad } F(0, 0) \neq 0$. Find the zero level set of F and the tangent to this set at the point $(0, 0)$.

3. Find all level sets of the function from the previous exercise.

4. Sketch the level sets and the graph of the function $f(x, y) = x^2 + (1 - x)^3 y^2$.

   HINT. To study the graph, it is useful to examine the dynamics of its cross sections cut by the planes $x = \text{const}$ in the cases where x varies from $-\infty$ to 1 and from 1 to $+\infty$.

5. Prove that a tangent plane to the graph of a differentiable function is unique.

6. Let f be a function defined on a subset $X \subset \mathbb{R}^m$. Show that the existence of an oblique tangent plane to the graph of f at a point $(a, f(a))$, where $a \in \text{Int}(X)$, implies that f is differentiable at a and the vector $(\text{grad } f(a), -1)$ is orthogonal to the tangent plane.

7. Show that a tangent in the plane is unique (if it exists).

8. Show that if a subset of $\mathbb{R}^m$ ($m \geq 3$) has more than one tangent plane at a point x, then it has infinitely many tangent planes at x.

9. Let φ be a function of several variables differentiable at a point a and f be a function that is constant on the graph of φ. Show that if f is differentiable at the point $(a, \varphi(a))$, then its derivative at this point in any direction tangent to the graph of φ (i.e., orthogonal to the normal to the tangent plane) vanishes.

5 Differentiable Maps

5.1 Definition of Differentiability

So far, we have only extended the notion of differentiability for functions of one variable to functions of several variables. The next important step is to extend it to arbitrary maps. The main motivation behind this is the very importance of the notion of a differentiable map. However, as we will see later, adopting a more general point of view, related to the notion of a differential map rather than a differentiable function, leads to a better understanding of the common nature of basic intrinsic properties of these objects.

In the following definition, and throughout this section, we assume that $X \subset \mathbb{R}^m$ and $T \colon X \to \mathbb{R}^n$.

Definition Let a be an interior point of X. A map $T \colon X \to \mathbb{R}^n$ is said to be *differentiable* at the point a if there exists a linear map $A \colon \mathbb{R}^m \to \mathbb{R}^n$ such that

$$T(a + h) - T(a) = A(h) + \alpha(h) \quad \text{where} \quad \alpha(h) \underset{h \to 0}{=} o(h) \tag{1}$$

(the last condition means that $\frac{\alpha(h)}{\|h\|} \underset{h \to 0}{\longrightarrow} 0$).

As in the definition of a differentiable function (see Section 2.1), we speak of differentiability only at interior points of the domain of definition of T. However, here we again keep this definition in the case where $m = 1$, X is an interval, and a is a boundary point of X. The reader can easily check that all subsequent theorems remain valid in this case.

We see from the definition that the differentiability of a map at a given point is a local property: it is not affected by changing the map outside an arbitrarily small neighborhood of this point.

As in Section 2.1, the infinitesimal function α can be written in the form

$$\alpha(h) = \|h\| \, \omega(h) \quad \text{where} \quad \omega(h) \underset{h \to 0}{\longrightarrow} 0$$

(we still assume that $\omega(0) = 0$).

Since a linear map is continuous, it follows from (1) that

$$T(a + h) - T(a) \underset{h \to 0}{\longrightarrow} 0$$

and, therefore, as in the case of functions, the continuity of T at a is a necessary condition for T to be differentiable at a.

The map A from (1) is uniquely determined. Indeed, if a linear map $B \colon \mathbb{R}^m \to \mathbb{R}^n$ satisfies

$$T(a + h) - T(a) = B(h) + \beta(h) \quad \text{where} \quad \beta(h) \underset{h \to 0}{=} o(h),$$

then, subtracting this equality from (1), we see that (for sufficiently small h) $(A - B)(h) + (\alpha(h) - \beta(h)) = 0$, or, setting $C = A - B$,

$$C(h) + o(h) = 0 \quad \text{for all sufficiently small } h.$$

Set $h = tx$ where x is an arbitrary vector and $t \in \mathbb{R}$. Dividing by $t \neq 0$ yields

$$C(x) + \frac{o(tx)}{t} = 0 \quad \text{for all sufficiently small } t.$$

Now, taking the limit as $t \to 0$ shows that $C(x) = 0$ for every x, i.e., $A = B$.

The above argument justifies the following definition.

Definition A linear map A satisfying (1) is called the *differential* of the map T at the point a. It is denoted by $d_a T$. The matrix of this map in the canonical bases of the spaces $\mathbb{R}^m$ and $\mathbb{R}^n$ is denoted by $T'(a)$. It will be called the *derivative* of T at a.

Note that the differential of a linear map coincides with this map.

For functions, the above definition of differentiability coincides with that given in Section 2. The definitions of differentials as linear functions representing the principal part of the increment coincide too.

5.2 The Jacobian Matrix

Here we establish a relation between the differentiability of a map and the differentiability of its coordinate functions and find the matrix representation of the differential.

Let $f_1, \ldots, f_n$ be the coordinate functions of a map T. Obviously, the coordinates of its increment $T(a + h) - T(a)$ coincide with the increments of the coordinate functions. If T is differentiable at a point a and its differential is given by a matrix with elements a_{jk}, then, writing (1) in coordinates, we see that for every j

$$f_j(a + h) - f_j(a) = \sum_{k=1}^{m} a_{jk} h_k + \alpha_j(h), \quad \alpha_j(h) \underset{h \to 0}{=} o(h), \tag{2}$$

where $h_1, \ldots, h_m$ and $\alpha_1(h), \ldots, \alpha_n(h)$ are the coordinates of h and $\alpha(h)$, respectively. The relation $\alpha_j(h) \underset{h \to 0}{=} o(h)$ follows from the inequality $|\alpha_j(h)| \leq \|\alpha(h)\|$. Thus, the differentiability of a map implies the differentiability of its coordinate functions.

Obviously, the converse is also true: if all coordinate functions are differentiable, then we have (2) for every j, which implies (1). Besides, it also follows from (2) that the rows of the matrix $T'(a)$ consist of the partial derivatives of the coordinate functions: $(a_{j1}, \ldots, a_{jm}) = f_j'(a)$ (see the end of Section 2.2). Thus,

$$T'(a) = \begin{pmatrix} \dfrac{\partial f_1}{\partial x_1}(a) & \cdots & \dfrac{\partial f_1}{\partial x_m}(a) \\ \cdots\cdots\cdots\cdots\cdots\cdots \\ \dfrac{\partial f_n}{\partial x_1}(a) & \cdots & \dfrac{\partial f_n}{\partial x_m}(a) \end{pmatrix}.$$

The matrix in the right-hand side is called the *Jacobian*[1] *matrix* of the map T, or the Jacobian matrix of the system of functions $f_1, \ldots, f_n$ (the latter term is also used in cases where the system of functions is not originated from any map). In the case $m = n$, the Jacobian matrix is square. Its determinant

$$\det T'(a) = \det\left(\frac{\partial f_j}{\partial x_k}(a)\right)_{j,k \le m}$$

is called the *Jacobian* of T.

If a map $T: O \to \mathbb{R}^n$ is differentiable on a set $E \subset O$ (i.e., differentiable at every point of this set), then with T we can associate a new map T' that sends each point $a \in E$ to the Jacobian matrix of T at this point: $a \mapsto T'(a)$. Since the set of $n \times m$ matrices can be regarded as a Euclidean space (see Section 0.13), we may, for example, speak of the continuity (on E) of T' (which, as we know, is equivalent to the coordinatewise continuity of T', meaning the continuity on E of the partial derivatives of the coordinate functions of T).

We draw the reader's attention to the difference between the gradient grad $f(a)$ of a function f at a point a and the matrix representation of its differential at a, which, according to the general definition, is denoted by $f'(a)$. In the special case under consideration (for $n = 1$), this matrix representation is the one-row Jacobian matrix

$$f'(a) = \left(\frac{\partial f}{\partial x_1}(a), \ldots, \frac{\partial f}{\partial x_m}(a)\right).$$

In terms of this matrix, the value of $d_a f$ at a vector h can be written as $d_a f(h) = f'(a) \cdot h$. The right-hand side of this equality is the product of the row vector $f'(a)$ and the vector h, which, according to the rules of matrix algebra, must be written as a column vector.

Meanwhile, according to our definition, the gradient grad $f(a)$, in terms of which the value of $d_a f$ at h can be written as $d_a f(h) = \langle \text{grad } f(a), h\rangle$, is a vector of the same space in which the function and its differential are defined. Thus, in calculations with the matrix representation of the differential, the gradient, as well as other vectors, must be written as a column vector (which is the transposed row vector $f'(a)$).

Note that if $T = \gamma$ is a path defined on an interval $[p, q]$ and differentiable at a point $a \in [p, q]$, then $\gamma'(a)$ is the vector whose coordinates are the derivatives of the coordinate functions of γ at a. It is called the *tangent vector* to the path γ at the point a.

Interpreting the argument of γ as time and the difference $\gamma(a + h) - \gamma(a)$ as the displacement of the point $\gamma(u)$ over time h, we may say that $\frac{1}{h}(\gamma(a + h) - \gamma(a))$ is

---

[1] Carl Gustav **Jacobi** (1804–1851) was a German mathematician.

the average velocity, while the limit

$$\gamma'(a) = \lim_{h \to 0} \frac{1}{h}(\gamma(a+h) - \gamma(a))$$

is the "instantaneous velocity" of the point $\gamma(u)$ at time a.

5.3 Differentiation of Composite Functions

Here we extend the basic differentiation rules to maps. First of all, note that the differential of a linear combination of maps coincides with the linear combination of the differentials, as for functions of one variable. We leave it to the reader to verify this. The following theorem shows that the ordinary rule for differentiating composite functions can be extended to maps in a natural way.

Theorem *Let $R = S \circ T$ be the composition of maps T and S. If T is differentiable at a point a and S is differentiable at the point $b = T(a)$, then R is differentiable at the point a and*

$$R'(a) = S'(b) \cdot T'(a). \tag{3}$$

Formula (3) can be rewritten in an equivalent form in terms of differentials:

$$d_a R = d_b S \circ d_a T.$$

Proof Let A and B be the differentials of the maps T and S at the points a and b, respectively. Then for sufficiently small h and η,

$$T(a+h) - T(a) = A(h) + \|h\|\,\omega(h) \qquad \text{where } \omega(h) \xrightarrow[h \to 0]{} 0,$$

$$S(b+\eta) - S(b) = B(\eta) + \|\eta\|\,\widetilde{\omega}(\eta) \qquad \text{where } \widetilde{\omega}(\eta) \xrightarrow[\eta \to 0]{} 0.$$

According to our convention, $\omega(0) = 0$ and $\widetilde{\omega}(0) = 0$.

Assuming that h is sufficiently small, we use the formula for the increment of S with $\eta = T(a+h) - T(a)$ to obtain

$$R(a+h) - R(a) = B(T(a+h) - T(a)) + \|T(a+h) - T(a)\|\,\widetilde{\omega}(T(a+h) - T(a))$$

$$= B \circ A(h) + \gamma(h),$$

where $\gamma(h) = \|h\|\,B(\omega(h)) + \|T(a+h) - T(a)\|\,\widetilde{\omega}(T(a+h) - T(a))$. To complete the proof, it remains to show that $\gamma(h)$ is an infinitesimal of higher order than h. Indeed,

$$\frac{\|\gamma(h)\|}{\|h\|} \le \|B\| \cdot \|\omega(h)\| + \frac{\|T(a+h) - T(a)\|}{\|h\|}\,\|\widetilde{\omega}(T(a+h) - T(a))\|.$$

Since

$$\|T(a+h) - T(a)\| \le \|A(h)\| + \|h\| \cdot \|\omega(h)\| = O(\|h\|) \quad \text{as } h \to 0$$

and $\widetilde{\omega}(T(a+h) - T(a)) \xrightarrow[h \to 0]{} 0$, we see that $\dfrac{\gamma(h)}{\|h\|} \xrightarrow[h \to 0]{} 0$. $\qquad\square$

Consider two special cases of (3).

I. Let $T : X \to \mathbb{R}^n$ be a map with coordinate functions $f_1, \ldots, f_n$ defined on a set X in $\mathbb{R}^m$ and differentiable at a point a, and $g = S$ be a function defined on a set Y containing $T(X)$ and differentiable at the point $b = T(a)$. Consider the new function

$$F(x) = g(T(x)) = g(f_1(x), \ldots, f_n(x)) \quad (x \in X).$$

By the above theorem, the function F is differentiable at a and $F'(a) = g'(b) \cdot T'(a)$. Writing this equation in coordinates, we see that for every $k = 1, \ldots, m$,

$$\frac{\partial F}{\partial x_k}(a) = \sum_{j=1}^{n} \frac{\partial g}{\partial y_j}(b) \cdot \frac{\partial f_j}{\partial x_k}(a) = \sum_{j=1}^{n} \frac{\partial g}{\partial y_j}(T(a)) \cdot \frac{\partial f_j}{\partial x_k}(a). \tag{4}$$

Formula (4) is extremely important, it is constantly used to compute partial derivatives. Note, however, that the general point of view leading to (3) better reveals an analogy with the simplest case of functions of one variable.

II. Now let $\gamma = T$ be a path defined on some interval. In this case, the composition $\Phi = S \circ \gamma$ is also a path. Under the assumptions of the theorem, it is differentiable at a and $\Phi'(a) = S'(\gamma(a)) \cdot \gamma'(a)$, i.e.,

$$\Phi'(a) = d_b S(\gamma'(a)), \quad \text{where } b = \gamma(a).$$

Thus, the differential of S sends the tangent vector to γ at the point $\gamma(a)$ to the tangent vector to Φ at the point $S(\gamma(a))$.

If S is a function, then Φ is a one-dimensional path, i.e., a function too. For its derivative at a, we obtain the formula

$$\Phi'(a) = S'(b) \cdot \gamma'(a) = \langle \operatorname{grad} S(b), \gamma'(a) \rangle. \tag{4'}$$

Let us consider two examples illustrating applications of the obtained results.

Example 1 (generalized fundamental theorem of calculus) Let f be a function with continuous partial derivatives in a domain O, and let a and b be points in O. We will show that, using a piecewise smooth path connecting these points, the corresponding increment of f can be represented as an integral. More exactly, if a, b are connected in O by a piecewise smooth path γ defined on an interval $[p, q]$ (i.e., $\gamma(p) = a$, $\gamma(q) = b$, $\gamma(t) \in O$ for all $t \in [p, q]$), then

$$F(b) - F(a) = \int_p^q \langle \operatorname{grad} F(\gamma(t)), \gamma'(t) \rangle \, dt.$$

Indeed, set $\Phi(t) = F(\gamma(t))$ for $t \in [p, q]$, so $\Phi(p) = F(a)$ and $\Phi(q) = F(b)$. Then, as follows from (4′) (with S replaced by F and a replaced by t), we obtain

$$\Phi'(t) = \langle \operatorname{grad} F(\gamma(t)), \gamma'(t) \rangle$$

for all points t except, possibly, finitely many. Hence, the fundamental theorem of calculus implies

$$F(b) - F(a) = \Phi(q) - \Phi(p) = \int\limits_p^q \Phi'(t)\, dt = \int\limits_p^q \langle \operatorname{grad} F(\gamma(t)), \gamma'(t) \rangle\, dt,$$

as required.

Example 2 (Euler's[2] homogeneous function theorem) Now we consider an example related to homogeneous functions. Recall that a set K is called a cone (with vertex at the origin) if it contains with every point x the points tx for all $t > 0$. A function f defined on a cone K is said to be *homogeneous of degree p* (where p is a fixed number) if $f(tx) = t^p f(x)$ for every $x \in K$ and every $t > 0$. If the cone K is open, $0 \notin K$, and F is a homogeneous function of degree p that is differentiable in K, then it satisfies Euler's homogeneous function theorem: for every $x \in K$,

$$x_1 f'_{x_1}(x) + \ldots + x_m f'_{x_m}(x) = p f(x).$$

To prove this, it suffices to differentiate the equality $f(tx) = t^p f(x)$ with respect to t at the point $t = 1$.

Clearly, the converse is also true: if a function is differentiable in a cone and satisfies Euler's equation, then it is homogeneous (since the derivative of the function $t \mapsto \frac{1}{t^p} f(tx)$ is identically zero).

5.4 Lagrange's Inequality

Here we prove two results generalizing the classical Lagrange's mean value theorem for differentiable functions.

Recall that for $x, y \in \mathbb{R}^m$ by $[x, y]$ we denote the closed interval (line segment) with endpoints x, y, i.e., the set $\{(1 - t)x + ty : 0 \le t \le 1\}$.

Theorem 1 *If a map T is differentiable on $[x, y]$, then there exists a point $z \in [x, y]$ such that*

$$\|T(y) - T(x)\| \le \|T'(z)\|\, \|y - x\|.$$

Proof Set $\Delta = T(y) - T(x)$. To estimate the length of this vector, we introduce an auxiliary function

---

[2] Leonhard **Euler** (1707–1783) was a Swiss mathematician.

$$\varphi(t) = \langle T(x + t(y - x)), \Delta \rangle, \quad t \in [0, 1].$$

Clearly, $\varphi(0) = \langle T(x), \Delta \rangle$, $\varphi(1) = \langle T(y), \Delta \rangle$ and, therefore,

$$\|\Delta\|^2 = \langle T(y) - T(x), \Delta \rangle = \varphi(1) - \varphi(0).$$

By Theorem 5.3, the function φ is differentiable and

$$\varphi'(t) = \langle T'(x + t(y - x))(y - x), \Delta \rangle.$$

By the mean value theorem, there exists $c \in (0, 1)$ such that $\varphi(1) - \varphi(0) = \varphi'(c)$. Hence, for $z = x + c(y - x)$ we have

$$\|\Delta\|^2 = \varphi(1) - \varphi(0) = \varphi'(c) = \langle T'(z)(y - x), \Delta \rangle$$
$$\leq \|T'(z)(y - x)\| \, \|\Delta\| \leq \|T'(z)\| \, \|y - x\| \, \|\Delta\|.$$

It follows immediately that the point z satisfies the required condition. $\quad\square$

Corollary 1 *Under the assumptions of the theorem,*

$$\|T(y) - T(x) - T'(x)(y - x)\| \leq \sup_{z \in [x,y]} \|T'(z) - T'(x)\| \, \|y - x\|.$$

To prove this, apply the theorem to the map $z \mapsto T(z) - T'(x)(z)$, using the fact that the differential of a linear map coincides with this map.

Corollary 2 *Let O be an open set, and assume that every coordinate function of a map $T: O \to \mathbb{R}^n$ has in O continuous partial derivatives with respect to all variables. Then on every compact set $K \subset O$, the map T satisfies the Lipschitz condition of order 1, i.e., for some C*

$$\|T(x) - T(y)\| \leq C \|x - y\| \quad \text{for any } x, y \in K.$$

Proof As observed at the end of Section 0.9, it suffices to check that T satisfies the local Lipschitz condition on K, and this property immediately follows from Lagrange's inequality, since the continuous function $z \mapsto \|T'(z)\|$ is bounded in every closed ball contained in O. $\quad\square$

Now we give another version of Lagrange's inequality. It is a generalization of Theorem 1 in which the interval $[x, y]$ is replaced with a piecewise smooth arc. The proof uses the fact that such an arc contains a point dividing it into two parts of equal length.

Theorem 2 *If T is a map differentiable at every point of a rectifiable arc K with endpoints x and y, then there is a point $z \in K$ such that*

$$\|T(y) - T(x)\| \leq \|T'(z)\| \, L,$$

where L is the length of K.

Proof For $a, b \in K$, let $K_{a,b}$ be the part of K between a and b. Using induction and the bisection method, one can easily construct a sequence of nested arcs K_{x_i, y_i} of length $L/2^i$ such that

$$\|T(y_i) - T(x_i)\| \geq \|T(y) - T(x)\|/2^i \quad \text{for all } i.$$

We claim that the common point z of these arcs is as required. Indeed, since the sum $\|x_i - z\| + \|y_i - z\|$ does not exceed the length of K_{x_i, y_i}, which is equal to $L/2^i$, we have $\|x_i - z\|, \|y_i - z\| \leq L/2^i$, and hence

$$\begin{aligned} T(y_i) - T(x_i) &= T(z) + T'(z)(y_i - z) - T(z) - T'(z)(x_i - z) + o(2^{-i}) \\ &= T'(z)(y_i - x_i) + o(2^{-i}). \end{aligned}$$

Since $\|y_i - x_i\| \leq L/2^i$, it follows that

$$\|T(y_i) - T(x_i)\| \leq \|T'(z)\| \, \|y_i - x_i\| + o(2^{-i}) \leq (\|T'(z)\| L + o(1))2^{-i}.$$

Thus,

$$\|T(y) - T(x)\|2^{-i} \leq \|T(y_i) - T(x_i)\| \leq (\|T'(z)\|L + o(1))2^{-i},$$

i.e., $\|T(y) - T(x)\| \leq \|T'(z)\| L + o(1)$. Passing to the limit completes the proof. $\square$

The next result refines the criterion of constancy obtained in Section 1.3.

Corollary *Let f be a function differentiable at every point of a rectifiable simple arc K. For f to be constant on K, it is sufficient that its gradient vanish on K.*

This is a special case of Theorem 2. Remarkably, the rectifiability condition is indispensable. As we will see in Section IV.6, there exists a function with continuous partial derivatives that is not constant on a simple curve consisting of points at which the gradient of the function vanishes.

5.5 Differentiability of the Inverse Map

As above, let $X \subset \mathbb{R}^m$ and a be an interior point of X. Let T be an invertible map from X to $\mathbb{R}^m$ and S be the inverse map. Under what conditions is S differentiable at the point $b = T(a)$? First of all, we establish an important necessary condition. If S is differentiable at b, then, by Theorem 5.3, the composition $S \circ T$ is differentiable at a. Now, since $S \circ T$ is the identity map, we have $S'(b) \cdot T'(a) = I_m$, where I_m is the $m \times m$ identity matrix. This provides a necessary condition for the differentiability of the inverse map: $\det(T'(a)) \neq 0$, i.e.,

$$\text{the matrix } T'(a) \text{ is invertible.} \tag{5}$$

Besides, the equality $S'(b) \cdot T'(a) = I_m$ implies that

$$S'(b) = (T'(a))^{-1}, \quad \text{i.e., } (T^{-1})'(b) = (T'(a))^{-1}. \tag{6}$$

In this connection, recall that if f is a strictly monotone function on an open interval that is differentiable at a point a, then the inverse function g is differentiable at the point $b = f(a)$ only if $f'(a) \neq 0$, and in this case $g'(b) = \frac{1}{f'(a)}$. Thus, condition (5) and identity (6) are a natural generalization of a classical result.

It is well known from algebra that the matrix $T'(a)$ is invertible if and only if $\det T'(a) \neq 0$. Thus, for the inverse map to be differentiable, it is necessary that the Jacobian of the original map be nonzero.

Before turning to the main theorem of this subsection, we prove an auxiliary result which gives a lower estimate for increments of a map.

Lemma *Let $T: X \to \mathbb{R}^m$. If T is differentiable at a point $a \in X$ and the derivative $T'(a)$ is invertible, then there exist $c > 0$, $\delta > 0$ such that $B(a, \delta) \subset X$ and $\|T(x) - T(a)\| \geq c\|x - a\|$ as long as $\|x - a\| < \delta$.*

Proof If T is linear and, consequently, $T = d_a T$, then the required estimate holds with coefficient $c = \frac{1}{\|T^{-1}\|}$, see inequality (4) from Section 0.13.

In the general case,

$$T(x) - T(a) = A(x - a) + \|x - a\|\, \omega(x),$$

where $A = T'(a)$ and $\omega(x) \xrightarrow[x \to a]{} 0$. Therefore,

$$\|T(x) - T(a)\| \geq \|A(x - a)\| - \|x - a\|\,\|\omega(x)\| \geq \left(\frac{1}{\|A^{-1}\|} - \|\omega(x)\| \right)\|x - a\|.$$

This proves the required estimate with coefficient $c = \frac{1}{2\|A^{-1}\|}$, since $\|\omega(x)\| < \frac{1}{2\|A^{-1}\|}$ for x sufficiently close to a. $\qquad\square$

Now we prove that under simple additional assumptions, condition (5) is not only necessary, but also sufficient for the inverse map to be differentiable.

Theorem *Assume that a map $T: X \to \mathbb{R}^m$ is differentiable at a point a and invertible, and let $S = T^{-1}$. If*
1) *$b = T(a)$ is an interior point of the set $T(X)$,*
2) *the map S is continuous at b,*
3) *the matrix $T'(a)$ is invertible,*
then S is differentiable at the point b, and, according to (6), we have $S'(b) = (T'(a))^{-1}$.

All three conditions are necessary for S to be differentiable at b (the first one, by the definition of differentiability).

Proof For brevity, set $A = T'(a)$. Then

$$T(x) - T(a) = A(x - a) + \|x - a\|\, \omega(x) \tag{7}$$

where $\omega(x) \xrightarrow[x \to a]{} 0$, $\omega(a) = 0$. Note that

$$\|T(x) - T(a)\| \geq c\|x - a\| \quad \text{for } x \in B(a, \delta), \tag{8}$$

where c and δ are from the lemma. Consider an arbitrary point $y \in T(X)$ and set $x = S(y)$. Substituting $x = S(y)$ and $a = S(b)$ into (7), we obtain

$$y - b = A(S(y) - S(b)) + \|S(y) - S(b)\|\omega(S(y)),$$

i.e.,

$$S(y) - S(b) = A^{-1}(y - b) - \|S(y) - S(b)\| A^{-1}(\omega(S(y))). \tag{9}$$

It remains to prove that the function

$$\beta(y) = -\|S(y) - S(b)\| A^{-1}(\omega(S(y)))$$

is an infinitesimal, as $y \to b$, of higher order than $\|y - b\|$. By the continuity of S, we may assume that y is so close to b that

$$\|x - a\| = \|S(y) - S(b)\| < \delta.$$

This allows us to use inequality (8) to estimate $\|\beta(y)\|$:

$$\|\beta(y)\| = \|x - a\| \cdot \|A^{-1}(\omega(S(y)))\|$$

$$\leq \frac{1}{c}\|T(x) - T(a)\| \cdot \|A^{-1}\| \cdot \|\omega(S(y))\| = \frac{\|A^{-1}\|}{c}\|y - b\| \cdot \|\omega(S(y))\|.$$

Since $x = S(y) \xrightarrow[y \to b]{} a$, we see that $\omega(S(y)) \xrightarrow[y \to b]{} 0$, and hence

$$\beta(y) \underset{y \to b}{=} o(\|y - b\|),$$

which completes the proof. $\square$

Exercises

1. Let $X \subset \mathbb{R}^m$, and let $f, g \colon X \to \mathbb{R}^n$ be maps differentiable at a point a. Show that the scalar product $\langle f, g \rangle$ is also differentiable at a and

$$d_a\langle f, g\rangle(h) = \langle d_a f(h), g(a)\rangle + \langle f(a), d_a g(h)\rangle \quad \text{for all } h \in \mathbb{R}^m.$$

   Find the gradient of the product $\langle f, g \rangle$ at a.

2. On the punctured plane $\mathbb{R}^2 \setminus \{0\}$, consider the maps $f = (f_1, f_2)$ and $g = (g_1, g_2)$ with coordinate functions

$$f_1(x, y) = g_1(x, y) = \frac{x^2 - y^2}{x^2 + y^2}, \quad f_2(x, y) = 2g_2(x, y) = \frac{2xy}{x^2 + y^2}.$$

Find the ranges of these maps and the ranks of the matrices f' and g'.

3. Show that for $x \neq 0$, the inversion $x \mapsto \frac{x}{\|x\|^2}$ is differentiable and find its differential.

4. Identifying the set of square $m \times m$ matrices with the space $\mathbb{R}^{m^2}$, compute the differentials of the maps $X \mapsto X^2$ and $X \mapsto X^3$. Find the Jacobian matrix of the former map for $m = 2$. Formulate a condition for the invertibility of this matrix.

5. Find the differential of the map $X \mapsto X^{-1}$ (defined on the set of invertible matrices)

   a) at the point I (where I is the identity matrix);

   b) at a point A (where A is an arbitrary invertible matrix).

6. Let O be an open subset in $\mathbb{R}^m$ and $F: O \to M_n$ be a map from O to the set of square $n \times n$ matrices that is differentiable at a point a. Show that

   a) the differential of F has the form

$$d_a F(h) = (\langle \operatorname{grad} f_{ij}(a), h \rangle)_{i,j \leq n},$$

where f_{ij} are the elements of the matrix F, i.e., the coordinate functions of this map;

   b) if $F(a)$ is invertible, then the differential of the map $\Phi(x) = (F(x))^{-1}$ at a has the form

$$d_a \Phi(h) = -(F(a))^{-1} \cdot d_a F(h) \cdot (F(a))^{-1}.$$

7. Let E be a subset in $\mathbb{R}^m$, and let $F: E \to \mathbb{R}^n$ be a map with coordinate functions $F_1, \ldots, F_n$ that is differentiable at a point $c \in \operatorname{Int}(E)$.

   For $m = p+q$ $(p, q \in \mathbb{N})$, we identify the space $\mathbb{R}^m$ with the product $\mathbb{R}^p \times \mathbb{R}^q$, and its points with pairs (x, y) where $x \in \mathbb{R}^p$, $y \in \mathbb{R}^q$. In particular, $c = (a, b)$ for some $a \in \mathbb{R}^p$ and $b \in \mathbb{R}^q$. Clearly, the Jacobian matrix $F'(c)$ has the form

$$\begin{pmatrix} \dfrac{\partial F_1}{\partial x_1}(c) & \cdots & \dfrac{\partial F_1}{\partial x_p}(c) & \dfrac{\partial F_1}{\partial y_1}(c) & \cdots & \dfrac{\partial F_1}{\partial y_q}(c) \\ \cdots\cdots\cdots\cdots\cdots\cdots\cdots\cdots\cdots\cdots\cdots\cdots \\ \dfrac{\partial F_n}{\partial x_1}(c) & \cdots & \dfrac{\partial F_n}{\partial x_p}(c) & \dfrac{\partial F_n}{\partial y_1}(c) & \cdots & \dfrac{\partial F_n}{\partial y_q}(c) \end{pmatrix}.$$

Its left and right parts

$$\begin{pmatrix} \dfrac{\partial F_1}{\partial x_1}(a, b) & \cdots & \dfrac{\partial F_1}{\partial x_p}(a, b) \\ \cdots\cdots\cdots\cdots\cdots\cdots \\ \dfrac{\partial F_n}{\partial x_1}(a, b) & \cdots & \dfrac{\partial F_n}{\partial x_m}(a, b) \end{pmatrix} \quad \text{and} \quad \begin{pmatrix} \dfrac{\partial F_1}{\partial y_1}(a, b) & \cdots & \dfrac{\partial F_1}{\partial y_q}(a, b) \\ \cdots\cdots\cdots\cdots\cdots\cdots \\ \dfrac{\partial F_n}{\partial y_1}(a, b) & \cdots & \dfrac{\partial F_n}{\partial y_q}(a, b) \end{pmatrix}$$

will be denoted by $F'_x(a, b)$ and $F'_y(a, b)$, respectively.

We also introduce the sets

$$E^y = \{ x \in \mathbb{R}^p : (x, y) \in E \} \quad (y \in \mathbb{R}^q)$$

and
$$E_x = \{y \in \mathbb{R}^q : (x, y) \in E\} \quad (x \in \mathbb{R}^p),$$

which will be called sections of E. On nonempty sections, we define maps F_x and F^y with values in $\mathbb{R}^n$ by the formulas

$$F^y(x) = F(x, y) \quad (\text{for } x \in E^y) \quad \text{and} \quad F_x(y) = F(x, y) \quad (\text{for } y \in E_x).$$

a) Check that the maps F^b and F_a are differentiable at the points a and b, respectively, and

$$(F^b)'(a) = F_x'(a, b), \quad (F_a)'(b) = F_y'(a, b).$$

b) By analogy with Theorem 2.3 on sufficient conditions of differentiability, show that if for all (x, y) from a sufficiently small cube centered at the point (a, b), the maps F^y and F_x are differentiable at the points x and y, respectively, with

$$(F^y)'(x) \xrightarrow[(x,y)\to(a,b)]{} (F^b)'(a) \quad \text{and} \quad (F_x)'(y) \xrightarrow[(x,y)\to(a,b)]{} (F_a)'(b),$$

then the map F is differentiable at the point $c = (a, b)$.

6 Higher Derivatives

6.1 Definitions and Notation

As in the case of functions of one variable, higher derivatives of a function of several variables are defined inductively. In Section 1 we have already defined partial derivatives of such a function, which will now be called first-order partial derivatives. Derivatives of order $n = 2, 3, \ldots$ are defined as partial derivatives of derivatives of order $n - 1$. In more detail: let f be a function defined on a subset X in $\mathbb{R}^m$, a be an interior point of X, and g be a partial derivative of f of order n in some neighborhood of a. If for some $k = 1, \ldots, m$ the function g has a partial derivative $g'_{x_k}(a)$ at a, then it is called a partial derivative of f of order $n + 1$ at the point a. Thus, we see from the definition that the existence of all partial derivatives of some order $n > 1$ automatically implies the existence of all partial derivatives of lower orders.

The notation for higher derivatives is as follows. Second-order partial derivatives of a function f at a point a, i.e., derivatives of the form $\frac{\partial}{\partial x_k}\left(\frac{\partial f}{\partial x_j}\right)(a)$, are denoted by $\frac{\partial^2 f}{\partial x_j \partial x_k}(a)$ or $f''_{x_j x_k}(a)$. For $j = k$, the notation $\frac{\partial^2 f}{\partial x_j^2}(a)$ or $f''_{x_j^2}(a)$ is used.

Correspondingly, third-order derivatives are denoted by $\frac{\partial^3 f}{\partial x_j \partial x_k \partial x_l}(a)$ or $f'''_{x_j x_k x_l}(a)$, and so on. The derivative of order n obtained by successively differentiating f with respect to $x_{k_1}, x_{k_2}, \ldots, x_{k_n}$, where the indices $k_1, \ldots, k_n$ range independently over

$1, 2, \ldots, m$, is denoted by

$$\frac{\partial^n f}{\partial x_{k_1} \ldots \partial x_{k_n}}(a) \quad \text{or} \quad f^{(n)}_{x_{k_1} \ldots x_{k_n}}(a).$$

Note that the coordinate corresponding to each subsequent differentiation is written *to the right* of the previous ones. Derivatives for which not all indices k_j coincide are said to be *mixed*; otherwise, they are said to be pure and denoted by

$$\frac{\partial^n f}{\partial x_k^n}(a) \quad \text{or} \quad f^{(n)}_{x_k^n}(a).$$

Clearly, if all derivatives of order n exist, then their number coincides with the number of all sequences $k_1, \ldots, k_n$ of n positive integers taking values from 1 to m, i.e., is equal to m^n.

If the coordinates of points are denoted by other letters, then the notation for derivatives changes correspondingly.

Now we consider several examples of computation of second-order derivatives.

Example 1 Let us find the partial derivatives of order 1 and 2 of the function $(x, y, z) \mapsto f(x, y, z) = xy^2z^3$. For any x, y, z, we have

$$f'_x(x, y, z) = y^2z^3, \quad f'_y(x, y, z) = 2xyz^3, \quad f'_z(x, y, z) = 3xy^2z^2.$$

Thus, we obtain

$$f''_{x^2}(x, y, z) = 0, \quad f''_{xy}(x, y, z) = 2yz^3, \quad f''_{xz}(x, y, z) = 3y^2z^2,$$
$$f''_{yx}(x, y, z) = 2yz^3, \quad f''_{y^2}(x, y, z) = 2xz^3, \quad f''_{yz}(x, y, z) = 6xyz^2,$$
$$f''_{zx}(x, y, z) = 3y^2z^2, \quad f''_{zy}(x, y, z) = 6xyz^2, \quad f''_{z^2}(x, y, z) = 6xy^2z.$$

Example 2 Now we verify that the function $U(x, t) = f(x + at) + g(x - at)$, where $a \in \mathbb{R}$ and f, g are arbitrary functions doubly differentiable on the real line, satisfies the equation $U''_{t^2} = a^2 U''_{x^2}$ (called the wave equation). Indeed,

$$U''_{x^2}(x, t) = f''(x + at) + g''(x - at),$$
$$U''_{t^2}(x, t) = a^2 f''(x + at) + a^2 g''(x - at),$$

which implies the required equality.

In many problems, in particular, those related to applications, an important role is played by the sum of all pure second-order derivatives

$$\Delta f = f''_{x_1^2} + \ldots + f''_{x_m^2}.$$

It is called the *Laplacian*[1] of f. Below, discussing the properties of Laplacians, we assume that the functions under consideration have continuous second-order derivatives, both pure and mixed. Clearly, $\Delta(f + g) = \Delta f + \Delta g$, and the function $f_c(x) = f(c\,x)$, $c \in \mathbb{R}$, satisfies the equation $\Delta f_c(x) = c^2 \Delta f(c\,x)$. Note also that an orthogonal change of variables preserves the Laplacian (see Exercise 5).

A function f is said to be *harmonic* in O if it satisfies the Laplace equation $\Delta f(x) = 0$ for $x \in O$. Clearly, linear functions are harmonic in the whole space $\mathbb{R}^m$. It is less obvious that the function $\|x\|^{2-m}$ is harmonic, which follows from the result of the next example.

Example 3 Given $p \in \mathbb{R}$, let us find the Laplacian of the function

$$f_p(x) = \|x\|^p, \quad x \in \mathbb{R}^m \setminus \{0\}.$$

Letting $x = (x_1, \ldots, x_m)$, we can easily obtain the equalities

$$\frac{\partial f_p}{\partial x_k}(x) = p\,x_k \|x\|^{p-2} \quad (k = 1, \ldots, m).$$

Therefore,

$$\frac{\partial^2 f_p}{\partial x_k^2}(x) = p\|x\|^{p-2} + p(p-2)x_k^2\|x\|^{p-4} \quad (k = 1, \ldots, m).$$

Adding these equalities, we obtain

$$\Delta f_p(x) = pm\|x\|^{p-2} + p(p-2)\|x\|^{p-2} = p(m + p - 2)\|x\|^{p-2}.$$

In particular, it follows that the function f_{2-m} is harmonic. For $m = 2$, this example is trivial ($f_{2-m} = f_0 = 1$ everywhere). For an important example of a harmonic function of two variables, which, in many respects, plays the same role as the function f_{2-m} for $m > 2$, see Exercise 4.

Example 4 Now we generalize the result of the previous example by finding the Laplacian of a *radial function* f, i.e., a function of the form $f(x) = F(\|x\|)$ where F is a function on the half-line $[0, +\infty)$ with continuous second-order derivatives. Since $\|x\|'_{x_k} = \frac{x_k}{\|x\|}$ for $\|x\| \neq 0$, it is clear that $f'_{x_k}(x) = \frac{x_k}{\|x\|} F'(\|x\|)$; therefore, for $k = 1, \ldots, m$ we have

$$f''_{x_k^2}(x) = \frac{\|x\| - \dfrac{x_k^2}{\|x\|}}{\|x\|^2} F'(\|x\|) + \frac{x_k^2}{\|x\|^2} F''(\|x\|).$$

Adding all these equalities, we obtain the required result (below, $r = \|x\|$):

$$\Delta f(x) = F''(r) + \frac{m-1}{r} F'(r).$$

---

[1] Pierre-Simon **Laplace** (1749–1827) was a French mathematician.

In particular, the Laplacian of a radial function is also a radial function. So, the problem of finding a radial function f satisfying the equation $\Delta f(x) = \varphi(\|x\|)$, where φ is a known function of one variable, is reduced to the ordinary differential equation

$$F''(r) + \frac{m-1}{r} F'(r) = \varphi(r),$$

which can easily be solved in quadratures by rewriting it in the form

$$(r^{m-1} F'(r))' = r^{m-1} \varphi(r).$$

The last equation makes it possible to describe all harmonic radial functions: if $\Delta F(\|x\|) = 0$, then $(r^{m-1} F'(r))' = 0$, which implies that $F(r) = \frac{C}{r^{m-2}} + \widetilde{C}$ for $m \geq 3$ and $F(r) = C \log r + \widetilde{C}$ for $m = 2$, where $C, \widetilde{C}$ are arbitrary real coefficients (as concerns the harmonicity of the latter function, see Exercise 4). This result shows that a harmonic radial function that is bounded in a punctured neighborhood of the origin is necessarily constant.

6.2 The Symmetry of Second Derivatives

In Example 1 at p. 57, the following feature of repeated differentiation attracts attention: derivatives obtained by differentiating with respect to the same variables but in different order coincide (the matrix composed of these derivatives is symmetric). Although this property is present in many other examples, it does not always hold (see the example at the end of this subsection). The following easy result shows that such a negative example is not possible if the function satisfies a natural additional condition.

Theorem *Let f be a function that is defined in a neighborhood of a point $(a, b) \in \mathbb{R}^2$ and has mixed derivatives f''_{xy} and f''_{yx} at this point. If they are continuous at (a, b), then $f''_{xy}(a, b) = f''_{yx}(a, b)$.*

Recall that, by definition, the assumption that a function has second-order partial derivatives automatically implies that it has first-order derivatives.

Proof Since differentiation commutes with translations, it suffices to consider the case where $a = b = 0$ and the mixed derivatives exist in a ball centered at the origin. With an arbitrary point (x, y) of this ball we associate the "second difference" of the function f:

$$D(x, y) = f(x, y) - f(x, 0) - f(0, y) + f(0, 0).$$

It can be expressed in two ways in terms of mixed derivatives. First,

$$D(x, y) = (f(x, y) - f(x, 0)) - (f(0, y) - f(0, 0)) = \varphi(x) - \varphi(0), \qquad (1)$$

where $\varphi(x) = f(x, y) - f(x, 0)$. Since $\varphi'(x) = f'_x(x, y) - f'_x(x, 0)$, applying the mean value theorem to the right-hand side of (1) yields (for some $\bar{x}$ with $|\bar{x}| \leq |x|$)

$$D(x, y) = \varphi(x) - \varphi(0) = x\varphi'(\bar{x}) = x\big(f_x'(\bar{x}, y) - f_x'(\bar{x}, 0)\big).$$

Again applying the mean value theorem (this time, to the function $y \mapsto f_x'(\bar{x}, y)$), we write the right-hand side of the last equality in the form $xy f_{xy}''(\bar{x}, \bar{y})$ for some $\bar{y}$ with $|\bar{y}| \le |y|$. Thus,

$$D(x, y) = xy f_{xy}''(\bar{x}, \bar{y}). \tag{2}$$

Interchanging the arguments of f, by a similar calculation (with $\varphi(x)$ replaced by $\psi(y) = f(x, y) - f(0, y)$ and so on), we arrive at a representation of $D(x, y)$ in terms of the other mixed derivative:

$$D(x, y) = yx f_{yx}''(\bar{\bar{x}}, \bar{\bar{y}}) \quad \text{where } |\bar{\bar{x}}| \le |x|, \ |\bar{\bar{y}}| \le |y|.$$

For $xy \ne 0$, it follows from (2) and the last equality that $f_{xy}''(\bar{x}, \bar{y}) = f_{yx}''(\bar{\bar{x}}, \bar{\bar{y}})$. Taking the limit as $x, y \to 0$ completes the proof. $\square$

The following example shows that the assumptions on partial derivatives in the statement of the theorem cannot be discarded completely (though they can be modified or weakened, see Exercises 1 and 2).

Example Let $f(x, y) = xy\dfrac{x^2 - y^2}{x^2 + y^2}$ if $x^2 + y^2 > 0$, $f(0, 0) = 0$. The reader can easily check that this function is differentiable everywhere, with $f_x'(0, y) = -y$ and $f_y'(x, 0) = x$ for any x and y (in particular, for $x = y = 0$). Hence,

$$f_{xy}''(0, 0) = -1, \quad f_{yx}''(0, 0) = 1.$$

6.3 Smooth Functions and Maps

Instead of considering the repeated differentiability of functions in full generality, below we restrict ourselves to the case where partial derivatives not only exist, but are continuous.

Definition Let n be an arbitrary positive integer. A function defined on an open subset O in $\mathbb{R}^m$ is said to be *n-smooth*, or a *function of class C^n*, if in O it has all partial derivatives up to order n and the derivatives of order n are continuous.

The set of all n-smooth functions on an open set O is denoted by $C^n(O)$. Obviously, $C^n(O) \supset C^{n+1}(O)$. A function that is n-smooth on O for every n is said to be *infinitely differentiable*, or a *function of class C^∞*, and the set of such functions is denoted by $C^\infty(O)$. Clearly, $C^\infty(O) = \bigcap_{n=1}^{\infty} C^n(O)$. For consistency of notation, we denote $C^0(O) = C(O)$.

A map from O to $\mathbb{R}^d$ is said to be *n-smooth* ($n = 1, 2 \ldots, \infty$), or a map of class C^n, if all its coordinate functions are of class C^n. The set of all such maps is denoted by $C^n(O; \mathbb{R}^d)$.

By Theorem 2.3, for a function f to be differentiable at a point a, it suffices that near a it have partial derivatives with respect to all variables and each of them be continuous at this point. Therefore, all coordinate functions of a map of class $C^1(O; \mathbb{R}^d)$ are differentiable in O, which implies (see Section 6.2) that the map itself is differentiable.

Below, we also use the following remark.

Remark A smooth map satisfies the local Lipschitz condition in its domain of definition. This immediately follows from Corollary 2 at p. 51.

6.4 Equality of Mixed Partial Derivatives of Arbitrary Order

Here we extend Theorem 6.2 to arbitrary partial derivatives that differ only by the order of differentiation. Let us explain the last term. Consider two partial derivatives of order n of an n-smooth function f. They have the form

$$\frac{\partial^n f}{\partial x_{j_1} \dots \partial x_{j_n}} \quad \text{and} \quad \frac{\partial^n f}{\partial x_{k_1} \dots \partial x_{k_n}},$$

where the indices $j_1, \dots, j_n$ and $k_1, \dots, k_n$ range independently over $1, 2, \dots, m$. We say that these derivatives differ only by the order of differentiation if the tuple $(k_1, \dots, k_n)$ is a permutation of the tuple $(j_1, \dots, j_n)$.

Theorem *If f is an n-smooth function, then its mixed partial derivatives of order n that differ only by the order of differentiation coincide.*

Proof We must prove that

$$\frac{\partial^n f}{\partial x_{j_1} \dots \partial x_{j_n}} = \frac{\partial^n f}{\partial x_{k_1} \dots \partial x_{k_n}}$$

provided that the tuples $(j_1, \dots, j_n)$ and $(k_1, \dots, k_n)$ can be obtained from each other by a permutation.

One can easily check by induction on n that every permutation can be expressed as a product of adjacent transpositions, i.e., permutations that interchange two adjacent indices. Hence, it suffices to show that such transpositions do not change the mixed derivative. Consider a transposition of two (nonequal) indices j_{i-1} and j_i in $(j_1, \dots, j_n)$. According to our notation,

$$\frac{\partial^i f}{\partial x_{j_1} \dots \partial x_{j_{i-1}} \partial x_{j_i}} = \frac{\partial^2}{\partial x_{j_{i-1}} \partial x_{j_i}} \left(\frac{\partial^{i-2} f}{\partial x_{j_1} \dots \partial x_{j_{i-2}}} \right).$$

By Theorem 6.2, we can change the order of differentiation with respect to x_{j_i} and $x_{j_{i-1}}$ in the right-hand side. Thus,

$$\frac{\partial^i f}{\partial x_{j_1} \dots \partial x_{j_{i-1}} \partial x_{j_i}} = \frac{\partial^2 f}{\partial x_{j_i} \partial x_{j_{i-1}}} \left(\frac{\partial^{i-2} f}{\partial x_{j_1} \dots \partial x_{j_{i-2}}} \right) = \frac{\partial^i f}{\partial x_{j_1} \dots \partial x_{j_i} \partial x_{j_{i-1}}}.$$

Now, differentiating both sides with respect to the other variables $x_{j_{i+1}}, \dots, x_{j_n}$, we see that a transposition of adjacent indices does not affect the partial derivative. As noted at the beginning of the proof, this implies that any permutation of indices does not change the partial derivative. □

Corollary *For a 2-smooth function f, the matrix*

$$H_f(x) = \begin{pmatrix} f''_{x_1^2}(x) & f''_{x_1 x_2}(x) & \cdots & f''_{x_1 x_m}(x) \\ f''_{x_2 x_1}(x) & f''_{x_2^2}(x) & \cdots & f''_{x_2 x_m}(x) \\ \cdots\cdots\cdots\cdots\cdots\cdots\cdots\cdots\cdots \\ f''_{x_m x_1}(x) & f''_{x_m x_2}(x) & \cdots & f''_{x_m^2}(x) \end{pmatrix}$$

composed of the values of the second-order derivatives of f at a point x is symmetric.

This matrix is called the *Hessian*[2] *matrix* of f. Note that its trace is equal to Δf, the Laplacian of f (see Section 6.1).

The above theorem gives a reason to discuss again the notation for partial derivatives. Since the partial derivative $\frac{\partial^n f}{\partial x_{j_1} \dots \partial x_{j_n}}$ does not depend on the order of differentiation, it suffices to indicate how many times the function is to be differentiated with respect to each coordinate. In other words, for a sufficiently smooth function, a partial derivative is uniquely determined by a vector with nonnegative integer elements indicating the number of differentiations with respect to the corresponding coordinate. Hence, it is convenient to use the notion of a multi-index, i.e., a vector of the form $\alpha = (\alpha_1, \dots, \alpha_m)$ whose all coordinates are nonnegative integers and α_i is equal to the number of differentiations with respect to the coordinate x_i (with $\alpha_i = 0$ if there is no differentiation with respect to x_i; in particular, $f^{(0)} = f$). The order of the partial derivative corresponding to a multi-index $\alpha = (\alpha_1, \dots, \alpha_m)$ is equal to the sum $\alpha_1 + \dots + \alpha_m$, which is denoted by $|\alpha|$. We call it the order of α, so the order of a multi-index is equal to the number of differentiations in the corresponding partial derivative. In what follows, it is also convenient to use the notation $\alpha!$ for the product $\alpha_1! \cdots \alpha_m!$.

In multi-index notation, a partial derivative of a function f is denoted as follows:

$$f^{(\alpha)}, \quad \frac{\partial^{|\alpha|} f}{\partial x^\alpha}, \quad \text{or, in more detail,} \quad \frac{\partial^{|\alpha|} f}{\partial x_1^{\alpha_1} \dots \partial x_m^{\alpha_m}}.$$

In some cases, especially when dealing with partial differential equations, one also writes $D^\alpha f$.

A monomial in $\mathbb{R}^m$ is a function of m variables of the form $x \mapsto x^\alpha = x_1^{\alpha_1} \cdots x_m^{\alpha_m}$, where $x = (x_1, \dots, x_m) \in \mathbb{R}^m$ and $\alpha = (\alpha_1, \dots, \alpha_m)$ is a multi-index (for $\alpha_j = 0$,

---

[2] Ludwig Otto **Hesse** (1811–1874) was a German mathematician.

the factor $x_j^{\alpha_j}$ equals 1 for all values of x_j). Obviously, this is an infinitely smooth function. The following result holds.

Proposition (derivatives of a monomial) *Given a multi-index* $\beta = (\beta_1, \ldots, \beta_m)$, *consider the partial derivative* $P_0^{(\beta)}$ *of a monomial* $P_0(x) = x^\alpha$. *If* $|\beta| \geq |\alpha|$, *then*

$$P_0^{(\beta)}(x) = \begin{cases} \alpha! = \alpha_1! \ldots \alpha_m! & \text{if } \alpha = \beta, \\ 0 & \text{if } \alpha \neq \beta; \end{cases}$$

if $|\alpha| > |\beta|$, *then* $P_0^{(\beta)}(0) = 0$.

Proof The first equality is obvious. To prove the second one, note that since $\alpha \neq \beta$ and $|\beta| \geq |\alpha|$, at least one of the numbers β_i is greater than α_i. Hence, differentiating the monomial β_i times with respect to x_i yields 0. We leave it to the reader to check the remaining equality. □

Since differentiation commutes with translations (see the end of Section 1.1), we obtain a similar result for polynomials of the form

$$P_a(x) = (x - a)^\alpha = (x_1 - a_1)^{\alpha_1} \ldots (x_m - a_m)^{\alpha_m} :$$

for every multi-index β,

$$P_a^{(\beta)}(x) = 0 \qquad \text{if } |\beta| \geq |\alpha|, \beta \neq \alpha,$$
$$P_a^{(\alpha)}(x) = \alpha!,$$
$$P_a^{(\beta)}(a) = 0 \qquad \text{if } |\beta| < |\alpha|.$$

Thus, the following corollary holds.

Corollary *The derivative* $P_a^{(\beta)}(a)$ *vanishes unless* $\alpha = \beta$; *in this case, it is equal to* $\alpha!$.

6.5 Algebraic Operations on Smooth Functions

Now we verify that algebraic operations on smooth functions result in functions of the same smoothness.

Theorem *The sum and the product of n-smooth functions* $(n = 1, 2, \ldots, \infty)$ *is again an n-smooth function. Their quotient is also an n-smooth function provided that the denominator does not vanish.*

Proof Clearly, it suffices to prove this for a finite n. We use induction and the following obvious relation between the classes $C^n(O)$ and $C^{n-1}(O)$ $(n \in \mathbb{N})$:

$$f \in C^n(O) \iff \frac{\partial f}{\partial x_k} \in C^{n-1}(O) \quad \text{for } k = 1, 2, \ldots, m. \tag{3}$$

We consider the case of a product, leaving that of a sum to the reader. Let $g, h \in C^n(O)$, $f = gh$. In the case $n = 1$, all partial derivatives of f are continuous, because $f'_{x_k} = g'_{x_k} h + g h'_{x_k}$. Therefore, the theorem is true for $n = 1$. In the case $n > 1$, the functions f'_{x_k} are $(n-1)$-smooth by the induction hypothesis, and applying (3) completes the proof.

In the case of a quotient, we can argue in a similar way using the formula

$$\frac{\partial}{\partial x_k} \left(\frac{g}{h} \right) = \frac{g'_{x_k} h - g h'_{x_k}}{h^2}. \qquad \Box$$

Remark The theorem can be supplemented by a formula for derivatives of order $2, 3, \ldots$ generalizing the well-known general Leibniz[3] rule for higher derivatives of the product of functions of one variable. Namely, for n-smooth functions f, g, we have

$$(fg)^{(\alpha)} = \sum_{\beta + \gamma = \alpha} \frac{\alpha!}{\beta! \, \gamma!} f^{(\beta)} g^{(\gamma)},$$

where α is a multi-index of order at most n. We leave it to the reader to prove this formula by induction. Another proof is given at the end of Section 8.4.

6.6 Composition of Smooth Maps

Let O, O' be open subsets in $\mathbb{R}^m$ and $\mathbb{R}^d$, respectively, and let $F \colon O \to \mathbb{R}^d$, $\Phi \colon O' \to \mathbb{R}^\ell$ be maps such that $F(O) \subset O'$, i.e., the composition $\Phi \circ F$ is well defined. In this case, the following theorem holds.

Theorem *The composition $\Phi \circ F$ of two n-smooth maps Φ and F is n-smooth too* $(n = 1, 2, \ldots, \infty)$.

Proof By definition, the n-smoothness of a map means the n-smoothness of its coordinate functions. Hence, we may restrict ourselves to the case where $\Phi = g$ is a function. As in the proof of the previous theorem, it suffices to consider the case of finite smoothness.

Let $\varphi = g \circ F$ and $f_1, \ldots, f_d$ be the coordinate functions of F. To prove that φ is n-smooth, we use induction on n. In the case $n = 1$, by the rule for differentiating composite functions (see formula (4) in Section 5.3), we have

$$\frac{\partial \varphi}{\partial x_k}(x) = \sum_{j=1}^{d} \frac{\partial g}{\partial y_j}(F(x)) \cdot \frac{\partial f_j}{\partial x_k}(x) \quad (x \in O)$$

for $k = 1, \ldots, m$ (a point in $\mathbb{R}^d$ is denoted by y). It is seen that the partial derivatives of φ are continuous and, therefore, $\varphi \in C^1(O)$. For $n > 1$, the same formula shows that $\varphi'_{x_k} \in C^{n-1}(O)$, since, obviously, $\frac{\partial f_j}{\partial x_k} \in C^{n-1}(O)$ for every j and

---

[3] Gottfried Wilhelm **Leibniz** (1646–1716) was a German mathematician.

$g'_{y_j} \circ F \in C^{n-1}(O)$ by the induction hypothesis. To complete the induction step, it remains to apply (3). $\square$

6.7 Fractional Smoothness

Our aim in this section is to embed the sequence of sets C^n in a family depending on a continuous numerical parameter characterizing the degree of smoothness of a function. First, recall the definition introduced in Section 0.9: a function f defined on a set E satisfies the Lipschitz condition of order α, $0 < \alpha \le 1$, on this set if there exists a constant L such that

$$|f(x) - f(x')| \le L\|x - x'\|^\alpha \quad \text{for all } x,\, x' \in E.$$

Clearly, if the set E is bounded, then a function satisfying on E the Lipschitz condition of some order α also satisfies the Lipschitz condition of a smaller order. Using the Lipschitz condition, one can define in a natural way the notion of "fractional smoothness."

Let $0 < \alpha < 1$, $r = 0, 1, 2, \ldots$, and let O be an open subset in $\mathbb{R}^m$. By the derivative of order 0 we mean the function itself and assume, as we agreed earlier, that $C^0(O) = C(O)$. Let $t = r + \alpha$; we say that a function $f \in C^r(O)$ belongs to the class $C^t(O)$ if all its derivatives of order r satisfy the Lipschitz condition of order α on every compact subset in O. The corresponding class for $\alpha = 1$ will be denoted by $LC^r(O)$. Thus, for all $r = 0, 1, 2, \ldots$ and $r \le t < r + 1$, we have

$$C^{r+1}(O) \subset LC^r(O) \subset C^t(O) \subset C^r(O).$$

A map belongs to the class C^t (or LC^r) if all its coordinate functions do.

One can easily check that algebraic operations on functions satisfying the Lipschitz condition of order α result in functions of the same class. Hence, Theorem 6.5 remains valid for functions of class C^t ($t > 1$) and LC^r.

However, it should be noted that for the composition of functions satisfying the Lipschitz condition of some order less than 1, this order may decrease. For instance, the function $x \mapsto f(x) = x^\alpha$ ($x \ge 0$) with $0 < \alpha < 1$ satisfies the Lipschitz condition of order α, while the composition $f \circ f$ satisfies the Lipschitz condition only of order α^2. But for functions and maps of classes C^t with $t \ge 1$ and LC^n, Theorem 6.6 remains valid with the same proof. We encourage the reader to verify this. Besides, it is easy to check that the composition (in any order) of maps of class C^α with $0 < \alpha < 1$ and LC again belongs to the class C^α.

Exercises

1. Prove the following version of Theorem 6.2 on the symmetry of second derivatives: if in a neighborhood of a point (a, b) a function f has first partial derivatives and they are differentiable at (a, b), then f has finite mixed derivatives at (a, b) and they coincide.
2. Check that Theorem 6.2 holds under less restrictive assumptions on the function: in addition to the existence of the first-order partial derivatives, it suffices to require that in a punctured neighborhood of (a, b) one of the mixed derivatives exist and have a finite limit at this point (Schwarz's[4] theorem).
3. Let $H_f(x)$ be the Hessian matrix of a function f of class $C^2(O)$ (where $O \subset \mathbb{R}^m$) at a point x, A be the matrix of a linear map from $\mathbb{R}^p$ to $\mathbb{R}^m$, and $g = f \circ A$. Show that the Hessian matrix of the function g is given by the formula

$$H_g(y) = A^T \cdot H_f(A(y)) \cdot A \quad (y \in A^{-1}(O)),$$

where A^T is the transpose of A.
4. Show that the function $(x, y) \mapsto \log(x^2 + y^2)$ is harmonic in $\mathbb{R}^2 \setminus \{0\}$.
5. Let $U : \mathbb{R}^m \to \mathbb{R}^m$ be an orthogonal transformation. Show that the Laplacians of the functions f and $f \circ U$ at points x and $U^{-1}(x)$ coincide.

To formulate Exercise 6, we need the following definition.

Definition Let $x \neq 0$. The point x' of the form $x' = \lambda x$ with $\lambda > 0$ (i.e., lying on the same ray as x) satisfying the condition $\|x\| \cdot \|x'\| = 1$ (so $x' = \frac{x}{\|x\|^2}$) is called the inversion of x with respect to the sphere $\|x\| = 1$.

6. Prove the following statement (Kelvin's[5] inversion theorem). Let U be a harmonic function in a domain O, $0 \notin O$, and O' be the inversion of O with respect to the unit sphere. Then the function V defined as

$$V(x) = \frac{1}{\|x\|^{m-2}} U\left(\frac{x}{\|x\|^2}\right)$$

is harmonic in O'.

   HINT. Since an orthogonal change of variables preserves harmonicity (see Exercise 5), it suffices to check that $\Delta V(x) = 0$ at the point $x = (1, 0, \ldots, 0)$.

7. (See Problem 1998-24 in [2].) Show that if a 2-smooth function U satisfies the equation

$$U''_{xx}(U'_t)^2 + U''_{tt}(U'_x)^2 - 2U''_{xt}U'_xU'_t = 0,$$

then the function $f \circ U$ with $f \in C^2(\mathbb{R})$ satisfies this equation too.
8. Show that the function

$$f(x, y) = \frac{1}{y - x}\left(\frac{\sin y}{y} - \frac{\sin x}{x}\right)$$

---

[4] Karl Hermann Amandus **Schwarz** (1843–1921) was a German mathematician.

[5] William **Thomson**, 1st Baron Kelvin (1824–1907) was an English physicist and mathematician.

extended to the coordinate axes and to the bisector $y = x$ by continuity belongs to the class $C^\infty(\mathbb{R}^2)$, and every its partial derivative of order n does not exceed $\frac{1}{(n+1)(n+2)}$ in absolute value.

HINT. Check that $f(x, y) = - \iint\limits_{T} \sin(xu + yv)\, du\, dv$ where T is the triangle with vertices at $(0, 0)$, $(1, 0)$, $(0, 1)$.

7 Polynomials in Several Variables and Higher Differentials

As we will see, higher differentials of functions of several variables are homogeneous polynomials depending on the increment of the argument. Before turning to them, we briefly discuss properties of polynomials in several variables.

7.1 Polynomials in Several Variables

In the previous section, we met the simplest example of a polynomial: a monomial $x_1^{\alpha_1} \cdots x_m^{\alpha_m}$, or, in abbreviated notation, x^α (hereafter, $x_1, \ldots, x_m$ are the coordinates of a vector $x \in \mathbb{R}^m$ in the canonical basis, α is an m-dimensional multi-index, its coordinates $\alpha_1, \ldots, \alpha_m$ are nonnegative integers); recall that, according to our convention, in the case $\alpha_j = 0$ the factor $x_j^{\alpha_j}$ is equal to 1 for all x_j).

The degree of a monomial x^α is the sum $|\alpha| = \alpha_1 + \ldots + \alpha_m$.

A *polynomial P in m variables* is a linear combination of monomials, i.e., a sum of the form

$$P(x) = \sum_\alpha c_\alpha x^\alpha = \sum_{\substack{\alpha_1, \ldots, \alpha_m \geq 0 \\ \text{are integres}}} c_{\alpha_1, \ldots, \alpha_m} x_1^{\alpha_1} \ldots x_m^{\alpha_m}$$

with finitely many nonzero coefficients c_α. The degree of a nonzero polynomial P is the largest degree of a monomial that appears in the sum with a nonzero coefficient. It is denoted by $\deg P$.

Lemma *Every polynomial can be uniquely represented as a sum of monomials.*

In other words, the system of all monomials in m variables is linearly independent.

Proof Let $P(x) = \sum_\alpha c_\alpha x^\alpha$. By Proposition 6.4, the derivative of $P^{(\beta)}$ at the origin is equal to $\beta! c_\beta$, i.e.,

$$c_\beta = \frac{P^{(\beta)}(0)}{\beta!}.$$

Thus, the coefficients of P are uniquely determined. $\qquad \square$

Another proof of the lemma can be obtained by induction on the number of variables, we leave it to the reader.

The above argument not only proves the uniqueness of the representation, but provides a formula for its coefficients:

$$P(x) = \sum_\alpha \frac{P^{(\alpha)}(0)}{\alpha!} x^\alpha. \tag{1}$$

A polynomial P is said to be *homogeneous of degree n* (where n is a nonnegative integer) if $P(tx) = t^n P(x)$ for every x and every real t. For instance, the monomial x^α is a homogeneous polynomial of degree $|\alpha|$. Homogeneous polynomials are also called *forms*; more exactly, a form of degree n is called an *n-form*. Forms of degree 1, 2, and 3 are also called linear, quadratic, and cubic forms, respectively.

An arbitrary polynomial P can be written as a sum of forms by grouping together the terms of the same homogeneity degree:

$$P(x) = \sum_i \sum_{|\alpha|=i} c_\alpha x^\alpha.$$

Example (multinomial theorem) Fix a nonnegative integer n and consider the homogeneous polynomial

$$P_{m,n}(x) = (x_1 + \ldots + x_m)^n$$

in $\mathbb{R}^m$ (for $m = 2$, this is the "binomial" $(x_1 + x_2)^n$ well known from high school algebra). Expanding the product, we obtain

$$P_{m,n}(x) = \sum_{j_1=1}^m \ldots \sum_{j_n=1}^m x_{j_1} \ldots x_{j_n}.$$

Of course, many terms in the sum coincide. Collecting like terms yields

$$P_{m,n}(x) = \sum_{|\alpha|=n} C_n^\alpha x^\alpha, \tag{2}$$

where C_n^α is the number of products $x_{j_1} \ldots x_{j_n}$ coinciding with x^α.

The coefficients C_n^α, which are similar to binomial coefficients, are called multinomial coefficients. Like binomial coefficients, they arise in a natural way in combinatorics, but we will not discuss this here and mention only a formula for calculating them. Since the variables $x_1, \ldots, x_m$ are interchangeable in $P_{m,n}(x)$, the partial derivatives of this polynomial depend only on the number of differentiations:

$$P_{m,n}^{(\alpha)}(x) = n(n-1) \ldots (n-s+1) P_{m,n-s}(x),$$

where $s = |\alpha| \leq n$. In particular, every partial derivative of order n is equal to $n!$. Hence, taking the derivative of order α of both sides of (2), for $|\alpha| = n$ we obtain

$$n! = C_n^\alpha \cdot \alpha!, \quad \text{i.e.,} \quad C_n^\alpha = \frac{n!}{\alpha!},$$

and thus we arrive at the *multinomial theorem*:

$$(x_1 + \ldots + x_m)^n = \sum_{|\alpha|=n} \frac{n!}{\alpha!} x^\alpha. \tag{3}$$

In particular, substituting $x_1 = \ldots = x_m = 1$ yields

$$\sum_{|\alpha|=n} \frac{n!}{\alpha!} = m^n.$$

We leave it to the reader to verify that in the case $m = 2$ formula (3) coincides with the binomial expansion well known from high school algebra.

To study the behavior of a polynomial P near some point $a \in \mathbb{R}^m$, it is more convenient to use an expansion in powers of $x - a$. It can be obtained by applying formula (1) to the polynomial $\widetilde{P}(x) = P(x + a)$. Since $\widetilde{P}^{(\alpha)}(0) = P^{(\alpha)}(a)$, we have $P(x + a) = \sum_\alpha \frac{P^{(\alpha)}(a)}{\alpha!} x^\alpha$, or

$$P(x) = \sum_\alpha \frac{P^{(\alpha)}(a)}{\alpha!} (x - a)^\alpha. \tag{4}$$

Thus, every polynomial $P(x) = \sum_\alpha c_\alpha x^\alpha$ can be "reexpanded in powers of $x - a$" by representing it as a linear combination of shifted monomials $(x - a)^\alpha$.

7.2 Higher Differentials

In what follows, the symbol O stands for an open subset in $\mathbb{R}^m$.

Defining differentials of order 2, 3, ... (or second, third, ... differentials), we assume that the function under consideration lies in the corresponding smoothness class. The definition is inductive, each differential is defined in terms of the (already defined) previous differential. The base case is the definition of the differential of a function, with which the reader is already familiar (see Section 2). Recall that the differential (which now will be also called the first differential) of a smooth function f at a point a is the linear homogeneous polynomial of degree 1

$$h \mapsto d_a f(h) = \langle \operatorname{grad} f(a), h \rangle = \sum_{k=1}^m f'_{x_k}(a) h_k, \quad h = (h_1, \ldots, h_m).$$

Now we define the second differential. Let $f \in C^2(O)$, $a \in O$. Fix $h = (h_1, \ldots, h_m) \in \mathbb{R}^m$ and consider the function

$$x \mapsto g(x) = d_x f(h) = \sum_{k=1}^{m} f'_{x_k}(x) h_k.$$

This function is, obviously, smooth and hence has a differential at a. Denoting the argument of this differential by $h' = (h'_1, \ldots, h'_m)$, we obtain

$$d_a g(h') = \sum_{j=1}^{m} g'_{x_j}(a) h'_j = \sum_{j=1}^{m} \left(\sum_{k=1}^{m} f''_{x_k x_j}(a) h_k \right) h'_j.$$

The right-hand side depends linearly both on h and on h'. Now set $h' = h$. The resulting quadratic form is called the *second differential* of the function f at the point a and denoted by $d_a^2 f(h)$. In other words,

$$d_a^2 f(h) = d_a(d_x f(h))(h')\big|_{h'=h} = d_a(d_x f(h))(h) \tag{5}$$

(to emphasize the inductive nature of this definition, we slightly abuse notation, replacing the function g in the right-hand side by its value at the point x). It is easy to see that

$$d_a^2 f(h) = \sum_{1 \le k, j \le m} f''_{x_k x_j}(a) h_k h_j. \tag{6}$$

In other words, the second differential is the quadratic form corresponding to the Hessian matrix: $d_a^2 f(h) = \langle H_f(a) h, h \rangle$ (see Corollary 6.4).

The third, fourth, ... differentials are defined in a similar way. Thus, we adopt the following definition.

Definition Let $f \in C^n(O)$ with $n \ge 2$ and $a \in O$. Assuming that the differentials of order $1, 2, \ldots, n-1$ are already defined, we define the *differential of f of order n at the point a* by the formula

$$d_a^n f(h) = d_a(d_x^{n-1} f(h))(h')\big|_{h'=h} = d_a(d_x^{n-1} f(h))(h) \quad \text{for } h \in \mathbb{R}^m$$

(here, as in (5), we do not use any notation for the function $x \mapsto d_x^{n-1} f(h)$, replacing it in the right-hand side by $d_x^{n-1} f(h)$).

Note that the above definition could be applied also to functions of class $C^{n-1}(O)$ whose all partial derivatives of order $n-1$ are differentiable at a.

The reader can easily verify by induction that

$$d_a^n f(h) = \sum_{j_1=1}^{m} \cdots \sum_{j_n=1}^{m} \frac{\partial^n f(a)}{\partial x_{j_1} \ldots \partial x_{j_n}} h_{j_1} \ldots h_{j_n}. \tag{7}$$

In view of Theorem 6.4 on the equality of mixed partial derivatives that differ only by the order of differentiation, we see that two terms in the right-hand side coincide if and only if the corresponding products $h_{j_1} \cdots h_{j_n}$ coincide. The number of such products coinciding with h^α is calculated in Example 7.1: it is equal to $C_n^\alpha = \frac{n!}{\alpha!}$. Hence, collecting like terms yields

$$d_a^n f(h) = \sum_{|\alpha|=n} C_n^\alpha \, f^{(\alpha)}(a) \, h^\alpha = \sum_{|\alpha|=n} \frac{n!}{\alpha!} f^{(\alpha)}(a) \, h^\alpha. \tag{8}$$

We see from (7) and (8) that the differential of order n is a homogeneous polynomial of degree n. For functions of one variable, a simple calculation gives $d_a^n f(h) = f^{(n)}(a) h^n$.

In the conventional notation for differentials, the vector h is denoted by dx and its coordinates, by $dx_1, \ldots, dx_m$. In this notation, formulas (7) and (8) take the following form (for brevity, we omit the indication of the point a at which the differential is being calculated):

$$d^n f = \sum_{j_1=1}^{m} \cdots \sum_{j_n=1}^{m} \frac{\partial^n f}{\partial x_{j_1} \ldots \partial x_{j_n}} \, dx_{j_1} \ldots dx_{j_n} \tag{7'}$$

and

$$d^n f = \sum_{|\alpha|=n} C_n^\alpha \, f^{(\alpha)} \, dx^\alpha. \tag{8'}$$

In particular, the second differential takes the form

$$d^2 f = \sum_{k,j=1}^{m} f''_{x_k x_j} \, dx_k \, dx_j.$$

For $m = 2$, we obtain

$$d^2 f = f''_{x^2} dx^2 + 2 f''_{xy} \, dx \, dy + f''_{y^2} dy^2.$$

In the right-hand side, the products $dx \, dx$ and $dy \, dy$ are traditionally denoted by dx^2 and dy^2.

When calculating higher differentials in specific problems, it often makes sense to directly use the definition, rather than formulas (7) or (8), i.e., to successively compute the "differential of a differential."

Example Let us find the second and third differentials of the function

$$f(x, y, z) = x^3 - xy^2 + 2yz^2 + 3x^2 - 5y + 7z + 4.$$

We obtain (not collecting like terms to save space)

$$df = 3x^2 dx - y^2 dx - 2xy \, dy + 2z^2 \, dy + 4 \, yz \, dz + 6x \, dx - 5 \, dy + 7 \, dz;$$
$$d^2 f = 6x \, dx^2 - 2y \, dx \, dy - 2(y \, dx + x \, dy) dy + 4z \, dy \, dz + 4(z \, dy + y \, dz) dz + 6 \, dx^2;$$
$$d^3 f = 6 \, dx^3 - 2 \, dx \, dy^2 - 4 \, dx \, dy^2 + 4 \, dy \, dz^2 + 8 \, dy \, dz^2.$$

The reader may verify that the calculations of these differentials via formulas (7') and (8') are not shorter.

Exercises

1. Find the fourth differential of the function $(x, y, z) \mapsto \log(x + 2y - 3z)$ at a point where $x + 2y - 3z = a > 0$.
2. Find the fourth differential of the polynomial

$$P(x, y, z) = x^4 - 5x^2y^2 + 2xyz^2 - x^3 - y^3 + 3xyz + 2x^2y - 10z.$$

3. Find the differentials $d_a^3 P$ and $d_a^4 P$ of the monomial $P(x, y, z) = xy^2z^3$ at the point $a = (1, 1, 0)$.
4. Find the first and second differentials of the function

$$(s, t, x, y) \mapsto \begin{vmatrix} s & t \\ x & y \end{vmatrix}.$$

   Extend the obtained result to determinants of matrices of arbitrary size.
5. Describe all functions of class C^2 in a domain O for which the second differential is identically zero. Extend this result to differentials of an arbitrary order.

8 Taylor's Formula

8.1 Preliminaries

The reader is already familiar with Taylor's[1] formula for functions of one variable. It allows one to obtain a local approximation of a sufficiently smooth function by a function of the simplest kind, a polynomial. The goal of such an approximation is twofold. First, it makes it possible to replace a local analysis of the function (for example, finding extrema) by an analysis of a simpler object (a polynomial). Second, the resulting polynomial can be used to approximate the function with controlled error and, in particular, to compute its values with any desired accuracy. Solving the latter problem requires additional information on the character of the approximation. Since in the case of many variables, polynomials are still functions of the simplest kind, the problem of approximating an arbitrary function (of a certain smoothness) by polynomials remains important. Before discussing this problem, recall some results related to Taylor's formula for functions of one variable. The reader is probably familiar with the following statement, which we will neither prove nor use.

   Let Δ be an interval, $a \in \Delta$, and φ be a function that is $(n - 1)$-smooth in Δ and has a (finite) derivative of order n at a. Then for $x \in \Delta$ we have

$$\varphi(x) = \sum_{k=0}^{n} \frac{\varphi^{(k)}(a)}{k!}(x - a)^k + \rho_n(\varphi, a; x), \tag{1}$$

---

[1] Brook **Taylor** (1685–1731) was an English mathematician.

where

$$\rho_n(\varphi, a; x) \underset{x \to a}{=} o((x-a)^n). \tag{1'}$$

The polynomial

$$\sum_{k=0}^{n} \frac{\varphi^{(k)}(a)}{k!}(x-a)^k$$

is called the *Taylor polynomial of order n* of the function φ at the point a (as usual, by the derivative of order 0 we mean the function itself, so the term with $k = 0$ is $\varphi(a)$) and denoted by $T_n(\varphi, a; x)$; the shorter notation $T_n(a; x)$ is also used.

Equality (1) is called *Taylor's formula*, and the term $\rho_n(\varphi, a; x)$ (in short, $\rho_n(a; x)$) is called its *n*th *remainder*. The form (1') of the remainder was obtained by Peano[2], and relations (1), (1') are called Taylor's formula *with Peano remainder term*.

The above result describes only the asymptotic behavior of the *n*th remainder and gives no estimates of this remainder. Such estimates can be obtained using additional information on the smoothness of the function. The first step in this direction is to assume that $\varphi \in C^n(\Delta)$ and to write the *n*th remainder as an integral, i.e., to obtain the *integral form of the remainder*.

Theorem (Taylor's formula with integral form of the remainder) *Let* $\varphi \in C^n(\Delta)$ *where* Δ *is an interval. Then for any* $a, x \in \Delta$,

$$\rho_{n-1}(\varphi, a; x) = \frac{1}{(n-1)!} \int_a^x \varphi^{(n)}(y)(x-y)^{n-1}\, dy$$

$$= \frac{(x-a)^n}{(n-1)!} \int_0^1 (1-t)^{n-1} \varphi^{(n)}(a + t(x-a))\, dt. \tag{2}$$

We will prove this theorem by integration by parts.

Proof It follows from the fundamental theorem of calculus that

$$\varphi(x) = \varphi(a) + \int_a^x \varphi'(y)\, dy = \varphi(a) - \int_a^x \varphi'(y)\, d(x-y)$$

$$= \varphi(a) - \varphi'(y)(x-y)\Big|_a^x + \int_a^x \varphi''(y)(x-y)\, dy$$

$$= T_1(a; x) + \int_a^x \varphi''(y)(x-y)\, dy.$$

Repeatedly integrating by parts (as long as the smoothness of φ allows), we obtain

---

[2] Giuseppe **Peano** (1858–1932) was an Italian mathematician.

$$\varphi(x) = T_1(a;x) - \frac{1}{2}\varphi''(y)(x-y)^2\Big|_a^x + \frac{1}{2!}\int_a^x \varphi'''(y)(x-y)^2\, dy$$

$$= T_2(a;x) + \frac{1}{2!}\int_a^x \varphi'''(y)(x-y)^2\, dy = \dots$$

$$= T_{n-1}(a;x) + \frac{1}{(n-1)!}\int_a^x \varphi^{(n)}(y)(x-y)^{n-1}\, dy.$$

This proves the first equality in (2). Making the change of variable $y = a + t(x - a)$, $0 \le t \le 1$, in the last integral, we obtain the second equality. $\square$

As we will see, in the case of functions of several variables there is a formula generalizing (2). However, in its proof we cannot use integration by parts. That is why, we give another proof of (2), which can be generalized to the case of several variables. In this proof, the idea related to integration by parts is realized in a somewhat different way by introducing an auxiliary function.

Proof Fixing a and x, set

$$U(y) = T_{n-1}(y;x) = \sum_{k=0}^{n-1} \frac{\varphi^{(k)}(y)}{k!}(x-y)^k \quad (y \in \Delta).$$

Clearly, $U \in C^1(\Delta)$, with $U(a) = T_{n-1}(a;x)$ and $U(x) = \varphi(x)$. Hence,

$$\rho_{n-1}(a;x) = \varphi(x) - T_{n-1}(a;x) = U(x) - U(a) = \int_a^x U'(y)\, dy.$$

Let us find $U'(y)$:

$$U'(y) = \sum_{k=0}^{n-1} \frac{\varphi^{(k+1)}(y)}{k!}(x-y)^k - \sum_{k=0}^{n-1} \frac{\varphi^{(k)}(y)}{k!}k(x-y)^{k-1}.$$

Discarding the (vanishing) term corresponding to $k = 0$ in the second sum and taking $k - 1$ for a new summation index (which we still denote by k), we obtain

$$U'(y) = \sum_{k=0}^{n-1} \frac{\varphi^{(k+1)}(y)}{k!}(x-y)^k - \sum_{k=0}^{n-2} \frac{\varphi^{(k+1)}(y)}{k!}(x-y)^k = \frac{\varphi^{(n)}(y)}{(n-1)!}(x-y)^{n-1}.$$

This implies the first equality in (2):

$$\rho_{n-1}(a;x) = U(x) - U(a) = \int_a^x U'(y)\, dy = \frac{1}{(n-1)!}\int_a^x \varphi^{(n)}(y)(x-y)^{n-1}dy.$$

The second equality can be obtained from it in the same way as in the previous version of the proof.   □

The theorem we have proved has important corollaries.

Corollary 1 (an estimate for the nth remainder term in Taylor's formula) *Let $\varphi \in C^n(\Delta)$ where Δ is an interval, and let ω be the modulus of continuity of the function $\varphi^{(n)}$ on Δ. Then for all $a, x \in \Delta$,*

$$|\rho_n(\varphi, a; x)| \leq \frac{|x - a|^n}{n!} \omega(|x - a|).$$

This quantitative estimate of the remainder refines the qualitative information which follows from the Peano form. Note also that if the highest derivative satisfies on Δ the Lipschitz condition of order α, $0 < \alpha \leq 1$, with constant L, then $|\rho_n(a; x)| \leq \frac{L}{n!} |x - a|^{n+\alpha}$ (since in this case $\omega(t) \leq Lt^\alpha$).

Proof It follows from (2) that

$$\rho_n(a; x) = \varphi(x) - T_n(a; x) = \rho_{n-1}(x) - \frac{\varphi^{(n)}(a)}{n!}(x - a)^n$$

$$= \frac{1}{(n-1)!} \int_a^x \varphi^{(n)}(y)(x - y)^{n-1} dy - \frac{1}{(n-1)!} \int_a^x \varphi^{(n)}(a)(x - y)^{n-1} dy$$

$$= \frac{1}{(n-1)!} \int_a^x (\varphi^{(n)}(y) - \varphi^{(n)}(a))(x - y)^{n-1} dy.$$

Estimating the difference in the integrand via the modulus of continuity and assuming for definiteness that $a < x$, we obtain the required inequality:

$$|\rho_n(a; x)| \leq \frac{1}{(n-1)!} \int_a^x |\varphi^{(n)}(y) - \varphi^{(n)}(a)|(x - y)^{n-1} dy$$

$$\leq \frac{1}{(n-1)!} \int_a^x \omega(x - a)(x - y)^{n-1} dy = \frac{(x - a)^n}{n!} \omega(x - a). \qquad □$$

We also mention another important result: Taylor's formula with Lagrange remainder term.

Corollary 2 (Taylor's formula with Lagrange remainder term) *Under the assumptions of the theorem, between the points a and x there is a point $\bar{x}$ such that*

$$\rho_{n-1}(\varphi, a; x) = \frac{\varphi^{(n)}(\bar{x})}{n!}(x - a)^n.$$

Proof It suffices to apply the mean value theorem to the second integral in (2). By this theorem, there is a point $\theta \in (0, 1)$ such that

$$\int\limits_0^1 (1-t)^{n-1} \varphi^{(n)}(a + t(x-a))dt = \varphi^{(n)}(a + \theta(x-a)) \int\limits_0^1 (1-t)^{n-1}dt$$

$$= \frac{1}{n}\varphi^{(n)}(a + \theta(x-a)).$$

Together with (2), this gives the Lagrange form of the remainder term with $\bar{x} = a + \theta(x-a)$. □

8.2 Taylor's Formula for Functions of Several Variables

To obtain an analog of Taylor's formula for a function of several variables, it is useful to rewrite (1) in a somewhat different form. With a view to subsequent generalization and since a function of several variables has many different partial derivatives, it makes sense to write Taylor's formula in terms of higher differentials (in the sum below, for consistency of notation, the term corresponding to $k = 0$, i.e., $\varphi(a)$, is written as the differential of order 0 of φ):

$$T_n(\varphi, a; x) = \sum_{k=0}^{n} \frac{d_a^k \varphi(x-a)}{k!}.$$

It is this equality that will serve as a foundation for a generalization of the notion of Taylor polynomial.

Recall that for a function f of several variables, we have (see formula (8) at p. 71)

$$d_a^k f(h) = \sum_{|\alpha|=k} \frac{k!}{\alpha!} f^{(\alpha)}(a) h^\alpha.$$

First, we establish a technical result.

In what follows, O is a nonempty open subset of $\mathbb{R}^m$.

Lemma *Let $f \in C^n(O)$, $a \in O$, and let $h = (h_1, \ldots, h_m)$ be a vector such that $a + th \in O$ for $0 \le t \le 1$. Let φ be the function defined on $[0, 1]$ as $\varphi(t) = f(a + th)$. Then for all $k = 1, 2, \ldots, n$,*

$$\varphi^{(k)}(t) = \sum_{j_1=1}^{m} \cdots \sum_{j_k=1}^{m} \frac{\partial^k f(a + th)}{\partial x_{j_1} \ldots \partial x_{j_k}} h_{j_1} \ldots h_{j_k} \quad (t \in [0, 1]).$$

One can see from formula (7) in the previous section that the right-hand side is nothing else than the kth differential of f at the point $a + th$. Thus, for $k = 1, \ldots, n$, we have

$$\varphi^{(k)}(t) = d_{a+th}^k f(h) = \sum_{|\alpha|=k} \frac{k!}{\alpha!} f^{(\alpha)}(a + th) h^\alpha$$

(the last equality holds by formula (8) at p. 71).

Proof of the lemma The n-smoothness of φ follows from Theorem 6.6, and the formula for $\varphi^{(k)}(t)$ can easily be obtained by induction (as in the proof of formula (7) at p. 70); we leave this to the reader. □

Now we are ready to obtain one of the most important results of the differential calculus of functions of several variables.

Theorem *Let $f \in C^n(O)$, and let a, x be points such that the segment*

$$[a, x] = \{a + t(x - a) : t \in [0, 1]\}$$

is contained in O. Then

$$f(x) = \sum_{k=0}^{n-1} \frac{1}{k!} d_a^k f(x - a) + \rho_{n-1}(f, a; x),$$

where

$$\rho_{n-1}(f, a; x) = \frac{1}{(n-1)!} \int_0^1 (1 - t)^{n-1} d_{a+t(x-a)}^n f(x - a) \, dt$$

$$= n \int_0^1 (1 - t)^{n-1} \sum_{|\alpha|=n} \frac{f^{(\alpha)}(a + t(x - a))}{\alpha!} (x - a)^\alpha \, dt. \qquad (3)$$

Proof We introduce the auxiliary function

$$t \mapsto \varphi(t) = f(a + t(x - a)), \quad \text{where } t \in [0, 1].$$

Then $\varphi \in C^n([0, 1])$, $\varphi(0) = f(a)$, $\varphi(1) = f(x)$. By Theorem 8.1 (with $a = 0$, $x = 1$),

$$\varphi(1) = \sum_{k=0}^{n-1} \frac{\varphi^{(k)}(0)}{k!} + \frac{1}{(n-1)!} \int_0^1 (1 - t)^{n-1} \varphi^{(n)}(t) \, dt.$$

To complete the proof, it suffices to replace the function φ and its derivatives by their values obtained in the lemma. □

Formula (3) together with other results motivates the following definition.

Definition Let $f \in C^n(O)$, $a \in O$. The polynomial[3]

$$T_n(f, a; x) = \sum_{k=0}^{n} \frac{1}{k!} d_a^k f(x - a)$$

---

[3] As before, by the term $d_a^0 f(x - a)$ corresponding to $k = 0$ we mean $f(a)$.

is called the nth *Taylor polynomial* of the function f at the point a (we will also use the shorter notation $T_n(a;x)$).

The equality

$$f(x) = T_n(f, a; x) + \rho_n(f, a; x)$$

is called Taylor's formula, and the difference

$$\rho_n(f, a; x) = f(x) - T_n(f, a; x)$$

(or, in short, $\rho_n(a;x)$) is called its nth remainder.

Substituting the expressions for the differentials of f into the definition of the Taylor polynomial, we obtain

$$T_n(f, a; x) = \sum_{k=0}^{n} \frac{1}{k!} \sum_{|\alpha|=k} \frac{k!}{\alpha!} f^{(\alpha)}(a) (x - a)^\alpha = \sum_{|\alpha| \le n} \frac{f^{(\alpha)}(a)}{\alpha!} (x - a)^\alpha.$$

Since a polynomial can be uniquely expanded in powers of $(x - a)^\alpha$ and in view of formula (4) from Section 7.1, all derivatives of the polynomial $T_n(f, a; x)$ up to order n at the point a coincide with the corresponding derivatives of the function f.

The integral form of the remainder obtained in the theorem implies corollaries similar to corollaries of Theorem 8.1.

Corollary 1 (an estimate for the nth remainder term in Taylor's formula) *Let $f \in C^n(O)$.*

1. As in the theorem, let a, x be points such that the segment $[a, x]$ is contained in O. If, besides, ω is a majorant of the moduli of continuity on $[a, x]$ of all n-order partial derivatives of f, then

$$|\rho_n(f, a; x)| \le \frac{(\sqrt{m}\, \|x - a\|)^n}{n!} \, \omega(\|x - a\|). \tag{4}$$

2. If A is a convex set contained in O (see Section 0.12) and ω is a majorant on A of the moduli of continuity of the highest derivatives of f, then (4) holds for any $a, x \in A$.

This corollary implies that at every point $a \in O$, Taylor's formula can be written with Peano remainder term:

$$\rho_n(f, a; x) \underset{x \to a}{=} o(\|x - a\|^n).$$

Indeed, let $A = \bar{B}(a, r) \subset O$ in the second assertion of the corollary. We may assume without loss of generality that

$$\omega(\|x - a\|) \xrightarrow[x \to a]{} 0,$$

hence the Peano form of the remainder immediately follows from (4).

Note also that if the highest derivatives of f satisfy on A the Lipschitz condition of order α, $0 < \alpha \le 1$, with constant L, then (4) takes the form

$$|\rho_n(a;x)| \le \frac{Lm^{n/2}}{n!} \|x - a\|^{n+\alpha} \quad \text{for } x \in A. \tag{4'}$$

Proof of Corollary 1 By (3), we have

$$f(x) = T_{n-1}(a;x) + n \int_0^1 (1-t)^{n-1} \sum_{|\alpha|=n} \frac{(x-a)^n}{\alpha!} f^{(\alpha)}(a + t(x-a))dt.$$

Besides, obviously,

$$T_n(a;x) = T_{n-1}(a;x) + \sum_{|\alpha|=n} \frac{(x-a)^\alpha}{\alpha!} f^{(\alpha)}(a)$$

$$= T_{n-1}(a;x) + n \int_0^1 (1-t)^{n-1} \sum_{|\alpha|=n} \frac{(x-a)^\alpha}{\alpha!} f^{(\alpha)}(a)\, dt.$$

Subtracting the last equality from the previous one yields

$$\rho_n(a;x) = f(x) - T_n(a;x)$$

$$= n \int_0^1 (1-t)^{n-1} \sum_{|\alpha|=n} \frac{(x-a)^\alpha}{\alpha!} \left(f^{(\alpha)}(a + t(x-a)) - f^{(\alpha)}(a)\right)dt.$$

Hence (below we use the multinomial theorem),

$$|\rho_n(a;x)| \le n \int_0^1 (1-t)^{n-1} \sum_{|\alpha|=n} \frac{|(x-a)^\alpha|}{\alpha!} |f^{(\alpha)}(a + t(x-a)) - f^{(\alpha)}(a)|dt$$

$$\le n\omega(\|x - a\|) \int_0^1 (1-t)^{n-1} \sum_{|\alpha|=n} \frac{|x_1 - a_1|^{\alpha_1} \ldots |x_m - a_m|^{\alpha_m}}{\alpha!} dt$$

$$= \frac{\omega(\|x - a\|)}{n!} \left(\sum_{j=1}^m |x_j - a_j|\right)^n \le \frac{(\sqrt{m}\,\|x - a\|)^n}{n!}\omega(\|x - a\|). \qquad \square$$

Let us mention another important result: Taylor's formula with Lagrange remainder term.

Corollary 2 (Taylor's formula with Lagrange remainder term) *Under the assumptions of the theorem, there exists a point $\bar{x} \in [a, x]$ such that*

$$\rho_{n-1}(f, a; x) = \frac{1}{n!} d_{\tilde{x}}^n f(x - a) = \sum_{|\alpha|=n} \frac{1}{\alpha!} f^{(\alpha)}(\bar{x}) (x - a)^\alpha.$$

Proof As in the case of one variable (see Corollary 1 at p. 75), it suffices to apply the mean value theorem to the integral from the integral representation of the remainder (see the first equality in (3)), which immediately implies the required result.   □

8.3 Another Estimate for the Remainder

In the theorem from the previous subsection, we obtained an integral representation of the remainder in Taylor's formula under the assumption that the segment between a and x is contained in O. Here we obtain a similar result in the case where these points are connected by an arbitrary piecewise smooth path. The proof relies on introducing a special auxiliary function and is similar to the second proof of Theorem 8.1.

Let F be a smooth function in a domain O. Recall (see Example 1 at p. 49) that the increment of F corresponding to points $a, x \in O$ can be written as an integral over a piecewise smooth path $\gamma : [0, 1] \to O$ connecting these points:

$$F(x) - F(a) = \int_0^1 \langle \operatorname{grad} F(\gamma(t)), \gamma'(t) \rangle \, dt. \tag{5}$$

This allows us to obtain an integral representation of the remainder in a case that is not covered by Theorem 8.2.

Theorem 1 *If $f \in C^n(O)$ and points a, x are connected by a piecewise smooth path γ defined on $[0, 1]$ and lying in O, then the remainders $\rho_{n-1}(f, a; x)$ and $\rho_n(f, a; x)$ can be written as*

$$\rho_{n-1}(f, a; x) = \int_0^1 \sum_{|\alpha|=n-1} \frac{(x - \gamma(t))^\alpha}{\alpha!} \langle \operatorname{grad} f^{(\alpha)}(\gamma(t)), \gamma'(t) \rangle \, dt,$$

$$\rho_n(f, a; x) = \int_0^1 \sum_{|\alpha|=n-1} \frac{(x - \gamma(t))^\alpha}{\alpha!} \langle \operatorname{grad} f^{(\alpha)}(\gamma(t)) - \operatorname{grad} f^{(\alpha)}(a), \gamma'(t) \rangle \, dt.$$

Proof First, we establish the first equality using the same idea as in the second proof of Theorem 8.1 in a more general situation. With this aim in mind, for $y \in O$ set

$$U(y) = T_{n-1}(y; x) = \sum_{|\alpha|<n} \frac{f^{(\alpha)}(y)}{\alpha!} (x - y)^\alpha.$$

Clearly, $U \in C^1(O)$. Besides, $U(a) = T_{n-1}(a; x)$, $U(x) = f(x)$, and hence

$$U(x) - U(a) = \rho_{n-1}(a; x). \tag{6}$$

Now, in order to make use of formula (5) (with $F = U$), we find the gradient of U. For this, we calculate its partial derivatives. For $j = 1, \ldots, m$, obviously,

$$\frac{\partial U}{\partial y_j}(y) = \sum_{|\alpha|<n} \left(\frac{1}{\alpha!} \frac{\partial f^{(\alpha)}(y)}{\partial y_j}(x-y)^\alpha - \frac{\alpha_j}{\alpha!} f^{(\alpha)}(y)(x-y)^{\alpha-e_j} \right).$$

Since for $\alpha_j = 0$ the subtrahends in the right-hand side are identically zero because of the factors α_j, we omit these terms and obtain

$$\frac{\partial U}{\partial y_j}(y) = \sum_{|\alpha|<n} \frac{1}{\alpha!} \frac{\partial f^{(\alpha)}(y)}{\partial y_j}(x-y)^\alpha - \sum_{\substack{|\alpha|<n, \\ \alpha_j>0}} \frac{1}{(\alpha-e_j)!} f^{(\alpha)}(y)(x-y)^{\alpha-e_j}$$

(here e_j is a canonical basis vector). Replacing in the second sum $\alpha - e_j$ by α and $f^{(\alpha)}$ by $f^{(\alpha+e_j)}$, i.e., by $\frac{\partial f^{(\alpha)}}{\partial y_j}$, we can rewrite the last equality as follows:

$$\frac{\partial U}{\partial y_j}(y) = \sum_{|\alpha|<n} \frac{1}{\alpha!} \frac{\partial f^{(\alpha)}(y)}{\partial y_j}(x-y)^\alpha - \sum_{|\alpha|<n-1} \frac{1}{\alpha!} \frac{\partial f^{(\alpha)}(y)}{\partial y_j}(x-y)^\alpha$$

$$= \sum_{|\alpha|=n-1} \frac{1}{\alpha!} \frac{\partial f^{(\alpha)}(y)}{\partial y_j}(x-y)^\alpha$$

for all $j = 1, \ldots, m$. So,

$$\operatorname{grad} U(y) = \sum_{|\alpha|=n-1} \frac{(x-y)^\alpha}{\alpha!} \operatorname{grad} f^{(\alpha)}(y).$$

Since the increment of U is equal to the remainder $\rho_{n-1}(a; x)$ (see (6)), formula (5) with U instead of F and the obtained expression for $\operatorname{grad} U$ give the required form of this remainder.

An integral representation of the remainder $\rho_n(a; x)$ can be obtained using formula (5) and the function $U + V$ instead of F, where

$$V(y) = \sum_{|\alpha|=n} \frac{f^{(\alpha)}(a)}{\alpha!}(x-y)^\alpha.$$

Indeed, since $V(x) = 0$ and

$$V(a) = T_n(a; x) - T_{n-1}(a; x) = \rho_{n-1}(a; x) - \rho_n(a; x),$$

we obtain $V(x) - V(a) = \rho_n(a; x) - \rho_{n-1}(a; x)$. Since, by (6), the increment of U is equal to $\rho_{n-1}(a; x)$, the increment of the sum $U + V$ is equal to $\rho_n(a; x)$. To make use of formula (5), it remains to find the gradient of this sum. Since $\operatorname{grad} U$ is already known, we must find only the gradient of the polynomial V. Repeating (with some simplifications) the calculation of the partial derivatives of U, we can easily obtain the equality

$$\operatorname{grad} V(y) = - \sum_{|\alpha|=n-1} \frac{(x-y)^\alpha}{\alpha!} \operatorname{grad} f^{(\alpha)}(a).$$

Therefore,

$$\operatorname{grad}(U+V)(y) = \sum_{|\alpha|=n-1} \frac{(x-y)^\alpha}{\alpha!} (\operatorname{grad} f^{(\alpha)}(y) - \operatorname{grad} f^{(\alpha)}(a)).$$

Now, to obtain an integral representation of $\rho_n(a;x)$, it remains to substitute this equality into (5) with $U+V$ instead of F. The details are left to the reader. $\qquad\square$

As in Sections 8.1, 8.2, we can derive from the theorem an important estimate for the remainder.

Theorem 2 (an estimate for the remainder) *Let $f \in C^n(O)$, let a, x be points from O connected by a piecewise smooth path γ of length S defined on $[0, 1]$ and lying in O, and let $L = \gamma([0, 1])$ be the support of γ. Besides, let ω be a majorant of the moduli of continuity on L of all n-order derivatives of f. Then the remainder $\rho_n(f, a; x) = f(x) - T_n(f, a; x)$ satisfies the following inequality:*

$$|\rho_n(f, a; x)| \le \frac{(\sqrt{m}\, S)^n}{n!} \omega(S).$$

Proof Using the representation of $\rho_n(a;x)$ established in Theorem 1, we obtain

$$|\rho_n(a;x)| \le \int_0^1 \sum_{|\alpha|=n-1} \left| \frac{(x-\gamma(t))^\alpha}{\alpha!} \langle \operatorname{grad} f^{(\alpha)}(\gamma(t)) - \operatorname{grad} f^{(\alpha)}(a), \gamma'(t) \rangle \right| dt$$

$$\le \int_0^1 \sum_{|\alpha|=n-1} \frac{|(x-\gamma(t))^\alpha|}{\alpha!} \left\| \operatorname{grad} f^{(\alpha)}(\gamma(t)) - \operatorname{grad} f^{(\alpha)}(a) \right\| \cdot \|\gamma'(t)\| \, dt.$$

Since for every α the increment (along L) of each derivative $\frac{\partial f^{(\alpha)}}{\partial x_j}$ $(j = 1, \dots, m)$ is majorized by its modulus of continuity $\omega_{\alpha,j}$ on L, we have

$$\left| \frac{\partial f^{(\alpha)}}{\partial x_j}(\gamma(t)) - \frac{\partial f^{(\alpha)}}{\partial x_j}(a) \right| \le \omega_{\alpha,j}(\|\gamma(t) - a\|) \le \omega_{\alpha,j}(S) \le \omega(S).$$

Hence, $\left\| \operatorname{grad} f^{(\alpha)}(\gamma(t)) - \operatorname{grad} f^{(\alpha)}(a) \right\| \le \sqrt{m}\, \omega(S)$, which yields the inequality

$$|\rho_n(a;x)| \le \sqrt{m}\, \omega(S) \int_0^1 \sum_{|\alpha|=n-1} \frac{|(x-\gamma(t))^\alpha|}{\alpha!} \|\gamma'(t)\| \, dt.$$

Using the multinomial theorem, we transform and estimate the arising sum, letting $x = (x_1, \ldots, x_m)$ and $\gamma(t) = (\gamma_1(t), \ldots, \gamma_m(t))$:

$$\sum_{|\alpha|=n-1} \frac{|(x - \gamma(t))^\alpha|}{\alpha!} = \sum_{|\alpha|=n-1} \frac{|x_1 - \gamma_1(t)|^{\alpha_1} \ldots |x_m - \gamma_m(t)|^{\alpha_m}}{\alpha!}$$

$$= \frac{1}{(n-1)!} (|x_1 - \gamma_1(t)| + \ldots + |x_m - \gamma_m(t)|)^{n-1}$$

$$\leq \frac{1}{(n-1)!} \left(\sqrt{m} \, \|x - \gamma(t)\| \right)^{n-1}.$$

Thus, the obtained inequality simplifies to

$$|\rho_n(a;x)| \leq \frac{m^{n/2}\omega(S)}{(n-1)!} \int_0^1 \|x - \gamma(t)\|^{n-1} \|\gamma'(t)\| \, dt.$$

Now, we introduce the function $\sigma(t) = \int_0^t \|\gamma'(u)\| \, du$, the length of the path γ corresponding to a segment $[0, t]$. Clearly, $\sigma'(t) = \|\gamma'(t)\|$ (for all points t except, possibly, finitely many), $\sigma(1) = S$, and $\|x - \gamma(t)\| \leq S - \sigma(t)$. This allows us to estimate the last integral and to obtain the required estimate:

$$|\rho_n(a;x)| \leq \frac{m^{n/2}\omega(S)}{(n-1)!} \int_0^1 (S - \sigma(t))^{n-1} \sigma'(t) \, dt = \frac{m^{n/2}}{n!} S^n \, \omega(S). \qquad \square$$

8.4 A Characteristic Property of Taylor Polynomials

Here we show that among all polynomials of given degree, Taylor polynomials are characterized by an important special property: they provide the best approximation to a smooth function near a fixed point. To make the corresponding result sufficiently general, first we refine Lemma 7.1 on the linear independence of monomials.

Lemma *Let $n \in \{0, 1, 2 \ldots\}$ and $a \in \mathbb{R}^m$. Among all polynomials P with $\deg P \leq n$, only the zero polynomial satisfies the condition*

$$P(x) \underset{x \to a}{=} o(\|x - a\|^n).$$

Proof It suffices to consider the case $a = 0$, since after that the result in full generality can be obtained by passing to the polynomial $\widetilde{P}(x) = P(x + a)$. In what follows, we assume that $a = 0$.

The assertion to be proved is obvious for $n = 0$. So, in what follows, $n \geq 1$. Assuming that the assertion is true for polynomials of degree less than n, we write a polynomial $P(x) = \sum\limits_{|\alpha| \leq n} c_\alpha x^\alpha$ in the form $P = P_0 + P_1$ where

$$P_0(x) = \sum_{|\alpha| < n} c_\alpha x^\alpha \quad \text{and} \quad P_1(x) = \sum_{|\alpha| = n} c_\alpha x^\alpha.$$

Clearly, $|x^\alpha| \leq \|x\|^{|\alpha|}$, and hence $P_1(x) = O(\|x\|^n) \underset{x \to 0}{=} o(\|x\|^{n-1})$. Therefore,

$$P_0(x) = P(x) - P_1(x) \underset{x \to 0}{=} o(\|x\|^{n-1}).$$

Since $\deg P_0 \leq n - 1$, it follows that $P_0 = 0$ by the induction hypothesis. Thus, $P = P_1$ is a homogeneous polynomial of degree n. Hence, for every $x \in \mathbb{R}^m$ we have

$$t^n P(x) = P(tx) = o(\|tx\|^n) \underset{t \to 0}{=} o(t^n) \quad (t \in \mathbb{R}).$$

Dividing by t^n and passing to the limit, we see that $P(x) = 0$. Since x is arbitrary, it follows that P is the zero polynomial.   □

Now we are ready to establish a characteristic property of Taylor polynomials.

Theorem *Let $f \in C^n(O)$, $a \in O$. A polynomial P of degree at most n is the nth Taylor polynomial of f at the point a if and only if*

$$f(x) - P(x) \underset{x \to a}{=} o(\|x - a\|^n). \tag{7}$$

Proof The fact that the nth Taylor polynomial has this property was observed after Corollary 1 in Section 8.2.

Now assume that a polynomial P satisfies condition (7) and, consequently,

$$P(x) - T_n(a; x) = (P(x) - f(x)) + (f(x) - T_n(a; x)) \underset{x \to a}{=} o(\|x - a\|^n).$$

Then the lemma applied to $P(x) - T_n(f, a; x)$ says that this difference is identically zero.   □

The above theorem sometimes allows one to find the Taylor polynomials $T_n(f, a; x)$ (and thus the derivatives of f at a) without repeated differentiation.

Example 1 Let us find the partial derivatives of the function

$$x \mapsto f(x) = \frac{1}{1 - \|x\|^2} \quad (\|x\| < 1)$$

at the origin.

Clearly, repeated differentiation leads to cumbersome calculations. However, the theorem allows us to avoid them and directly find the Taylor polynomials of arbitrary order. For this, we use the equality

$$f(x) = 1 + \|x\|^2 + \|x\|^4 + \ldots + \|x\|^{2n} + \frac{\|x\|^{2n+2}}{1 - \|x\|^2}$$

(valid for every $n \in \mathbb{N}$). Since the last term is $o(\|x\|^{2n})$ as $x \to 0$, it follows from the theorem (with n replaced by $2n$) that

$$T_{2n}(f, 0; x) = \sum_{k=0}^{n} \|x\|^{2k} = \sum_{k=0}^{n} (x_1^2 + \ldots + x_m^2)^k = \sum_{k=0}^{n} \sum_{|\beta|=k} \frac{k!}{\beta!} x^{2\beta}$$

(here we have used the multinomial theorem). Hence,

$$\frac{f^{(2\beta)}(0)}{(2\beta)!} = \frac{k!}{\beta!} \quad \text{for } k = |\beta|, \quad \text{i.e., } f^{(2\beta)}(0) = \frac{(2\beta)!}{\beta!}(|\beta|)!,$$

and $f^{(\alpha)}(0) = 0$ if the multi-index α has an odd coordinate.

Example 2 Now we give a simple proof of Leibniz's rule which we discussed in the remark after Theorem 6.5.

Multiplying the Taylor formulas for n-smooth functions f, g, one can easily check that

$$f(x)g(x) = \sum_{|\alpha| \le n} \frac{1}{\alpha!} \left(\sum_{\beta+\gamma=\alpha} \frac{\alpha!}{\beta!\gamma!} f^{(\beta)}(a) g^{(\gamma)}(a) \right) (x-a)^{\alpha} + o(\|x-a\|^n)$$

as $x \to a$. By the above theorem, the outer sum is the Taylor polynomial of order n of the function fg, and the inner sum is exactly the derivative $(fg)^{(\alpha)}(a)$, which proves the formula stated in Section 6.5.

Exercises

1. Find
   a) the derivative $\frac{\partial^4 f(0)}{\partial x_i^2 \partial x_j^2}$ of the function $x \mapsto f(x) = \frac{1-\|x\|^2}{1+\|x\|^2}$;

   b) the derivative $\frac{\partial^6 f(0,0)}{\partial x^4 \partial y^2}$ of the function $(x, y) \mapsto f(x, y) = \log \frac{1+xy^2}{1-x^2y}$.
2. Show that condition (7) from Theorem 8.4 can be satisfied for every n for a function that is discontinuous everywhere except the point a (and, consequently, that has no partial derivatives of order greater than 1 at this point).
3. A function f defined on a convex subset E (see Section 0.12) in $\mathbb{R}^m$ is said to be *convex* if

$$f(tx + (1-t)y) \le tf(x) + (1-t)f(y) \quad \text{for any } x, y \in E, \ t \in [0, 1].$$

Show that a function of class $C^2(O)$ is convex if and only if at every point from O its second differential is a nonnegative quadratic form (the set O is assumed to be convex).

4. Show that if a function f in a neighborhood of a point a has all partial derivatives of order $n - 1$ and they are differentiable at a, then

$$f(x) - T_n(f, a; x) = f(x) - \sum_{|\alpha| \le n} \frac{f^{(\alpha)}(a)}{\alpha!} (x - a)^\alpha \underset{x \to a}{=} o(\|x - a\|^n).$$

9 Extrema of Functions of Several Variables

Finding the extrema of a function is one of the most important problems whose solution was made possible by the discovery of differential calculus. The reader is undoubtedly familiar with this problem and methods for solving it in the case of functions of one variable. In this section, our aim is to learn how to find extrema of functions of several variables.

9.1 A Necessary Condition for an Extremum

The central notion of this section, that of an extremum of a function of several variables, is defined in essentially the same way as in the one-dimensional case.

Definition We say that a function f defined on a set $X \subset \mathbb{R}^m$ has a *local minimum* at a point $a \in X$ if there exists a neighborhood U of this point such that

$$f(x) \ge f(a) \quad \text{for all } x \in X \cap U.$$

If, besides, $f(x) > f(a)$ for $x \neq a$, then the local minimum is said to be *strict*.

A *local maximum* and a *strict local maximum* are defined in a similar way. Local minima and maxima are included under the common term *local extremum*.

For a function that is differentiable at a point of extremum, there is a necessary condition quite similar to that established for a function of one variable.

Theorem *Let a be an interior point of a set $X \subset \mathbb{R}^m$ at which a function $f : X \to \mathbb{R}$ has a local extremum. If f is differentiable at this point, then* $\operatorname{grad} f(a) = 0$, *i.e.*,

$$f'_{x_1}(a) = \ldots = f'_{x_m}(a) = 0. \tag{1}$$

Points at which condition (1) is satisfied are called *critical points* of f.

Note that to find the critical points, one must solve a system (1) of m equations with the same number of unknowns (the coordinates of critical points).

Proof Let e be an arbitrary unit vector. Consider the function

$$t \mapsto \varphi(t) = f(a + te).$$

Clearly, it is defined in a neighborhood of 0 and has a local extremum at this point. Besides, φ is differentiable at 0, and hence $\varphi'(0) = 0$, which is equivalent to $\frac{\partial f}{\partial e}(a) = 0$. Taking for e the canonical basis vectors, we arrive at (1). □

Looking at the function $(x, y) \mapsto xy$, for which the obtained necessary condition is satisfied at the origin, we see that it is not sufficient. That is why, at critical points further study of the behavior of the function is necessary (see Section 9.3 below).

9.2 A Refined Necessary Condition for an Extremum

Below, we consider only C^2-smooth functions defined on an open subset of $\mathbb{R}^m$. For such functions, the necessary condition (1) can be refined using the second differential. Since it is a quadratic form, we need some terminology from algebra. Recall the following definition.

Definition A quadratic form Q is said to be *positive* (*negative*) *definite* if $Q(h) > 0$ (respectively, $Q(h) < 0$) for all $h \neq 0$. A quadratic form that assumes both positive and negative values is said to be *indefinite*.

A quadratic form that is either positive definite or negative definite is said to be definite. If Q is the form given by a symmetric $m \times m$ matrix A, i.e.,

$$Q(h) = \langle Ah, h \rangle,$$

then the main diagonal of A contains the values of Q on the basis vectors. Hence, for Q to be definite, it is necessary that the diagonal elements of A have the same sign. If they include both positive and negative values, then the form is indefinite.

Our definition does not cover all possible cases, since it may happen that a quadratic form does not change sign but has nontrivial zeros (other than 0), as, for instance, the squared sum of the coordinates.

Theorem *Let a be a point of extremum of a C^2-smooth function f. Then the first differential of f at a vanishes, and the second differential does not change sign. More exactly, for every $h \in \mathbb{R}^m$, the following assertions hold:*
1) $d_a f(h) = 0$ *and* $d_a^2 f(h) \geq 0$ *if f has a local minimum at a;*
2) $d_a f(h) = 0$ *and* $d_a^2 f(h) \leq 0$ *if f has a local maximum at a.*

Proof The equality $d_a f(h) = 0$ for all h follows from Theorem 9.1. Below, we assume without loss of generality that f has a local minimum at a. By assumption, for every vector h, the difference $f(a + th) - f(a)$ is nonnegative for sufficiently small $t > 0$. On the other hand, by Taylor's formula with Peano remainder term (see Section 8.2),

$$f(a + th) - f(a) \underset{t \to 0}{=} d_a f(th) + \frac{1}{2} d_a^2 f(th) + o(\|th\|^2) \underset{t \to 0}{=} \frac{t^2}{2} d_a^2 f(h) + o(t^2)$$

(we have used the fact that $d_a f = 0$). Since the left-hand side is nonnegative for small t, dividing by t^2 and taking the limit as $t \to 0$ yields $\frac{1}{2} d_a^2 f(h) \geq 0$. This completes the proof, since h is an arbitrary vector.   □

This theorem immediately implies the following important corollary.

Corollary *If f is a function of class C^2 and the differential $d_a^2 f$ is an indefinite quadratic form, then f does not have a local extremum at a.*

In particular, this is the case if among the numbers $\frac{\partial^2 f}{\partial x_1^2}(a), \ldots, \frac{\partial^2 f}{\partial x_m^2}(a)$ there are both positive and negative values.

The above corollary has no counterpart for a function of one variable, since the second differential of such a function is proportional to $(dx)^2$ and cannot change sign.

We conclude this section by considering examples which show that the refined necessary condition is not sufficient for a function to have an extremum.

Example 1 The functions

$$(x, y) \mapsto x^2 + y^4 \quad \text{and} \quad (x, y) \mapsto x^2 - y^4$$

have the same (nonnegative) second differential at the common critical point $(0, 0)$; however, the first one has an extremum at this point, while the second one has not.

Example 2 Let $\varphi(y) = e^{-\frac{1}{y^2}}$ if $y \neq 0$, $\varphi(0) = 0$. The infinitely smooth functions

$$(x, y) \mapsto x^2 + \varphi(y) \quad \text{and} \quad (x, y) \mapsto x^2 - \varphi(y)$$

have the same differentials of all orders at the common critical point $(0, 0)$; however, the first one has an extremum at this point, while the second one has not.

Example 3 The refined necessary condition is not sufficient for a function to have an extremum even if its restriction to every line passing through the critical point has a strict extremum at this point. To see this, look at the function

$$f(x, y) = (y - x^2)(y - 2x^2).$$

Since $f(x, y) = y^2 + O((x^2 + y^2)^{3/2})$, the origin is a critical point at which the second differential is nonnegative. Drawing the zero level set of f and indicating the signs of f at this figure, one can easily see that its restriction to every line passing through the origin has a strict minimum at 0, though f changes sign in an arbitrarily small neighborhood of the origin (and hence does not have an extremum at this point).

9.3 Sufficient Conditions for an Extremum

In the previous subsections, we have found necessary conditions for a smooth function to have an extremum. As already mentioned, they are not sufficient. Now, we will use the second differential to obtain conditions that guarantee the existence of an extremum at a critical point. First, we prove a lemma.

Lemma *If Q is a positive definite quadratic form, then there exists $c > 0$ such that $Q(x) \geq c\|x\|^2$ for all x.*

Proof Assume that the smallest value of Q on the unit sphere of the space where it is defined is attained at a point x_* (by Weierstrass's theorem, such a point does exist). We claim that the number $c = Q(x_*)$ satisfies the required condition. Clearly, $c > 0$, since $x_* \neq 0$. Represent an arbitrary point x in the form $\|x\| x'$ where $\|x'\| = 1$. Then, using the homogeneity of the quadratic form, we obtain

$$Q(x) = Q(\|x\|x') = \|x\|^2 Q(x') \geq \|x\|^2 Q(x_*) = c\|x\|^2. \qquad \square$$

Theorem *Let f be a function of class C^2, and assume that the second differential of f at a critical point a is a definite quadratic form. Then f has an extremum at a, which is a strict minimum if $d_a^2 f$ is positive definite, and a strict maximum if $d_a^2 f$ is negative definite.*

Remark The above theorem, together with Corollary 9.2, solves the problem of the existence and classification of an extremum in the case where the critical point is *nondegenerate* (or, as one sometimes says, nonsingular). This means that the second differential at this point is nondegenerate, i.e., that the Hessian matrix is invertible. The existence or nonexistence of an extremum at such a critical point depends on whether or not the Hessian matrix has both positive and negative eigenvalues.

Proof Let us prove that f has a minimum at a if the second differential is positive definite. By Taylor's formula, for all sufficiently small h we have

$$f(a + h) - f(a) = \frac{1}{2} d_a^2 f(h) + \|h\|^2 \omega(h) \tag{2}$$

where $\omega(h) \xrightarrow[h \to 0]{} 0$. By the lemma, there exists $c > 0$ such that $d_a^2 f(h) \geq c\|h\|^2$. Together with (2), this shows that

$$f(a + h) - f(a) \geq \frac{c}{2}\|h\|^2 + \omega(h)\|h\|^2 = \left(\frac{c}{2} + \omega(h)\right)\|h\|^2.$$

Since $\omega(h) \xrightarrow[h \to 0]{} 0$, the right-hand side is positive for all $h \neq 0$ with sufficiently small norm, which implies the assertion.

If the second differential is negative definite, then we may argue similarly or consider the function $-f$ instead of f. $\qquad \square$

Example 1 Let us find the local extrema of the function

$$f(x, y, z) = 2x^3 yz - (x^2 + y^2 + z^2) \quad (x, y, z \in \mathbb{R}).$$

To find the critical points, we set the partial derivatives equal to zero and consider the resulting system of equations

$$\begin{cases} 6x^2yz - 2x = 0, \\ 2x^3z - 2y = 0, \\ 2x^3y - 2z = 0. \end{cases}$$

One can easily see that it has five solutions:

$$(0,0,0), \ \left(1, \frac{1}{\sqrt{3}}, \frac{1}{\sqrt{3}}\right), \ \left(1, -\frac{1}{\sqrt{3}}, -\frac{1}{\sqrt{3}}\right), \ \left(-1, \frac{1}{\sqrt{3}}, -\frac{1}{\sqrt{3}}\right), \ \left(-1, -\frac{1}{\sqrt{3}}, \frac{1}{\sqrt{3}}\right).$$

At the first point, the second differential is a negative definite quadratic form:

$$d^2_{(0,0,0)}f(h) = -2\|h\|^2.$$

Therefore, at the origin f has a strict local maximum.

At the other critical points, we have $xyz = \frac{1}{3}$, as follows from the first equation of the system. Hence, at these points the derivative f''_{xx} is equal to 2, while the derivatives f''_{yy} and f''_{zz} are equal to -2. Therefore, at the remaining four points the second differential is an indefinite quadratic form and the function does not have a local extremum.

Example 2 Let us find the local extrema of the function

$$f(x, y) = x^2 - xy^2 + ay^3 + \frac{5}{4}y^4$$

where a is a real parameter.

Setting the partial derivatives equal to zero, one can easily see that there are two critical points: $x = y = 0$ and $x = \frac{9}{32}a^2$, $y = -\frac{3}{4}a$.

Simple calculations show that at the point $\left(\frac{9}{32}a^2, -\frac{3}{4}a\right)$ the second differential of f is positive definite for all $a \neq 0$. Thus, $\left(\frac{9}{32}a^2, -\frac{3}{4}a\right)$ is a strict minimum of f.

To analyze the behavior of f near the origin, we must proceed differently, since at this point the sufficient condition for an extremum is not satisfied:

$$d^2_{(0,0)}f(h_1, h_2) = 2h_1^2.$$

So far, it is clear only that $f(0,0)$ is not a local maximum (this follows from Theorem 9.2). To obtain the answer, it is useful to transform the formula for f:

$$f(x, y) = \left(x - \frac{1}{2}y^2\right)^2 + ay^3 + y^4.$$

Now, it is clear that for $a = 0$ the function f has a strict minimum at the origin. For $a \neq 0$, it does not have an extremum at this point, since the function $f(0, y) = ay^3 + \frac{5}{4}y^4$ changes sign at 0.

9.4 The Absolute Maximum and Minimum Values of a Function of Several Variables

So far, we have discussed problems related only to the local behavior of functions. However, it is of great importance, in particular, in various applications, to have information about their global properties. First and foremost, we mean the absolute maximum and minimum values of a function on a given set. If this set is compact and the function is continuous on it, then the existence of these values is ensured by the Weierstrass extreme value theorem (Section 0.8). The absolute maximum and minimum values of a function can often be found without analyzing the second differentials, simply by comparing its values at the critical points or using its specific properties. For example, if a nonnegative function vanishes at the boundary of a compact set and a critical point is unique, then, clearly, the function attains an absolute maximum at this point.

Example 1 (Huygens's[1] problem) Consider two elastic balls of the same size but with different masses whose centers can move along a fixed straight line. Assume that a ball of mass M moving with velocity V collides with a motionless ball of mass m. It follows from the laws of conservation of energy and momentum that the latter acquires the velocity $\frac{2M}{M+m} V$. If $M > m$, then this velocity is greater than the velocity of the first ball. Placing a ball of intermediate mass between the two balls, we can ensure that after two collisions the ball of mass m acquires a velocity greater than $\frac{2M}{M+m} V$. How should we choose the masses of n intermediate balls in order to maximize the velocity of the last ball after the $(n + 1)$th collision? What is the limit of this velocity as $n \to \infty$ (the masses m and $M > m$ are assumed fixed)?

It is easy to see by induction that the velocity acquired by the last ball after successive collisions with balls of masses $x_1, \ldots, x_n$ is equal to

$$V_n(x) = V_n(x_1, \ldots, x_n) = \frac{2M}{M + x_1} \cdot \frac{2x_1}{x_1 + x_2} \cdots \frac{2x_n}{x_n + m} \cdot V.$$

Thus, the problem reduces to that of finding the absolute maximum of the function V_n on the set $\mathbb{R}_+^n$. Note that V_n vanishes at the boundary of this set. Besides, the limit $\lim\limits_{\|x\| \to \infty} V_n(x)$ vanishes too; indeed, removing all factors $\frac{x_i}{x_i + x_{i+1}}$ that do not contain the maximum coordinate, we only increase the limit, which gives the bound

$$V_n(x) \le \frac{\text{const}}{\max\{x_1, \ldots, x_n\}} \le \frac{\text{const}}{\|x\|} \sqrt{n}.$$

Therefore, the absolute maximum of V_n is attained at some point of $\mathbb{R}_+^n$. It is a critical point of V_n. Setting the partial derivatives of V_n equal to zero, simple calculations give

$$\frac{x_1}{M} = \frac{x_2}{x_1} = \ldots = \frac{x_n}{x_{n-1}} = \frac{m}{x_n}.$$

---

[1] Christiaan **Huygens** (1629–1695) was a Dutch physicist and mathematician.

Multiplying these fractions, we see that $q^{n+1} = \frac{m}{M}$ where q is the common value of the fractions. Thus, V_n has a unique critical point in $\mathbb{R}^n_+$. Its coordinates form a geometric sequence with common ratio $q = \left(\frac{m}{M}\right)^{1/(n+1)}$. It is at this point that V_n attains an absolute maximum value V_n^*, which can be easily calculated: $V_n^* = \left(\frac{2}{1+q}\right)^{n+1} V$. It is not difficult to see that

$$V_n^* < V_{n+1}^* \quad \text{and} \quad V_n^* \xrightarrow[n\to\infty]{} \sqrt{\frac{M}{m}} V.$$

Example 2 Let us find the absolute maximum and minimum values of the function

$$f(x, y) = \left(\frac{1}{2} + \cos x + \cos y\right) \sin x \sin y$$

on the square $[0, \pi]^2$.

Obviously, the function changes sign inside the square and vanishes at its boundary. Therefore, it attains an absolute maximum and an absolute minimum at interior (and hence critical) points. To find them, we must solve the system of equations

$$\begin{cases} \cos x \sin y \left(\frac{1}{2} + \cos x + \cos y\right) - \sin^2 x \sin y = 0, \\ \sin x \cos y \left(\frac{1}{2} + \cos x + \cos y\right) - \sin x \sin^2 y = 0. \end{cases}$$

Since $\sin x \sin y \neq 0$, it can be simplified:

$$\begin{cases} \cos x \left(\frac{1}{2} + \cos x + \cos y\right) = \sin^2 x, \\ \cos y \left(\frac{1}{2} + \cos x + \cos y\right) = \sin^2 y. \end{cases}$$

Subtracting the second equation multiplied by $\cos x$ from the first equation multiplied by $\cos y$, we see that either $\cos x \cos y = -1$, or $\cos x = \cos y$. The first case cannot occur inside the square, while the second one leads to the equation $\cos x \left(\frac{1}{2} + 2 \cos x\right) = 1 - \cos^2 x$. Solving this quadratic (with respect to $\cos x$) equation, we see that either $\cos x = \cos y = \frac{1}{2}$, or $\cos x = \cos y = -\frac{2}{3}$. In the first case, $x = y = \frac{\pi}{3}$ and $\sin x = \sin y = \frac{\sqrt{3}}{2}$, so the value of f is equal to $\frac{9}{8}$. In the second case, $\sin x = \sin y = \frac{\sqrt{5}}{3}$ and the value of f is equal to $-\frac{25}{54}$. Since there are no other critical points, we have

$$\max_{[0,\pi]^2} f = f\left(\frac{\pi}{3}, \frac{\pi}{3}\right) = \frac{9}{8} \quad \text{and}$$

$$\min_{[0,\pi]^2} f = f\left(\arccos\left(-\frac{2}{3}\right), \arccos\left(-\frac{2}{3}\right)\right) = -\frac{25}{54}.$$

Along the way we obtain (since the function is even and periodic in each variable)

$$\max_{\mathbb{R}^2} f = \max_{[-\pi,\pi]^2} f = \max_{[0,\pi]^2} f = \frac{9}{8} \quad \text{and} \quad \min_{\mathbb{R}^2} f = \min_{[-\pi,\pi]^2} f = -\max_{[-\pi,\pi]^2} f = -\frac{9}{8}.$$

Example 3 Let us find the absolute maximum and minimum values of the function

$$f(x, y) = x^3 + y^3 - 9xy + 27$$

on the square $Q_a = [0, a]^2$, $a > 0$.

Clearly,

$$\text{grad } f(x, y) = 3(x^2 - 3y, y^2 - 3x),$$

and there are only two critical points: $(0,0)$ and $(3,3)$. Thus, two cases arise.

If $a > 3$, then the critical point $(3,3)$ lies inside the square Q_a. Obviously, $f(3,3) = 0$. Due to symmetry, analyzing the restriction of f to the boundary of Q_a, we can confine ourselves to considering only its restriction to the horizontal sides of the square. On the lower side, where $y = 0$, we obtain the function of one variable

$$g(x) = f(x, 0) = x^3 + 27, \quad 0 \le x \le a,$$

which is, obviously, increasing. Hence, the values of f at the lower side of the square lie between its values at the vertices $(0,0)$ and $(a,0)$, i.e., between 27 and $a^3 + 27$. The restriction of f to the upper side of the square yields the function

$$h(x) = f(x, a) = x^3 + a^3 - 9ax + 27 \quad (0 \le x \le a)$$

with a critical point $x_0 = \sqrt{3a}$, at which it attains a minimum. Since

$$f(x_0, a) = h(x_0) = a^3 - 6a\sqrt{3a} + 27 = \left(a^{3/2} - 3^{3/2}\right)^2 > 0 = f(3,3),$$

the absolute minimum of f on the square Q_a is attained at the point $(3,3)$ and equal to zero, i.e., f is nonnegative. The absolute maximum value is either $f(a, a) = 2a^3 - 9a^2 + 27$ or $f(a, 0) = f(0, a) = a^3 + 27$. Comparing these values, we see that $f(a, a) - f(0, a) = a^2(a - 9)$. Thus, for $a > 3$ we have

$$\min_{Q_a} f = f(3,3) = 0$$

and

$$\max_{Q_a} f = \max_{\partial Q_a} f = \begin{cases} f(0, a) = f(a, 0) = a^3 + 27 & \text{if } 3 < a \le 9, \\ f(a, a) = 2a^3 - 9a^2 + 27 & \text{if } a \ge 9. \end{cases}$$

For $a \le 3$, there are no critical points inside the square. Hence, the absolute maximum and minimum values are attained at the boundary. On the sides of Q_a that lie on the coordinate axes, the function f is increasing, while on the other sides, it is decreasing. It is easy to see that $f(a, a) < f(0, 0)$. Hence, in the case $a \in (0, 3]$ we have

$$\max_{Q_a} f = \max_{\partial Q_a} f = f(a,0) = f(0,a) = a^3 + 27,$$

$$\min_{Q_a} f = \min_{\partial Q_a} f = f(a,a) = 2a^3 - 9a^2 + 27.$$

In the last example, we have encountered a situation where points at which a function can attain an extreme value lie at the boundary of the set. This difficulty did not arise in the study of a function of one variable, which is defined on an interval, whose boundary consists of only two points. Now we discuss in more detail the problem of finding boundary points at which a function can attain extreme values.

Assume that we study a function f defined on a compact subset E in $\mathbb{R}^m$ and differentiable at its interior points. Also, assume that the boundary ∂E of E consists of finitely many parts each of which is a (wide-sense) graph of a function of $m - 1$ variables. Such a situation (with a very simple boundary) appeared in Example 3. Assume that our task is to find an extreme value of f on a part of ∂E that coincides with the graph of a function φ of $m - 1$ variables defined on a compact subset K in $\mathbb{R}^{m-1}$. This problem is equivalent to finding an extreme value of the composition

$$u = (u_1, \ldots, u_{m-1}) \mapsto g(u) = f(u_1, \ldots, u_{m-1}, \varphi(u)) \quad (u \in K).$$

Thus, the analysis of the function under consideration reduces to that of a function of a smaller number of variables. Note that in order to find an extreme value of g on the boundary of K, one may have to repeat the technique to further reduce the number of variables.

Consider an example illustrating the above considerations.

Example 4 Let us find the absolute maximum and minimum values of the function $f(x, y, z) = x(y^2 + 1)z^3$ on the ball $x^2 + y^2 + z^2 \leq 5$.

Since f is linear in the first variable, we have $\max f = -\min f$, and these values are attained at the boundary sphere. The form of the function suggests that this sphere should be regarded as the union of the graphs of the functions $y = \pm\sqrt{5 - x^2 - z^2}$. Thus, the original problem reduces to a simpler one, that of finding the absolute maximum value of the function

$$g(x, z) = xz^3(6 - x^2 - z^2)$$

on the disk $x^2 + z^2 \leq 5$. Setting its partial derivatives equal to zero and solving the resulting system of equations, one can easily find four critical points of the function g: $(\pm 1, \pm\sqrt{3})$. They lie inside the disk, and the values of g at these points are equal to $\pm 6\sqrt{3}$. It remains to find out whether these are the absolute extrema of g on the closed disk. Analyzing the behavior of g on the boundary circle, we may assume that $z \geq 0$. Then the problem reduces to the study of the function of one variable

$$h(x) = f\left(x, 0, \sqrt{5 - x^2}\right) = x(5 - x^2)^{3/2}$$

on the interval $\left[-\sqrt{5}, \sqrt{5}\right]$. Since $h(\pm\sqrt{5}) = 0$, the absolute maximum is attained at an interior point of the interval. It is easy to check that there are only two critical points: $x = \pm\frac{1}{2}\sqrt{5}$. The corresponding values of f are equal to $\pm\frac{75}{16}\sqrt{3}$. Since $\frac{75}{16}\sqrt{3} < 6\sqrt{3}$, the absolute maximum and minimum values of f are attained at the points $(\pm1, \pm1, \pm\sqrt{3})$ lying on the sphere $x^2 + y^2 + z^2 = 5$. More exactly, the absolute maximum value $6\sqrt{3}$ is attained at the points for which the signs of the first and last coordinates coincide, while the absolute minimum value $(-6\sqrt{3})$ is attained when their signs are opposite.

To conclude, note that the described method of analyzing a function on the boundary of its domain of definition is often impeded by the fact that the functions whose graphs form the boundary are unknown or given by cumbersome formulas. A method that allows one to overcome this difficulty in some cases will be considered in Section III.3.

Exercises

1. Let f be a function of class C^3. Show that if at a critical point

$$d^2 f = 0 \quad \text{and} \quad d^3 f \neq 0,$$

then f does not have an extremum at this point.

2. Let f be an n-smooth function that at a critical point a satisfies the conditions

$$d_a^2 f = \ldots = d_a^{n-1} f = 0.$$

Show that if the differential $d_a^n f$ changes sign (in particular, this is the case for odd n unless $d_a^n f$ is identically zero), then f does not have an extremum at a; and if n is even and $d_a^n f(h) \neq 0$ for all $h \neq 0$ (this condition is automatically satisfied for functions of one variable if $f^{(n)}(a) \neq 0$), then f does have an extremum at a.

3. Find the absolute maximum value of the function

$$f(x, y, z) = x^2 y(z^3 + 2)$$

on the ball $x^2 + y^2 + z^2 \leq 4$.

4. Show that

$$\max_{\mathbb{R}^3} xyz \frac{x+y+z}{e^{x^2+y^2+z^2}} = \frac{4}{3e^2} \quad \text{and} \quad \min_{\mathbb{R}^3} xyz \frac{x+y+z}{e^{x^2+y^2+z^2}} = -\frac{1+\sqrt{2}}{4e^2}.$$

5. Generalizing the result of the previous exercise, show that all points at which the function

$$x = (x_1, \ldots, x_m) \mapsto f(x) = x_1 \ldots x_m (x_1 + \ldots + x_m) e^{-\|x\|^2}$$

has an extremum lie on the sphere $\|x\| = \sqrt{\frac{m+1}{2}}$, and its absolute maximum value is attained at a point with equal coordinates and is equal to $m\left(\frac{m+1}{2em}\right)^{(m+1)/2}$. Show that if m is odd, then the absolute minimum value of f is attained at a critical point whose all coordinates but one are equal. Find this point.

6. Does there exist a function $f \in C^\infty(\mathbb{R}^2)$ such that $f(\mathbb{R}^2) = \mathbb{R}$ and f has a unique critical point at which it has a strict extremum (this is obviously impossible for a function of one variable).

   HINT. Consider the function from Exercise 4 in Section 1.

7. Find a polynomial in two variables that has no critical points in the upper half-plane except for two points of maximum.

8. Let f be a smooth function in an open disk B that is continuous in the closure of B and vanishes on the boundary circle. Show that if f has two local maxima in B, then it has at least one other critical point in this disk. Give an example showing that the boundary condition cannot be dropped.

10 Implicit Function

In this section, we deal with a problem which will be discussed in a more general setting in Section III.1. Here we consider a special case of this problem, with only one equation rather than a system of equations. This allows us, on the one hand, to avoid a number of technical issues and, on the other hand, to make fuller use of the geometric interpretation of the results. Besides, in this simple case we may benefit from monotonicity of functions.

10.1 Statement of the Problem

The reader is undoubtedly familiar with the problem of solving equations and systems of equations with parameters from high school. Now we will look at this problem from a slightly different angle. We do not seek to obtain any formulas describing the dependence of the solution on the known parameters, which is usually impossible to do. Instead, we are interested in conditions ensuring the existence of a solution and, if a solution does exist, its uniqueness and the properties of its dependence on the parameters (continuity, smoothness, etc.). Besides, we ask ourselves whether the solution is stable: is the solvability of the equation preserved under small perturbations of the parameters? In particular, this question is of importance for applications, where the values of parameters cannot, in principle, be determined exactly.

   As we have already mentioned, systems of equations will be considered in Chapter III, and now we restrict ourselves to the case of one equation with one unknown (but with an arbitrary number of parameters). The uniqueness problem in this case turns out to be relatively simple, since we can use the notion of monotonicity. It is also useful to bear in mind geometric considerations, interpreting a solution of an equation as the coordinate of a point belonging to a level set of the function under consideration.

In what follows, we assume that $\mathbb{R}^{m+1} = \mathbb{R}^m \times \mathbb{R}$, i.e., identify points of the space $\mathbb{R}^{m+1}$ with pairs (x, y) where $x \in \mathbb{R}^m$, $y \in \mathbb{R}$. We are interested in the solvability and properties of solutions of an equation $F(x, y) = 0$ (the function F is defined on some subset of $\mathbb{R}^{m+1}$) for a scalar variable y, where x is a vector from $\mathbb{R}^m$ regarded as a parameter. To make the statement of the problem more precise, we introduce the following definition.

Definition Let F be a function defined on a subset E of $\mathbb{R}^{m+1}$ and $X \subset \mathbb{R}^m$. A function $f : X \to \mathbb{R}$ is called an *implicit function defined by the equation*

$$F(x, y) = 0 \tag{1}$$

if

$$(x, f(x)) \in E \quad \text{and} \quad F(x, f(x)) = 0 \quad \text{for every } x \in X.$$

Besides, if $f(X) \subset \Delta$ where Δ is a (possibly infinite) interval and $X \times \Delta \subset E$, we also say that f is an *implicit function defined by equation* (1) *in the product* $X \times \Delta$.

Clearly, the graph of an implicit function is contained in the zero level set of the function F. Note also that equation (1) defines a unique implicit function in $X \times \Delta$ if and only if for every $x \in X$ there is only one point $y \in \Delta$ such that $F(x, y) = 0$.

Example Let $F(x, y) = x^2 + y^2 - 1$ for $(x, y) \in \mathbb{R}^2$. Then (1) takes the form

$$x^2 + y^2 - 1 = 0, \tag{1\,a}$$

so the zero level set of F is the unit circle. Since its projection to the x-axis coincides with the interval $[-1, 1]$, equation (1 a) can define an implicit function only on a set contained in $[-1, 1]$. On this interval, an implicit function can be defined in many ways. Indeed, set

$$f(x) = \pm\sqrt{1 - x^2},$$

choosing the sign arbitrarily for each $x \in [-1, 1]$. In this way we obtain infinitely many implicit functions. If, however, we are interested in implicit functions defined by equation (1 a) in the semi-strip $[-1, 1] \times [0, +\infty)$, then the answer is quite different: in this semi-strip, equation (1 a) defines a unique implicit function $x \mapsto \sqrt{1 - x^2}$ ($|x| \le 1$), and this function is continuous.

The very simple example considered above can serve as an illustration of other problems arising in the study of implicit functions. For example, it is clear that equation (1 a) defines two continuous implicit functions on $[-1, 1]$. Obviously, uniqueness cannot be achieved by replacing this interval with an arbitrarily small neighborhood of some value of the parameter and considering continuous implicit functions defined only on this neighborhood: there will still be two of them. The situation changes if instead of a small neighborhood (on the line) of some value x_0 of the parameter, we consider a neighborhood (on the plane) of a point (x_0, y_0) satisfying equation (1 a).

For example, take the neighborhood $P = (-1, 1) \times (0, 2)$ of the point $(0, 1)$ of the circle. Then, considering equation $(1\,a)$ in P, we impose restrictions not only on the parameter x, but also on possible solutions of the equation. The result is that in P equation $(1\,a)$, obviously, defines a unique implicit function $x \mapsto \sqrt{1 - x^2}$ $(|x| < 1)$.

However, in some cases one can guarantee neither the existence, nor the uniqueness of an implicit function even in an arbitrarily small neighborhood of a point satisfying equation (1). Clearly, in our example both points $(1, 0)$ and $(0, 1)$ lie on the circle. However, equation $(1\,a)$ does not define an implicit function in any neighborhood of the first point, since it is not solvable for $x > 1$. Hence, an implicit function can be defined only in a one-sided neighborhood of the point 1. Besides, however small a neighborhood of the point $(1, 0)$ may be, an implicit function is never unique, since there are two continuous implicit functions whose graphs lie in an arbitrarily small rectangle $(1 - \delta^2, 1] \times (-2\delta, 2\delta)$ (as to discontinuous functions with this property, there are, obviously, infinitely many of them). Thus, we see that a small perturbation of the parameter can lead, on the one hand, to a loss of solvability of equation (1) and, on the other hand, to a loss of uniqueness of an implicit function, even if we restrict ourselves only to solutions from an arbitrarily small neighborhood of a point lying on the zero level set. We have yet to find out what analytic properties of the function F are responsible for this phenomenon.

10.2 The Existence and Uniqueness of an Implicit Function

Due to the intermediate value theorem, the existence of an implicit function can be proved quite easily under natural assumptions.

Theorem *Let $a, b \in \mathbb{R}$, $a < b$, and let X be a nonempty subset in $\mathbb{R}^m$. Assume that a function F is continuous at least on the product $X \times [a, b]$ and for every $x \in X$ satisfies the following conditions*:
1) *the values $F(x, a)$ and $F(x, b)$ have opposite signs $(F(x, a)F(x, b) < 0)$*;
2) *the function $y \mapsto F(x, y)$ is strictly monotone on $[a, b]$.*
Then the equation $F(x, y) = 0$ defines a unique implicit function on $X \times [a, b]$, and this function is continuous on X.

Proof It follows from the first condition that for every $x \in X$ the (continuous) function defined in the second condition takes values of opposite sign at the endpoints of the interval $[a, b]$, which guarantees that in (a, b) there exists a solution of the equation $F(x, y) = 0$. Clearly, by the second condition, this solution is unique, and we denote it by $f(x)$. Thus, we obtain a function $f \colon x \mapsto f(x)$ defined on X with values in (a, b).

We claim that f is continuous at an arbitrary point $x_0 \in X$. Setting $y_0 = f(x_0)$, let us check that $|f(x) - y_0| < \varepsilon$ for every $\varepsilon > 0$ as long as $x \in X$ and the norm $\|x - x_0\|$ is sufficiently small. Here we may assume ε to be so small that the interval $[y_0 - \varepsilon, y_0 + \varepsilon]$ is contained in (a, b). Since $F(x_0, y_0) = 0$, the strictly monotone function $y \mapsto F(x_0, y)$ changes sign at the point $y_0 = f(x_0)$, hence

$$F(x_0, y_0 - \varepsilon)F(x_0, y_0 + \varepsilon) < 0.$$

By the continuity of F, this inequality is preserved under small perturbations of x_0: there exists a small disk B centered at x_0 such that

$$F(x, y_0 - \varepsilon)F(x, y_0 + \varepsilon) < 0 \quad \text{for all } x \in X \cap B.$$

Since $F(x, f(x)) = 0$, all values assumed by the function f in $X \cap B$ belong to the interval $(y_0 - \varepsilon, y_0 + \varepsilon) = (f(x_0) - \varepsilon, f(x_0) + \varepsilon)$, which proves that f is continuous at x_0.   □

Remark The first assumption of the theorem can be weakened by replacing the product $X \times [a, b]$ with a "curvilinear trapezoid," i.e., a set of the form

$$T_{U,V} = \{(x, y) : x \in X, \ U(x) \le y \le V(x)\},$$

where U and V are continuous functions on X (and $U < V$). If for every $x \in X$ the values $F(x, U(x))$ and $F(x, V(x))$ have opposite signs, then, under the assumption that F is strictly monotone with respect to the second argument, the theorem remains valid: an implicit function defined on X exists and is continuous. Besides, its graph lies between the graphs of U and V, i.e.,

$$U(x) < f(x) < V(x) \quad \text{for all } x \in X.$$

To illustrate this remark, consider an example.

Example Let $F(x, y) = y - xe^{-xy}$. Obviously, $F'_y > 0$, and hence the function F is strictly increasing with respect to the second argument. We leave it to the reader to verify that the equation $F(x, y) = 0$ defines in $\mathbb{R}^2$ a unique implicit function f, which is continuous and odd. Besides, for $x > 2$ and $U(x) = \frac{1}{2}V(x) = \frac{\log x}{x}$, we have $F(x, U(x)) < 0$, $F(x, V(x)) > 0$. Hence, by the above remark,

$$\frac{\log x}{x} < f(x) < 2\frac{\log x}{x} \quad \text{for } x > 2.$$

For a refined asymptotics of $f(x)$ as $x \to +\infty$, see Exercise 2.

10.3 The Smoothness of an Implicit Function

Having proved the existence and continuity of an implicit function, we still cannot regard the problem as fully resolved. Indeed, in the examples above the function F was not only continuous, but also smooth (and even infinitely differentiable). To what extent is the smoothness of F inherited by the implicit function?

Of course, posing this question, we must be sure that equation (1) defines (under some assumptions) a unique implicit function, since otherwise, as we have seen in the discussion of the example in Section 10.1, there may exist infinitely many discontinuous implicit functions. Since smoothness is a local property, it suffices to have a theorem that would allow us to establish it in a neighborhood of a given point. As we will see, the assumptions that will be made in this theorem imply also the existence of an implicit function "in the small." These assumptions are quite natural. To see this, consider equation (1) near a point (x_0, y_0) satisfying it, assuming for simplicity that F is a function of two variables. Since this function is smooth, with equation (1) we can associate an approximate equation, replacing $F(x, y)$, i.e., the difference $F(x, y) - F(x_0, y_0)$, by the differential of F at (x_0, y_0). Then instead of (1) we obtain the equation

$$F'_x(x_0, y_0)(x - x_0) + F'_y(x_0, y_0)(y - y_0) = 0, \qquad (1')$$

which is, obviously, (uniquely) solvable with respect to y if $F'_y(x_0, y_0) \neq 0$. As we will see, this condition, which is sufficient for equation $(1')$ to be solvable, remains sufficient for the solvability of equation (1) "in the small."

Theorem *Let F be a smooth function in an open subset O in $\mathbb{R}^{m+1}$, and let $(x_0, y_0) \in O$. If $F(x_0, y_0) = 0$ and*

$$\frac{\partial F}{\partial y}(x_0, y_0) \neq 0, \qquad (2)$$

then there exists a number $\delta > 0$ and a ball B centered at x_0 such that in the neighborhood $B \times (y_0 - \delta, y_0 + \delta)$ of the point (x_0, y_0), the equation $F(x, y) = 0$ defines a unique implicit function f. This function is smooth, and for every $j = 1, \ldots, m$ and all $x \in B$ we have

$$\frac{\partial f}{\partial x_j}(x) = -\frac{F'_{x_j}(x, f(x))}{F'_y(x, f(x))} \qquad (3)$$

(in particular, $f'(x) = -\frac{F'_x(x, f(x))}{F'_y(x, f(x))}$ for $m = 1$).

In other words, the zero level set $\{(x, y) \in O : F(x, y) = 0\}$ of the function F, which contains the point (x_0, y_0), coincides near this point with the graph of a smooth function of m variables.

Note that the uniqueness of an implicit function implies that $f(x_0) = y_0$.

Special emphasis should be placed upon the role of condition (2). In the example at p. 97, it is violated for $x_0 = \pm 1$, which is the reason why equation (1 a) ceases to be solvable for arbitrarily small perturbations of the parameter. Other examples related to the violation of this condition will be considered after the proof of the theorem.

Turning to equation $(1')$, from formula (3) with $m = 1$ we see that the slope of the line defined by this equation is nothing else than $f'(x_0)$. Thus, passing from (1) to $(1')$ gives an equation for the tangent to the graph of the implicit function at the point x_0.

Proof Since (x_0, y_0) is an interior point of O, for a sufficiently small ball B centered at x_0 and sufficiently small $\delta > 0$ the direct product $B \times [y_0 - \delta, y_0 + \delta]$ is contained in O. By assumption, $\frac{\partial F}{\partial y}(x_0, y_0) \neq 0$, hence a ball B and a number δ can be chosen so small that

$$\frac{\partial F}{\partial y}(x, y) \neq 0 \quad \text{as long as } x \in B \text{ and } |y - y_0| \leq \delta.$$

Then for every $x \in B$, the function $y \mapsto F(x, y)$ is strictly monotone on $[y_0 - \delta, y_0 + \delta]$, and since $F(x_0, y_0) = 0$, we obtain that

$$F(x_0, y_0 - \delta)F(x_0, y_0 + \delta) < 0.$$

By the continuity of F, this inequality remains valid if x_0 is replaced by a point x sufficiently close to x_0. Shrinking the radius of B if necessary, we may assume that

$$F(x, y_0 - \delta)F(x, y_0 + \delta) < 0 \quad \text{if } x \in B.$$

Thus, we are under the assumptions of the previous theorem with $X = B$ and (a, b) replaced by $(y_0 - \delta, y_0 + \delta)$. It guarantees the existence of a unique implicit function $f : B \to (y_0 - \delta, y_0 + \delta)$, as well as its continuity in B.

Now we calculate the partial derivatives of f at an arbitrary point of B. Recall that, by construction,

$$F_y'(x, y) \neq 0 \quad \text{for} \quad (x, y) \in B \times (y_0 - \delta, y_0 + \delta).$$

Calculating the partial derivative with respect to one coordinate, we fix all the other coordinates, hence we may assume without loss of generality that F is a function of two variables and, correspondingly, B is an interval containing x_0. Then the partial derivative of f becomes an ordinary derivative.

Consider points x and $x+u$ from B, with $u \neq 0$, and set $y = f(x)$, $y+v = f(x+u)$. Since $F(x, y) = F(x + u, y + v) = 0$, the mean value theorem (see formula (1′) in Section 1) implies that

$$0 = F(x + u, y + v) - F(x, y) = F_x'(x + \theta u, y + v)u + F_y'(x, y + \widetilde{\theta}v)v$$

for some $\theta, \widetilde{\theta} \in [0, 1]$. Therefore,

$$\frac{f(x + u) - f(x)}{u} = \frac{v}{u} = -\frac{F_x'(x + \theta u, y + v)}{F_y'(x, y + \widetilde{\theta}v)}.$$

Since $v = f(x + u) - f(x) \xrightarrow[u \to 0]{} 0$ by the continuity of f, the right-hand side has a finite limit $-\frac{F_x'(x,y)}{F_y'(x,y)}$ as $u \to 0$ (recall that the functions F_x' and F_y' are continuous by assumption, and $F_y' \neq 0$ in $B \times (y_0 - \delta, y_0 + \delta)$). This means that the implicit function f has a derivative at x, and

$$f'(x) = -\frac{F_x'(x, f(x))}{F_y'(x, f(x))} \quad (x \in B).$$

Besides, we have proved formula (3) for partial derivatives. It implies that each of them is continuous in B (as a composition of continuous functions). □

Corollary *Under the assumptions of the theorem, if $F \in C^t(O)$ for some $t \geq 1$, then the implicit function f also belongs to the class $C^t(B)$.*

For $t \in \mathbb{N}$, the proof proceeds by induction, with the base case ($t = 1$) proved in the theorem. The induction step relies on formula (3) and Theorem 6.6 on compositions of smooth maps. The case of fractional smoothness follows from the complement to Theorem 6.6 given in Section 6.7.

Now we use the theorem proved above to show that simple roots of an algebraic equation smoothly depend on its coefficients for small perturbations.

Example Consider an algebraic equation

$$y^m + a_1 y^{m-1} + \ldots + a_{m-1} y + a_m = 0$$

with real coefficients $a_1, \ldots, a_m$. Let $y = b$ be a real root of this equation. How does it change if we slightly perturb the coefficients of the equation? Clearly, this is a special case of the implicit function problem. To make use of the results we have already obtained, consider in $\mathbb{R}^{m+1}$ the infinitely differentiable function

$$F(x, y) = y^m + x_1 y^{m-1} + \ldots + x_{m-1} y + x_m, \quad x = (x_1, \ldots, x_m) \in \mathbb{R}^m, \ y \in \mathbb{R}.$$

By assumption, the equation $F(a, y) = 0$, where $a = (a_1, \ldots, a_m)$, has a real root $y = b$: $F(a, b) = 0$. If $\frac{\partial F}{\partial y}(a, b) \neq 0$, then, by Theorem 10.3, for coefficients $(x_1, \ldots, x_m)$ sufficiently close to $(a_1, \ldots, a_m)$, the equation $F(x, y) = 0$ has a unique root $y = f(x)$ near b, and the function f is of class C^∞ near a (a root of an equation smoothly depends on its parameters). The condition $\frac{\partial F}{\partial y}(a, b) \neq 0$ means that b is a simple root of the original algebraic equation. Thus, a simple root is stable under small perturbations of the coefficients: in a neighborhood of this root, the perturbed equation still has a unique root, which is an infinitely smooth function of the coefficients.

In the case of a multiple root, the situation is much more complicated. To get an idea of possible situations, compare, for instance, the equations $y^2 = 0$ or $y^3 = 0$ with the slightly modified equations $y^2 + a = 0$, $y^2 + ay = 0$ or $y^3 + ay^2 = 0$, $y^3 + ay = 0$, and $y^3 + a = 0$, where the coefficient a is small. Sketching the graphs of these polynomials, one can easily see that a small perturbation of the coefficients can cause a root both to vanish and to be replaced by several new roots. If b is a root of even multiplicity, then, changing only the constant term of the equation, we can obtain both a polynomial without roots near b and a polynomial with a pair of roots near b. Therefore, in this case an implicit function that calculates a root of the equation $F(x, y) = 0$ does not exist in any neighborhood of a (for the simplest case of such a situation, see Exercise 1).

If, however, b is a root of odd multiplicity, then a small perturbation of the coefficients cannot cause the root to vanish. It can only cause the appearance of new roots near b. But every implicit function f that calculates a root of the equation $F(x, y) = 0$ close to b in a neighborhood of a is necessarily nonsmooth (see Exercise 11).

10.4 The Role of the Gradient of F

As we see from the example in Section 10.1, if condition (2) is violated, then Theorem 10.3 is no longer true. Indeed, at the point $(1, 0)$ the derivative F'_y vanishes, and, as we have already observed, equation (1 a) does not define an implicit function in any two-sided neighborhood of this point. Besides, in an arbitrarily small rectangle $(1 - \delta^2, 1] \times (-2\delta, 2\delta)$ equation (1 a) defines two continuous implicit functions. At the same time, the part of the level set (circle) contained in a sufficiently small neighborhood of $(1, 0)$ is congruent to the graph of a smooth function and coincides with such a graph up to a permutation of coordinates (in Section 4, we have agreed to call such a set a wide-sense graph). Hence, if at least one of the partial derivatives of a function of several variables (e.g., with respect to the first variable) does not vanish at the point under consideration, then, interchanging the first and last coordinates makes it possible to use the theorem.

Since the equation $F(x, y) = F(x_0, y_0)$ is equivalent to equation (1) for the function $\widetilde{F} = F - F(x_0, y_0)$, we see that the following result holds.

Proposition *Let $F \in C^1(O)$, $z_0 \in O$. If* grad $F(z_0) \neq 0$, *then the intersection of a sufficiently small neighborhood of z_0 with the level set $\{z \in O \colon F(z) = F(z_0)\}$ is a (wide-sense) graph of a smooth function.*

This result shows that a "true" singularity of a level set of F, when it is not congruent to a graph even "in the small," may arise only if the gradient of f vanishes at the point under consideration, i.e., if it is a critical point of f.

Let us consider some examples illustrating the types of singularities of level sets that can arise at critical points. In all cases, we consider an equation $F(x, y) = 0$ with $F \in C^\infty$ and study the point $(0, 0)$.

Example 1 Let $F(x, y) = (x^2 + y^2)(x^2 + y^2 - 1)$. Obviously, $(0, 0)$ is an isolated point of the level set.

Example 2 Let $F(x, y) = x^2 - y^2$. The corresponding level set is the union of the lines $y = \pm x$. Clearly, its intersection with any neighborhood of the origin is not congruent to a graph. Among the four continuous implicit functions defined by the equation $F(x, y) = 0$ in the square $(-1, 1)^2$, only two are smooth.

Example 3 Let $F(x, y) = y(x^2 + y^2 - 2y)$. The corresponding level set is the union of the x-axis and the circle $x^2 + (y - 1)^2 = 1$. Its intersection with any neighborhood of the origin is not congruent to a graph. Among the four continuous implicit functions defined by the equation $F(x, y) = 0$ in the square $(-1, 1)^2$ (find them), there are smooth functions of class C^1, but not C^2.

Exercises

1. As we observed after Definition 10.1, the domain of definition of every implicit function $(x_1, x_2) \mapsto y = f(x_1, x_2)$ generated by an equation $F(x_1, x_2, y) = 0$ is

contained in the projection of the zero set level of f to the plane $y = 0$. Find this projection in the case $F(x_1, x_2, y) = y^2 + x_1 y + x_2$. Verify that on this set one can define exactly two continuous implicit functions.

2. Show that in the example from Section 10.2, the implicit function f is odd and has the following asymptotic behavior as $x \to +\infty$:

$$f(x) = 2\frac{\log x}{x} - \frac{\log(2\log x)}{x} + O\left(\frac{1}{x}\right).$$

3. Show that the equation $F(x, y) = 0$ defines a unique implicit function in the first quadrant $\mathbb{R}_+^2$ in the following cases:
   a) $F(x, y) = \cos x + e^{-xy} + x^2 - y - 2$;
   b) $F(x, y) = x + 2y - \arctan(x + y)$;
   c) $F(x, y) = e^{xy} + x^2 + y - 1$.
   In what cases does the equation $F(x, y) = 0$ define a unique implicit function on the entire real line? In which of these cases is the implicit function odd?

4. Verify that all functions from Exercises 2 and 3 satisfy the assumptions of Theorem 10.3 at the point $(0, 0)$ and find the derivatives of the implicit functions at $x_0 = 0$. Check that the implicit functions are infinitely differentiable. In all these cases, find the first three nonzero terms of the Taylor series of the implicit function at the point $x_0 = 0$.

   HINT. Write the implicit function in the form $f(x) = c_1 x + c_2 x^2 + \ldots + c_n x^n + o(x^n)$ with undetermined coefficients (the coefficient c_1 is already known from Theorem 10.3, but you need not necessarily use this fact) and substitute this expansion into (1). This method is faster than a direct calculation of the derivatives of the implicit function. The existence of such a representation follows from the infinite differentiability of the implicit function.

5. Show that the implicit functions from Exercise 3 have the following asymptotic behavior as $x \to +\infty$:
   a) $x^2 + \cos x - 2 + O(e^{-(x-2)^2})$;
   b) $-\frac{x}{2} + \frac{\pi}{4} - \frac{2}{x} + O\left(\frac{1}{x^2}\right)$;
   c) $-x^2 + 1 - e^{x-x^3} + O(xe^{2(x-x^3)})$.

6. Show that the equation $x^3 - x^2 + y^3 + y = 0$ defines a unique implicit function in $\mathbb{R}^2$. Find its Taylor expansion at the origin up to $o(x^6)$. Find the asymptotic expansion of this function as $x \to +\infty$ ($x \to -\infty$) up to $O\left(\frac{1}{x}\right)$.

7. Check that the equation $y^7 + xy - x^5 = 0$ defines a unique implicit function in the half-plane $(0, +\infty) \times \mathbb{R}$. Find its representation of the form $y = Ax^\alpha + Bx^\beta + \ldots$ ($\alpha > \beta > \ldots$) as $x \to +\infty$ up to $O(x^{-4})$.

8. Find a parallelepiped in $\mathbb{R}^3$ centered at the point $x_1 = -3$, $x_2 = -2$, $y = 2$ in which the equation $y^3 + x_1 y + x_2 = 0$ defines a unique implicit function. Estimate the norm of its gradient from above.

9. Consider the equation $x e^{-y} - y e^x = 0$.
   a) Find a rectangle centered at the origin in which this equation defines a unique implicit function. Find the first several terms of its Taylor expansion near the origin and sketch its graph.

   b) Show that in the first quadrant $\mathbb{R}_+^2$ this equation defines a unique implicit function f. Find the first two terms of its asymptotic expansion as $x \to +\infty$.

10. Consider an equation $T(\varphi) = 0$ where T is a trigonometric polynomial, i.e., a function of the form

$$T(\varphi) = c_0 + c_1 \cos \varphi + c_2 \sin \varphi + \ldots + c_{2m-1} \cos m\varphi + c_{2m} \sin m\varphi \quad (\varphi \in \mathbb{R}),$$

where $c_0, c_1, \ldots, c_{2m}$ are real coefficients. Let φ_0 be a simple root of the equation, i.e., $T(\varphi_0) = 0$, but $T'(\varphi_0) \neq 0$. Show that for a sufficiently small perturbation of the coefficients, the equation still has a simple root near φ_0, and it smoothly depends on the coefficients.

11. Complement the result of the example from Section 10.3 by considering the case where b is a root of the equation $F(a, y) = 0$ of multiplicity $k \geq 2$. Show that

   a) there exists a sequence of points $(a^{(n)}, y_n)$ converging to (a, b) such that $F(a^{(n)}, y_n) = 0$ for all n and $|y_n - b| \asymp \|a^{(n)} - a\|^{1/k}$ (hence, the implicit function f cannot be smooth if it is defined in a neigborhood of a and $f(a) = b$);

   b) if k is odd, then in some neighborhood of a there exists an implicit function f that is continuous at a and such that $f(a) = b$.

11 *Whitney's Extension Theorem

This section deals with the problem of extending a function f defined on a closed subset E of a Euclidean space to an r-smooth function in a neighborhood of this set. Multiplying such an extension by a smooth function approximating the characteristic function of E within this neighborhood (see Corollary V.1.2), we obtain a function of the same smoothness that is an extension of f to the whole space. That is why, below by a smooth extension of a function from a closed set we mean an extension to the whole space.

11.1 Statement of the Problem and a Preliminary Result

As follows from the Heine–Cantor theorem, a necessary condition for a function to have a continuous extension to the closure of a set is that it is uniformly continuous on this set. At the same time, as observed in Remark 0.9, this condition is also sufficient for the existence of such an extension. When considering problems related to extensions of functions, it is naturally to assume that this condition is satisfied and the corresponding extension has already been constructed. That is why, in what follows we assume from the very beginning that the continuous function under consideration is defined on a closed set. It is well known that such a function (as well as a function satisfying the Lipschitz condition) can be extended as a continuous

(Lipschitz) function to the whole space, but we do not dwell on this result, referring the reader, e.g., to [9, Appendix II]. In what follows, we are interested in the existence of a smooth extension of a given function. We reiterate that the original function is always assumed to be defined on a closed set. Since this problem is by no means always solvable, first of all we describe the simplest necessary conditions on the original function f for the existence of an r-smooth ($r \in \mathbb{N}$) extension of f.

Obviously, besides being continuous on E, the function f must be r-smooth inside E, i.e., on the set $G = \text{Int}(E)$ (for the time being, we assume that $G \neq \varnothing$). Moreover, it is necessary for the existence of an r-smooth extension of f to a neighborhood of E that all derivatives of f up to order r are uniformly continuous on every bounded subset in G. This property makes it possible to extend each partial derivative $f^{(\alpha)}$ (of order at most r) to the closure $\bar{G}$ of G, obtaining a function g_α continuous on $\bar{G}$. These conditions (the r-smoothness of the original function in G and the uniform continuity of its derivatives on bounded subsets of G) will be called the natural necessary conditions for the existence of a smooth extension.

Assuming for the moment that $\bar{G} = E$, we are faced with the main difficulty of the problem under consideration, which is clearly visible already in the case $r = 1$. Having extended the partial derivatives $\frac{\partial f}{\partial x_1}, \ldots, \frac{\partial f}{\partial x_m}$ to functions $g_1, \ldots, g_m$ continuous on $\bar{G}$, we can, by the theorem mentioned above, extend them by continuity to the whole space. Denote these extensions by $h_1, \ldots, h_m$, respectively. Now, the question is whether these functions, obtained by independently extending $g_1, \ldots, g_m$, are the partial derivatives of a smooth function. Indeed, for this to be the case, they must also satisfy some necessary conditions, for example, the integral of the 1-form $h_1 dx_1 + \ldots + h_m dx_m$ over every smooth closed path must vanish. Obviously, for arbitrary extensions this is not necessarily the case. We must somehow make the extensions of $g_1, \ldots, g_m$ agree with each other. But what conditions must be satisfied for such an agreement (especially for higher derivatives) is not at all clear. It turns out useful to consider from the very beginning not separate partial derivatives, but all of them together, and deal with Taylor polynomials of the function f. These polynomials, constructed for a sufficiently smooth function f at points $c \in G = \text{Int}(E)$, will be denoted, as before (see Section 8), by $T_r(f, c)$, and their values at a point x, by $T_r(f, c; x)$. Correspondingly, $T_r^{(\alpha)}(f, c; x)$ stands for the value at x of the derivative of order α of this polynomial, i.e.,

$$T_r^{(\alpha)}(f, c; x) = \frac{\partial^\alpha T_r(f, c)}{\partial x^\alpha}(x);$$

in particular, $T_r^{(\alpha)}(f, x; x) = T_r^{(\alpha)}(f, x; y)\big|_{y=x}$. By Corollary 1 at p. 75, we have

$$f(x) - T_r(f, c; x) \underset{x \to c}{=} o(\|x - c\|^r).$$

Since the derivatives of Taylor polynomials are the Taylor polynomials of the derivatives, this statement can be substantially refined: for all multi-indices α of order at most r,

$$f^{(\alpha)}(x) - T_r^{(\alpha)}(f, c; x) \underset{x \to c}{=} o(\|x - c\|^{r - |\alpha|})$$

(as usual, $f^{(0)} = f$). Since $f^{(\alpha)}(x) = T_r^{(\alpha)}(f, x; x)$, these relations can be rewritten as

$$T_r^{(\alpha)}(f, x; x) - T_r^{(\alpha)}(f, c; x) \underset{x \to c}{=} o\left(\|x - c\|^{r - |\alpha|}\right). \tag{1}$$

Thus, near points where f is r-smooth, its Taylor polynomials with $|\alpha| \leq r$ satisfy (1). If f can be extended to a function of class $C^r(\mathbb{R}^m)$, then there necessarily exists a family of polynomials $\{P_c\}_{c \in \mathbb{R}^m}$ (namely, the Taylor polynomials of the extension) that satisfy relations of the form (1) not only at the interior points of E, but also at its boundary points (and, in fact, everywhere in $\mathbb{R}^m$). As we will see, this condition is not only necessary for the existence of an r-smooth extension, but also, with some clarifications, sufficient.

We begin with a simple case where a family of polynomials is defined on an open set and prove that the above condition is sufficient: if it is satisfied, then the family coincides with the family of Taylor polynomials of a smooth function. Let us state this result more precisely.

Lemma *Let $r \in \mathbb{N}$, and let U be an open subset in $\mathbb{R}^m$. Assume that each point $c \in U$ is assigned a polynomial P_c of degree at most r such that*

$$P_x^{(\alpha)}(x) - P_c^{(\alpha)}(x) \underset{x \to c}{=} o\left(\|x - c\|^{r - |\alpha|}\right) \quad \textit{for every } \alpha \textit{ with } |\alpha| \leq r. \tag{2}$$

Then the function $g \colon x \mapsto P_x(x)$ belongs to the class $C^r(U)$, and $P_c = T_r(g, c)$ for every point $c \in U$.

Here and in what follows, the symbol $P_x^{(\alpha)}(x)$ is understood by analogy with our convention for Taylor polynomials:

$$P_x^{(\alpha)}(y) = \frac{\partial^\alpha P_x}{\partial y^\alpha}(y), \quad \text{in particular,} \quad P_x^{(\alpha)}(x) = P_x^{(\alpha)}(y)\big|_{y=x}.$$

Proof We use the equality (hereafter, $x, c \in U$)

$$g(x) - g(c) = P_x(x) - P_c(c) = P_x(x) - P_c(x) + P_c(x) - P_c(c). \tag{3}$$

Since

$$P_x(x) - P_c(x) \underset{x \to c}{=} o(\|x - c\|^r),$$

condition (2) with $\alpha = 0$ implies

$$g(x) - g(c) \underset{x \to c}{=} P_c(x) - P_c(c) + o(\|x - c\|).$$

The polynomial P_c is differentiable, so it follows that the function g is differentiable and $d_c g = d_c P_c$.

To prove the continuity of the first-order partial derivatives, we write relations similar to (3) with $|\alpha| = 1$:

$$g^{(\alpha)}(x) - g^{(\alpha)}(c) = P_x^{(\alpha)}(x) - P_c^{(\alpha)}(c)$$
$$= P_x^{(\alpha)}(x) - P_c^{(\alpha)}(x) + P_c^{(\alpha)}(x) - P_c^{(\alpha)}(c)$$
$$= P_c^{(\alpha)}(x) - P_c^{(\alpha)}(c) + o(\|x - c\|^{r-1}).$$

They imply the continuity of the derivatives of g at the point c. Since this is true for every point $c \in U$, we see that $g \in C^1(U)$.

We can now apply induction. Let $1 \le i < r$, and assume that (for a given r) we have already proved that $g \in C^i(U)$ and $d_c^j g = d_c^j P_c$ for $1 \le j \le i < r$ for all $c \in U$. Fix an arbitrary multi-index α of order i and consider the family of polynomials $\{Q_c\}_{c \in U}$ where $Q_c = P_c^{(\alpha)}$. Then for every m-dimensional multi-index β with $|\beta| \le r - i$, we have

$$Q_x^{(\beta)}(x) - Q_c^{(\beta)}(x) = P_x^{(\alpha+\beta)}(x) - P_c^{(\alpha+\beta)}(x) \underset{x \to c}{=} o(\|x - c\|^{r-|\alpha|-|\beta|}),$$

i.e., the polynomials of the constructed family satisfy condition (2) with r replaced by $r - i$ and α replaced by β. Setting $h(x) = Q_x(x)$, we see that

$$h(x) = P_x^{(\alpha)}(x) = g^{(\alpha)}(x) \quad \text{for } x \in U.$$

As we have already proved, h is a smooth function and, consequently, g has continuous partial derivatives up to order $i + 1$, which completes the proof of the induction step.   □

11.2 Whitney's[1] Extension Theorem

As we have already mentioned in the previous subsection, when solving the problem of constructing a smooth extension, it is impractical to consider separate extensions of the partial derivatives of the original function (if they exist at all) and then make them agree with each other. An alternative approach is suggested by Lemma 11.1, which shows that to construct a smooth extension of a function, one can use, instead of separate extensions of its derivatives, whole "aggregates" that can approximate it in the same way as Taylor polynomials approximate a smooth function. Thus, the extension theorem will be based on the assumption that for a given function f defined on a closed set E there exists a family of polynomials $\{P_c\}_{c \in E}$ "similar" to the family of its Taylor polynomials. Of course, the polynomials P_c are not exactly the Taylor polynomials of f (if only because c is not necessarily an interior point of E), but they will have the same approximation properties as these polynomials and will be, in fact, nothing else than the Taylor polynomials of the extended function. Besides, obviously, the constant term of P_c coincides with the value of the original function f at the point c, as befits a Taylor polynomial: $P_c(c) = f(c)$. It should be added that in order to obtain sufficient conditions, we must take into account that the

---

[1] Hassler **Whitney** (1907–1989) was an American mathematician.

remainder term in Taylor's formula can be uniformly estimated in a domain where the function is smooth (see Section 8.2) and impose on P_c restrictions similar to properties (1) and (2).

If there exists such a family, Whitney's theorem allows one to establish the existence of a smooth extension of f even in cases where E has no interior points at all and, consequently, one cannot speak of the smoothness and Taylor polynomials of the original function.

Theorem *Let* $r \in \mathbb{N}$ *and* E *be a nonempty closed subset in* $\mathbb{R}^m$. *Assume that each point* $c \in E$ *is assigned a polynomial* P_c *of degree at most* r, *and* f *is the function defined on* E *by the formula* $f(x) = P_x(x)$. *If for every compact set* $K \subset E$ *there exists a nonincreasing function* ω_K *on* $[0, \infty)$ *with* $\lim_{t \to 0} \omega_K(t) = \omega_K(0) = 0$ *such that*

$$|P_x^{(\alpha)}(x) - P_c^{(\alpha)}(x)| \le \|x - c\|^{r - |\alpha|} \omega_K(\|x - c\|) \quad \text{for } x, c \in K \qquad (4)$$

for all multi-indices α *with* $|\alpha| \le r$, *then there exists an extension* $\widetilde{f}$ *of* f *from the class* $C^r(\mathbb{R}^m)$ *such that*

$$T_r(\widetilde{f}, c) = P_c \quad \text{for all } c \in E,$$

i.e., $\widetilde{f}^{(\alpha)}(x) = P_x^{(\alpha)}(x)$ *if* $x \in E$ *and* $|\alpha| \le r$.

For $E = \mathbb{R}^m$, the conclusion of the theorem is a special case of Lemma 11.1. Hence, in what follows we assume that $E \ne \mathbb{R}^m$.

Inequality (4) can be refined by replacing ω_K by the functions $\widetilde{\omega}_K$ (with $\widetilde{\omega}_K \le \omega_K$) where

$$\widetilde{\omega}_K(t) = \sup_{\substack{|\alpha| \le r; x, c \in K; \\ 0 < \|x - c\| \le t}} \frac{|P_x^{(\alpha)}(x) - P_c^{(\alpha)}(x)|}{\|x - c\|^{r - |\alpha|}}.$$

These functions depend monotonically on K: to a larger set there corresponds a larger function. In what follows, we assume (without loss of generality) that all functions ω_K have this property too.

11.3 Proof of Whitney's Theorem

The proof of the theorem entails overcoming substantial technical difficulties. It will proceed along the following lines. First, we construct a function g infinitely smooth in the complement $O = \mathbb{R}^m \setminus E$ of the set E. This will be done in such a way that the values of g at points $x \in O$ close to the boundary depend only on the polynomials P_c with indices close to x. It will turn out later that it is this function that is an extension of f to O. Therefore, it is natural that we extend the original family $\{P_c\}_{c \in E}$ by using the family of Taylor polynomials (of order r) of the function g, i.e., the family $\{T_r(g, c)\}_{c \in O}$. After this, we must overcome the main difficulty: verify that these two families agree, i.e., that their union satisfies (in the whole space $\mathbb{R}^m$) condition (2) at p. 107.

Proof First, following the outlined plan, we define a function g in O. For this, we need nonnegative infinitely smooth functions $\{\varphi_j\}_{j\in\mathbb{N}}$ in $\mathbb{R}^m$ constructed in the partition of unity theorem (see Section V.1.2). Recall that

$$\varphi_1 + \varphi_2 + \ldots = 1 \quad \text{everywhere in } O$$

and every function φ_j takes nonzero values only in a ball B_j contained in O whose radius R_j is comparable with the distance from B_j to ∂O:

$$R_j \leq \text{dist}(B_j, \partial O) \leq 2R_j. \tag{5}$$

Besides, a sufficiently small neighborhood of every point from O has a nonempty intersection with a bounded number (at most 10^{2m}) of the balls $B_1, B_2, \ldots$.

Set

$$g(x) = \sum_{j=1}^{\infty} P_{z_j}(x)\varphi_j(x) \quad (x \in O)$$

where z_j is a point of E closest to B_j (in general, it is not necessarily unique; in this case, we arbitrarily fix one of such points throughout the argument). The function g is infinitely smooth in O, since near every point of O the series defining g reduces to a finite sum.

As we have already mentioned, for $c \in O$ we take P_c to be the rth Taylor polynomial of g at the point c, so $g^{(\alpha)}(c) = P_c^{(\alpha)}(c)$ for $|\alpha| \leq r$. Combining the families $\{P_c\}_{c\in E}$ and $\{P_c\}_{c\in O}$, we obtain an "extension" of the original family. It is our task now to prove that these families agree, in the sense that their union satisfies condition (2) of Lemma 11.1 at every point of $\mathbb{R}^m$. Then the lemma will imply that the function $\widetilde{f}$ corresponding to the extended family, i.e., the function

$$\widetilde{f}(x) = \begin{cases} f(x) = P_x(x) & \text{if } x \in E, \\ g(x) = T_r(g, x; x) & \text{if } x \in O, \end{cases}$$

which is an extension of f, belongs to the class $C^r(\mathbb{R}^m)$, which completes the proof of the theorem.

At points of the set O, condition (2) is satisfied by the properties of Taylor polynomials; at interior points of the set E (if there are any), it follows from condition (4). So, we must verify it only at boundary points. If $c \in \partial E = \partial O$, this condition is satisfied "along" E by (4), and it remains to check that it is satisfied at points $c \in \partial O$ "along" the set O.

To begin with, assuming that K is a compact subset in E and $|\alpha| \leq r$, we estimate the difference $P_b^{(\alpha)}(x) - P_a^{(\alpha)}(x)$ for $a, b \in K$ and $x \in \mathbb{R}^m$ using the function ω_K. To this end, we expand the polynomial $P_w^{(\alpha)}$, where $w = a$ or b, in powers of $x - a$ (recall that P_w is a polynomial of degree at most r, so the degree of $P_w^{(\alpha)}$ does not exceed $r - |\alpha|$; below, β is an m-dimensional multi-index):

$$P_w^{(\alpha)}(x) = \sum_{|\beta| \leq r - |\alpha|} \frac{1}{\beta!} P_w^{(\alpha+\beta)}(a)(x-a)^\beta.$$

Set $\Delta = \|x - a\|$. Then

$$|P_b^{(\alpha)}(x) - P_a^{(\alpha)}(x)| \leq \sum_{|\beta| \leq r - |\alpha|} \frac{1}{\beta!} |P_b^{(\alpha+\beta)}(a) - P_a^{(\alpha+\beta)}(a)| \Delta^{|\beta|}.$$

Hence, by condition (4) applied to each term at the points a, b instead of x, c, we obtain (below $t = \|a - b\|$)

$$|P_b^{(\alpha)}(x) - P_a^{(\alpha)}(x)| \leq \omega_K(t) \sum_{|\beta| \leq r - |\alpha|} \frac{1}{\beta!} t^{r-|\alpha|-|\beta|} \Delta^{|\beta|}.$$

It is not difficult to estimate the resulting sum:

$$\sum_{|\beta| \leq r - |\alpha|} \frac{1}{\beta!} t^{r-|\alpha|-|\beta|} \Delta^{|\beta|} = \sum_{i=0}^{r-|\alpha|} t^{r-|\alpha|-i} \Delta^i \sum_{|\beta|=i} \frac{1}{\beta!} \leq (t + m\Delta)^{r-|\alpha|}$$

(the inner sum is equal to $\frac{m^i}{i!}$, this follows from the multinomial theorem). So, for any $a, b \in K$, $x \in \mathbb{R}^m$, and $\alpha \in \mathbb{Z}_+^m$, $|\alpha| \leq r$, we have

$$|P_a^{(\alpha)}(x) - P_b^{(\alpha)}(x)| \leq (\|a - b\| + m\|x - a\|)^{r-|\alpha|} \omega_K(\|a - b\|). \tag{6}$$

Now we verify that relation (2) holds for the extended family of polynomials at every point $c \in \partial O$ "along" the set O. To this end, we estimate the difference of derivatives

$$P_x^{(\alpha)}(x) - P_c^{(\alpha)}(x) = g^{(\alpha)}(x) - P_c^{(\alpha)}(x),$$

where $x \in O \cap B(c, 1)$ and $|\alpha| \leq r$, using inequality (6) with $K = E \cap \bar{B}(c, 4)$.

Let $\bar{x}$ be a point of E closest to x. Since $\|\bar{x} - x\| \leq \|x - c\| < 1$, we conclude that $\bar{x} \in K$. Split the difference $g^{(\alpha)}(x) - P_c^{(\alpha)}(x)$ into two parts:

$$g^{(\alpha)}(x) - P_c^{(\alpha)}(x) = (g - P_{\bar{x}})^{(\alpha)}(x) + (P_{\bar{x}} - P_c)^{(\alpha)}(x). \tag{7}$$

Consider the first of them:

$$(g - P_{\bar{x}})^{(\alpha)}(x) = \left(\sum_{j=1}^\infty P_{z_j} \varphi_j - P_{\bar{x}} \sum_{j=1}^\infty \varphi_j \right)^{(\alpha)}(x)$$

$$= \left(\sum_{j=1}^\infty (P_{z_j} - P_{\bar{x}}) \varphi_j \right)^{(\alpha)}(x) = \sum_{j=1}^\infty ((P_{z_j} - P_{\bar{x}}) \varphi_j)^{(\alpha)}(x) \tag{8}$$

(the term-by-term differentiation is valid, since in a sufficiently small neighborhood of x the series reduces to a finite sum). Below, estimating the jth term of this sum,

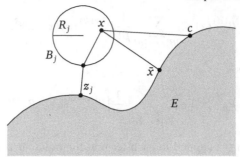

Fig. 4

we assume that $x \in B_j$, because $\varphi_j = 0$ outside the ball B_j. Since $x \in B_j$ and $\bar{x} \in \partial E = \partial O$, the left inequality in (5) implies

$$R_j \leq \operatorname{dist}(B_j, \partial O) \leq \|\bar{x} - x\| < 1.$$

Then the set K contains not only the point $\bar{x}$ (as mentioned after the definition of this point), but also z_j. Indeed, it is obvious that z_j belongs to E, and, since $R_j \leq \|\bar{x} - x\|$, it satisfies the inequality (see Fig. 4)

$$\|z_j - x\| \leq \operatorname{dist}(B_j, E) + \operatorname{diam}(B_j) \leq \|x - \bar{x}\| + 2R_j \leq 3\|x - \bar{x}\|.$$

It remains to observe that

$$\|z_j - c\| \leq \|z_j - x\| + \|x - c\| < 3\|x - \bar{x}\| + 1 < 4,$$

and hence $z_j \in E \cap B(c, 4) \subset K$.

We also need the inequalities

$$\|z_j - \bar{x}\| \leq \|z_j - x\| + \|x - \bar{x}\| \leq 4\|x - \bar{x}\|. \tag{9}$$

By Leibniz's rule (see Remark 6.5), each term $((P_{z_j} - P_{\bar{x}})\varphi_j)^{(\alpha)}(x)$ in the sum (8) is a linear combination of products of the form $(P_{z_j} - P_{\bar{x}})^{(\beta)}(x)\, \varphi_j^{(\alpha - \beta)}(x)$ (where the multi-index β is "not greater than" α, i.e., $\alpha - \beta \in \mathbb{Z}_+^m$).

Now we estimate the factor $(P_{z_j} - P_{\bar{x}})^{(\beta)}(x)$ for $|\beta| \leq r$ using inequality (6) (which can be applied, since the points $\bar{x}$ and z_j belong to K):

$$\left| (P_{z_j} - P_{\bar{x}})^{(\beta)}(x) \right| \leq \left(\|z_j - \bar{x}\| + m\|x - \bar{x}\| \right)^{r - |\beta|} \omega_K \left(\|z_j - \bar{x}\| \right).$$

Together with (9), this yields the inequality

$$\left| (P_{z_j} - P_{\bar{x}})^{(\beta)}(x) \right| \leq \left((m + 4)\|x - \bar{x}\| \right)^{r - |\beta|} \omega_K \left(4\|x - \bar{x}\| \right). \tag{10}$$

By the partition of unity theorem, the derivatives of φ_j satisfy the inequalities

$$\left|\varphi_j^{(\alpha-\beta)}(x)\right| \le \frac{C_{\alpha-\beta}}{\rho_x^{|\alpha|-|\beta|}},$$

where $\rho_x = \text{dist}(x, E) = \|x - \bar{x}\|$. Hence (see (10)),

$$\left|(P_{z_j} - P_{\bar{x}})^{(\beta)}(x)\varphi_j^{(\alpha-\beta)}(x)\right| \le C_1 \|x - \bar{x}\|^{r-|\alpha|}\omega_K(4\|x - \bar{x}\|)$$

(hereafter, $C_1, C_2 \ldots$ are constants independent of x and j). Due to Leibniz's rule, this allows us to estimate the terms of the sum (8):

$$\left|((P_{z_j} - P_{\bar{x}})\varphi_j)^{(\alpha)}(x)\right| \le C_2 \|x - \bar{x}\|^{r-|\alpha|}\omega_K(4\|x - \bar{x}\|).$$

Since this sum has at most 10^{2m} nonzero terms, we obtain

$$\left|(g - P_{\bar{x}})^{(\alpha)}(x)\right| \le C_3 \|x - \bar{x}\|^{r-|\alpha|}\omega_K(4\|x - \bar{x}\|). \tag{11}$$

The differences $P_{\bar{x}}^{(\alpha)}(x) - P_c^{(\alpha)}(x)$ in the right-hand side of (7) are much easier to estimate. Indeed, since $c, \bar{x} \in K$, we can apply (6) at once:

$$\left|P_{\bar{x}}^{(\alpha)}(x) - P_c^{(\alpha)}(x)\right| \le (\|\bar{x} - c\| + m\|x - c\|)^{r-|\alpha|}\omega_K(\|\bar{x} - c\|).$$

Since

$$\|\bar{x} - c\| \le \|x - c\| + \|x - \bar{x}\| \le 2\|x - c\|,$$

this implies the inequality

$$\left|P_{\bar{x}}^{(\alpha)}(x) - P_c^{(\alpha)}(x)\right| \le C_4 \|x - c\|^{r-|\alpha|}\omega_K(2\|x - c\|).$$

Together with (11) (where $\|x - \bar{x}\|$ must be replaced by the greater value $\|x - c\|$) and (7), it shows that for $c \in \partial O$ and $x \in O \cap B(c, 1)$ we have

$$\left|P_x^{(\alpha)}(x) - P_c^{(\alpha)}(x)\right| = |g^{(\alpha)}(x) - P_c^{(\alpha)}(x)| \le C_5 \|x - c\|^{r-|\alpha|}\omega_K(4\|x - c\|). \tag{12}$$

Thus, for the extended family of polynomials, relation (2) is satisfied at all points $c \in \mathbb{R}^m$. Hence, by Lemma 11.1 (with $U = \mathbb{R}^m$), the function $\tilde{f}$ that coincides with f on E and with g on O belongs to the class $C^r(\mathbb{R}^m)$, and $P_c = T_r(\tilde{f}, c)$ for $c \in E$.   □

Remark As one can see from the proof, inequality (11) is valid for all α, not only for $|\alpha| \le r$. In particular, if $|\alpha| = r + 1$, then (since $P_{\bar{x}}^{(\alpha)}$ is identically zero) for $x \in B(c, 1) \cap O$ we have

$$\left|\tilde{f}^{(\alpha)}(x)\right| = |g^{(\alpha)}(x)| \le C \frac{\omega_K(4\|x - \bar{x}\|)}{\|x - \bar{x}\|} = C \frac{\omega_K(4\rho_x)}{\rho_x}, \tag{13}$$

where $c \in \partial O$, $K = E \cap \bar{B}(c, 4)$, and the coefficient C does not depend on the choice of c.

11.4 Preserving the Degree of Smoothness of the Highest Derivatives

As we have established in Whitney's theorem, the family of polynomials $\{P_c\}_{c \in \mathbb{R}^m}$ is the family of rth Taylor polynomials of the extended function $\widetilde{f}$. For $|\alpha| = r$, we rewrite condition (4) of this theorem in a slightly different form. Since

$$P_x^{(\alpha)}(x) = \widetilde{f}^{(\alpha)}(x) \quad \text{and} \quad P_c^{(\alpha)}(c) = \widetilde{f}^{(\alpha)}(c)$$

by Whitney's theorem, while $P_c^{(\alpha)}(x) = P_c^{(\alpha)}(c)$ (because $P_c^{(\alpha)}$ is a polynomial of degree 0), we see that for every compact subset K in E,

$$\left| \widetilde{f}^{(\alpha)}(x) - \widetilde{f}^{(\alpha)}(c) \right| = \left| P_x^{(\alpha)}(x) - P_c^{(\alpha)}(x) \right| \le \omega_K (\|x - c\|) \quad \text{for all } x, c \in K. \quad (14)$$

Thus, for $|\alpha| = r$ condition (4) is a bound on the moduli of continuity of the restriction to K of the highest derivatives of the extended function. Our goal here is to find out whether similar bounds hold for arbitrary compact subsets in $\mathbb{R}^m$. Since in some cases (for example, if the set E is countable) the functions ω_K may not have the typical properties of moduli of continuity, we will assume that the functions ω under consideration not only are increasing and infinitesimal at the origin, but also do not grow too rapidly when we double the argument:

$$\omega(2t) \le 2\omega(t) \quad \text{for all } t > 0.$$

Such (nonzero) functions will be called *estimation functions*. It is easy to see that

$$\omega(at) \le 2a\,\omega(t) \quad \text{for } a > 1.$$

In particular, taking $a = \frac{1}{t}$, we see that on $(0, 1]$ the lower bound $Ct \le \omega(t)$ holds with $C = \frac{1}{2}\omega(1)$. Thanks to this property, every estimation function is a majorant (with some coefficient) for the modulus of continuity of every bounded function satisfying the Lipschitz condition. Hence, difficulties in estimating the increments of derivatives of the extended function may arise only near the boundary of E.

An example of an estimation function is the power function $t \mapsto At^\theta$ ($t \ge 0$) where $A > 0, 0 < \theta \le 1$. Our goal is to show that if all functions ω_K in Whitney's theorem are estimation functions, then the rth derivatives of the extended function satisfy (up to constant factors in the right-hand side) inequalities similar to (14) on any compact subsets in $\mathbb{R}^m$ (and not only on subsets of E). Thus, the extension method used in the proof preserves the "degree of smoothness" of the highest derivatives. In particular, if all functions ω_K in Whitney's theorem have the form $A_K t^\theta$ (with fixed $\theta \in (0, 1)$), then the extended function belongs to the class $C^{r+\theta}$. In other words, if the original family of polynomials satisfies the conditions necessary for the extended function to belong to the class $C^{r+\theta}$ (see inequality (14) with $\omega_K(t) = A_K t^\theta$), then the constructed extension does indeed belong to this class.

We now turn to implementing this program.

In the following statement, we assume that all assumptions of Whitney's theorem are satisfied and keep the notation of this theorem, assuming that all majorants of moduli of continuity arising below are estimation functions.

Proposition *Let Q be a compact subset in $\mathbb{R}^m$, $Q_0 = Q \cap E$,*

$$R = \{x \in E : \text{dist}(x, Q_0) \le 5\},$$

and ω_R be the corresponding estimation function. The derivatives of order r of the function $\tilde{f}$ constructed in Whitney's theorem satisfy, for all $x, y \in Q$, the inequality

$$|\tilde{f}^{(\alpha)}(x) - \tilde{f}^{(\alpha)}(y)| \le A \omega_R(\|x - y\|), \tag{15}$$

where A is a sufficiently large coefficient (depending on Q).

We assume without loss of generality that $Q_0 \ne \varnothing$ (otherwise we can replace Q with an ambient compact set that has a nonempty intersection with E). Recall also that, according to our convention, the functions ω_K in Whitney's theorem depend monotonically on the set: to a larger set there corresponds a larger function.

Proof First of all, note that since the left-hand side of (15) is bounded on Q, this inequality holds for

$$\|x - y\| \ge \delta > 0 \quad \text{if } A \ge \frac{M}{\omega_R(\delta)},$$

where M is an upper bound on the left-hand side. Hence, below we take $\delta = \frac{1}{4}$ and prove inequality (15) assuming that $\|x - y\| < \frac{1}{4}$. Under this assumption, we consider successively several cases.

I. Both points x, y belong to Q_0. Then inequality (15) holds just by the assumptions of Whitney's theorem (since $\omega_{Q_0} \le \omega_R$).

II. At least one of the points x, y (say, x) is "far" from Q_0, more exactly, $\text{dist}(x, Q_0) \ge \frac{1}{2}$. Then both points belong to the compact set

$$Q' = \left\{ u \in Q \mid \text{dist}(u, Q_0) \ge \frac{1}{4} \right\}$$

contained in O. Hence, inequality (15) is valid for them, because the extended function is infinitely differentiable in O and all its derivatives satisfy the Lipschitz condition on Q'.

III. One of the points x, y (for definiteness, y) belongs to Q_0 while the other one does not: $x \notin Q_0$, $y \in Q_0$. As in the proof of Whitney's theorem, let $\bar{x}$ be a point of E closest to x. It may not belong to Q_0, but is not far away from this set, because

$$\|\bar{x} - y\| \le \|\bar{x} - x\| + \|x - y\| \le 2\|x - y\| < \frac{1}{2}.$$

In particular, $\bar{x} \in R$ (see Fig. 5). Clearly,

$$|\tilde{f}^{(\alpha)}(x) - \tilde{f}^{(\alpha)}(y)| \le |\tilde{f}^{(\alpha)}(x) - \tilde{f}^{(\alpha)}(\bar{x})| + |\tilde{f}^{(\alpha)}(\bar{x}) - \tilde{f}^{(\alpha)}(y)|. \tag{16}$$

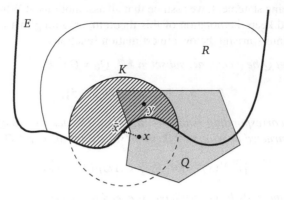

Fig. 5

The second term in the right-hand side satisfies (15), because $\bar{x}, y \in R$ and $\|\bar{x} - y\| \leq 2\|x - y\|$, which implies (by condition (4) of Whitney's theorem) that

$$|\widetilde{f}^{(\alpha)}(\bar{x}) - \widetilde{f}^{(\alpha)}(y)| \leq \omega_R(\|\bar{x} - y\|) \leq \omega_R(2\|x - y\|). \tag{17}$$

Now we estimate the difference $\widetilde{f}^{(\alpha)}(x) - \widetilde{f}^{(\alpha)}(\bar{x})$, writing it in the form

$$\widetilde{f}^{(\alpha)}(x) - \widetilde{f}^{(\alpha)}(\bar{x}) = \widetilde{f}^{(\alpha)}(x) - P_{\bar{x}}^{(\alpha)}(\bar{x}) = \widetilde{f}^{(\alpha)}(x) - P_{\bar{x}}^{(\alpha)}(x) = g^{(\alpha)}(x) - P_{\bar{x}}^{(\alpha)}(x)$$

(the middle equality holds because $P_{\bar{x}}^{(\alpha)}$ is a polynomial of degree 0). To this end, we apply inequality (11) with $|\alpha| = r$ and $c = \bar{x}$ (the conditions under which this inequality is valid are satisfied, since $K = E \cap \bar{B}(\bar{x}, 4) \subset R$).

In our case, inequality (11) takes the form

$$|\widetilde{f}^{(\alpha)}(x) - \widetilde{f}^{(\alpha)}(\bar{x})| = |g^{(\alpha)}(x) - P_{\bar{x}}^{(\alpha)}(x)| \leq C_3\, \omega_K(4\|x - \bar{x}\|). \tag{17'}$$

Substituting this estimate together with (17) into (16), we obtain

$$|\widetilde{f}^{(\alpha)}(x) - \widetilde{f}^{(\alpha)}(y)| \leq C_3\, \omega_K(4\|x - \bar{x}\|) + \omega_R(2\|x - y\|) \leq (4C_3 + 2)\omega_R(\|x - y\|).$$

This proves inequality (15) in case III.

IV. It remains to consider the case where both points x, y do not belong to Q_0 but are close to this set:

$$0 < \text{dist}(x, Q_0) < \frac{1}{2}, \quad 0 < \text{dist}(y, Q_0) < \frac{1}{2}.$$

There are two possible subcases (as before, $\bar{x}$ is a point of E closest to x):

$$\text{a)}\ \|x - y\| \geq \frac{1}{2}\|x - \bar{x}\| \quad \text{and} \quad \text{b)}\ \|x - y\| < \frac{1}{2}\|x - \bar{x}\|.$$

In the first case, $\|x - \bar{x}\| \le 2\|x - y\|$ and $\|y - \bar{x}\| \le 3\|x - y\|$. Repeating the derivation of inequalities (17') (for the points $x, \bar{x}$) and (17) (for $y, \bar{x}$), we obtain

$$|\widetilde{f}^{(\alpha)}(x) - \widetilde{f}^{(\alpha)}(\bar{x})| \le C_3 \omega_R(4\|x - \bar{x}\|) \le C_3 \omega_R(8\|x - y\|) \le 8C_3 \,\omega_R(\|x - y\|),$$
$$|\widetilde{f}^{(\alpha)}(y) - \widetilde{f}^{(\alpha)}(\bar{x})| \le \omega_R(\|y - \bar{x}\|) \le \omega_R(3\|x - y\|) \le 4\,\omega_R(\|x - y\|),$$

which implies (15) (with an appropriate constant).

Finally, consider the case IV (b), where $\|x - y\| < \frac{1}{2}\|x - \bar{x}\| = \frac{1}{2}\rho_x$. Then

$$y \in B\left(x, \frac{1}{2}\rho_x\right) \subset O.$$

Since the segment $[x, y]$ between x, y lies in O, the difference $\widetilde{f}^{(\alpha)}(x) - \widetilde{f}^{(\alpha)}(y)$ can be estimated using Lagrange's inequality:

$$|\widetilde{f}^{(\alpha)}(x) - \widetilde{f}^{(\alpha)}(y)| \le \max_{z \in [x,y]} \left\|\operatorname{grad} \widetilde{f}^{(\alpha)}(z)\right\| \cdot \|x - y\|. \tag{18}$$

To estimate $\operatorname{grad} f^{(\alpha)}(z)$, we may use inequality (13) (since $|\alpha| = r$). It implies that for $c = \bar{x}$ and $K = E \cap \bar{B}(\bar{x}, 4)$, all coordinates of the gradient satisfy the inequality

$$\left|\frac{\partial f^{(\alpha)}}{\partial x_i}(z)\right| \le C \frac{\omega_K(4\rho_z)}{\rho_z} \quad (i = 1, \ldots, m).$$

Further, all points of the segment $[x, y]$ are at distance at least $\frac{1}{2}\rho_x$ from E, so

$$\rho_z \ge \frac{1}{2}\rho_x > \|x - y\|.$$

Hence, taking into account that $\omega_K(at) \le 2a\omega_K(t)$, for $a = 4\frac{\rho_z}{\|x-y\|} > 1$ and $t = \|x - y\|$ we obtain the following inequality:

$$\|\operatorname{grad} \widetilde{f}^{(\alpha)}(z)\| \le \sqrt{m}\, C \frac{\omega_K(4\rho_z)}{\rho_z} \le 8\sqrt{m}\, C \frac{\omega_K(\|x - y\|)}{\|x - y\|}.$$

Now, due to (18), we have

$$|\widetilde{f}^{(\alpha)}(x) - \widetilde{f}^{(\alpha)}(y)| \le 8\sqrt{m}\, C\, \omega_K(\|x - y\|). \tag{19}$$

The function ω_K depends on $K = E \cap \bar{B}(\bar{x}, 4)$, and hence on x. To get rid of this dependence, observe that $K \subset R$, and hence $\omega_K \le \omega_R$ for all x under consideration. This allows us to obtain from (19) the final inequality:

$$|\widetilde{f}^{(\alpha)}(x) - \widetilde{f}^{(\alpha)}(y)| \le 8\sqrt{m}\, C\, \omega_R(\|x - y\|).$$

Increasing the constants obtained in different cases, we get a constant A for which inequality (15) holds for any $x, y \in Q$. $\square$

Let us mention a corollary which complements the proposition proved above. We keep the notation introduced in Whitney's theorem.

Corollary *Let* $\theta \in (0, 1]$, *and assume that all assumptions of Whitney's theorem are satisfied. If all functions* ω_K *have the form* $\omega_K(t) = A_K t^\theta$, *then the function* f *has an extension of class* $C^{r+\theta}$ *(for* $\theta < 1$*) or* LC^r *(for* $\theta = 1$*).*

The existence of such an extension follows from Whitney's theorem, and the fact that it belongs to the required smoothness class follows from the proposition, since in our case all functions ω_K are estimation functions.

11.5 Extension from Totally Connected Sets

Since condition (4) of Whitney's theorem is not only sufficient but also necessary for the existence of an r-smooth extension, this theorem basically resolves the problem under consideration. However, by its very generality, this condition is quite difficult to verify even in cases of relatively simple sets E. For instance, it is not at all obvious that for one of the curvilinear trapezoids (see Fig. 6)

$$\{(x, y): |x| \le 1, \ 0 \le y \le 2 - \sqrt{|x|}\}$$

and

$$\{(x, y): |x| \le 1, \ 0 \le y \le 1 + \sqrt{|x|}\},$$

any continuous function with uniformly continuous (inside the trapezoid) partial derivatives has a smooth extension, while for the other trapezoid, this is not the case. We will return to these examples later.

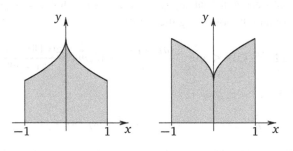

Fig. 6

In this subsection, we introduce a simple yet sufficiently general geometric property of sets which can be used to prove that every function defined on a set with this property and satisfying simple necessary conditions has a sufficiently smooth extension to the whole space.

Definition A subset X of a Euclidean space is said to be *totally connected* if for every bounded subset A of X there exists a number C such that any two points $x, y \in A$

can be connected by a path in X of length at most $C \|x - y\|$. A path satisfying this condition will be called a *C-shortest path*.

Here are several examples of totally connected sets, the first of which is obvious.

Example 1 A convex set is totally connected.

Note that the union of two convex sets with a nonempty intersection is not necessarily totally connected. An example is provided by two externally tangent disks.

Example 2 The hypograph of a positive Lipschitz function f defined on a convex set $A \subset \mathbb{R}^{m-1}$ is totally connected.

Recall that the *hypograph* of a nonnegative function f defined on a subset A in $\mathbb{R}^{m-1}$ is the set (below, the product $\mathbb{R}^{m-1} \times \mathbb{R}$ is identified with $\mathbb{R}^m$)

$$\{(x, y) \in \mathbb{R}^m : x \in A, \ 0 \le y \le f(x)\}.$$

Proof Assume without loss of generality that A is bounded. Let P be the hypograph of f and M be the Lipschitz constant of f.

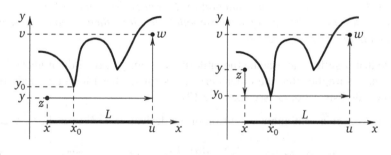

Fig. 7

Consider points $z = (x, y)$, $w = (u, v)$ of P and the segment L between x and u. Besides, let y_0 be the absolute minimum of f on L, $y_0 = f(x_0)$ for some $x_0 \in L$. If $\min\{y, v\} \le y_0$ (for definiteness, $y \le y_0$), then a path satisfying the condition from the definition of a totally connected set (with $C = 2$) can be obtained by moving from (x, y) first horizontally over L, and then upwards (or downwards) to (u, v). Otherwise, first descend from (x, y) to (x, y_0), taking into account that

$$0 \le y - y_0 \le f(x) - y_0 = f(x) - f(x_0) \le M \|x - x_0\| \le M \|x - u\|,$$

and then repeat the procedure used in the first case (see Fig. 7).

The result is a path connecting z and w whose length does not exceed

$$2M \|x - u\| + \|x - u\| \le (2M + 1) \|z - w\|. \qquad \square$$

Not only the hypograph itself, but also its interior is totally connected, which can be checked by repeating the above argument.

Here is another simple example showing that the interior of a totally connected set E is not necessarily totally connected even if it is connected and dense in E.

Example 3 Consider the set $E = \{(x, y) \in \mathbb{R}^2 : x \le x^2 + y^2 \le 2x\}$ bounded by two internally tangent circles. It is easy to see that E is totally connected. At the same time, its interior

$$G = \{(x, y) \in \mathbb{R}^2 : x < x^2 + y^2 < 2x\}$$

does not have this property. Indeed, for $x \in (0, 1)$ both points $(x, \pm\sqrt{x})$ belong to G. The distance between them is infinitesimal as $x \to 0$, while the lengths of paths connecting these points and lying in G are bounded away from zero.

The next theorem contains the main result of this subsection. Recall that necessary conditions for a continuous function f defined on a set E to have an r-smooth extension are that f is r-smooth inside E and its partial derivatives up to order r are uniformly continuous on every bounded subset of $\text{Int}(E)$. We call them the natural necessary conditions for the existence of an r-smooth extension.

Theorem *Let E be a closed set in $\mathbb{R}^m$ whose interior G is dense in E and totally connected. If f is a continuous function on E that satisfies the natural necessary conditions for the existence of an r-smooth extension, then it has an r-smooth extension to the whole space.*

Note that it suffices to assume the uniform continuity only of the highest derivatives, since it implies that the other derivatives satisfy the Lipschitz condition on every bounded subset of G (see Exercise 12).

Proof Let g_α be a continuous extension of the derivative $f^{(\alpha)}$ to E. For $c \in E$ set

$$P_c(x) = f(c) + \sum_{1 \le |\alpha| \le r} \frac{g_\alpha(c)}{\alpha!}(x - c)^\alpha \qquad (x \in \mathbb{R}^m).$$

It follows immediately from this definition that $P_x(x) = f(x)$ for $x \in E$ and $P_c = T_r(f, c)$ for $c \in G$. Besides, obviously,

$$P_c^{(\alpha)}(x)\big|_{x=c} = g_\alpha(c) \quad \text{for all } c \in E, |\alpha| \le r. \tag{20}$$

To prove the theorem, it remains to check that the family of polynomials $\{P_c\}_{c \in E}$ satisfies condition (4) of Whitney's theorem. Let us specify what it means in our case.

Replacing c in (20) by $x \in E$, we write the left-hand side of (4) in the form

$$\left|P_x^{(\alpha)}(x) - P_c^{(\alpha)}(x)\right| = \left|g_\alpha(x) - P_c^{(\alpha)}(x)\right|, \quad \text{where } c, x \in E. \tag{21}$$

Let K be an arbitrary compact subset in E (if E is compact, we may assume that $K = E$). Without loss of generality, we assume that K is the closure of its interior W

(otherwise we can replace K by the closure of a bounded relative neighborhood of K). Let C be the constant corresponding to W according to the definition of the total connectedness of G, and let U be the intersection of G with the Δ-neighborhood of W where $\Delta = C \cdot \mathrm{diam}(K)$. It is clear that every C-shortest path connecting points of W is contained in U.

Finally, let ω be a majorant (a function that is increasing on $[0, \infty)$ and infinitesimal at the origin) of all moduli of continuity in U of the functions g_α for $|\alpha| \leq r$. We may and will assume that ω is continuous[2]. To check condition (4), it suffices to verify that for some M we have

$$\left| P_x^{(\alpha)}(x) - P_c^{(\alpha)}(x) \right| \leq M\omega(C\|x - c\|)\|x - c\|^{r-|\alpha|}$$

for $|\alpha| \leq r$ and all $c, x \in K$; due to (21), these inequalities can be written in the form

$$\left| g_\alpha(x) - P_c^{(\alpha)}(x) \right| \leq M\omega(C\|x - c\|)\|x - c\|^{r-|\alpha|} \quad \text{for } c, x \in E \qquad (22)$$

(the role of the function ω_K from (4) will be played by the function $t \mapsto M\omega(Ct)$). So, to complete the proof, we must check inequality (22).

First, let $c, x \in W$, γ be a C-shortest path connecting these points in G, and S be the length of γ. As we have noted, this path is contained in U. By Theorem 2 at p. 82 (for $n = r - |\alpha|$, with O replaced by U and f replaced by $f^{(\alpha)}$), since $g_\alpha = f^{(\alpha)}$, $P_c^{(\alpha)} = T_{r-|\alpha|}(f^{(\alpha)}, c)$ in U, we have

$$\left| g_\alpha(x) - P_c^{(\alpha)}(x) \right| = \left| f^{(\alpha)}(x) - T_{r-|\alpha|}(f^{(\alpha)}, c; x) \right| \leq \frac{(\sqrt{m}\, S)^{r-|\alpha|}}{(r - |\alpha|)!}\, \omega(S).$$

But $S \leq C\|x - c\|$ for any $c, x \in W$, hence this inequality implies (22) with $M = (\sqrt{m}\, C)^r$:

$$|g_\alpha(x) - P_c^{(\alpha)}(x)| \leq M\omega(C\|x - c\|)\, \|x - c\|^{r-|\alpha|}.$$

Since both sides depend continuously on x and c and W is dense in K, passing to the limit shows that this inequality holds for any pair of points from K. This completes the verification of condition (4) of Whitney's theorem, and hence the proof of the existence of a smooth extension. $\qquad\qquad\square$

The assumption that $\mathrm{Int}(E)$ is totally connected cannot be dropped in the theorem, since otherwise the natural necessary conditions may not be sufficient even if the set E and the function f are quite simple. Consider the following example.

---

[2] To achieve this, one can replace $\omega(t)$ with the average value of ω on $[t, 2t]$.

Example 4 Let
$$E = \{(x, y) : |x| \leq 1, \ -1 \leq y \leq \sqrt{|x|}\}.$$

Obviously, every boundary point of E except the origin has a convex relative neighborhood, while the origin has no totally connected relative neighborhood.

Define a continuous function f on E by the formula
$$f(x, y) = \begin{cases} y^2 \operatorname{sign} x & \text{if } y \geq 0, \\ 0 & \text{if } y \leq 0 \end{cases} \quad \text{for } (x, y) \in E.$$

It is easy to see that its first-order derivatives exist and are uniformly continuous in $\operatorname{Int}(E)$. Now, the increment of f corresponding to the points $(\pm y^2, y)$ is equal to the distance between them. Hence, if f had a smooth extension to some neighborhood of the origin, for sufficiently small $y > 0$ we would have $\frac{\partial f}{\partial x}(\bar{x}, y) = 1$ at some point $(\bar{x}, y)$, $|\bar{x}| \leq y^2$. This contradicts the continuity of $\frac{\partial f}{\partial x}$ at the origin, since this derivative vanishes for $y \leq 0$. Therefore, the function f (which satisfies the natural necessary conditions) has no smooth extension even to an arbitrarily small neighborhood of the origin.

In conclusion, note if the function whose hypograph we consider fails to satisfy the Lipschitz condition at some point, this does not necessarily preclude the smooth extendability of a function defined on this hypograph and satisfying the natural necessary conditions (for example, if the interior of the hypograph remains totally connected). An example of such a situation is the hypograph of the function $1 - \sqrt{|x|}$ on the interval $[-1, 1]$ and the point $(0, 1)$ (see also Exercise 1 b)).

11.6 Sets with Minimally Smooth Boundaries

Here we consider an interesting class of closed sets for which the necessary conditions for the existence of a smooth extension turn out to be sufficient too. We need the following result.

Lemma *If every boundary point of a connected open set G has a neighborhood whose intersection with G is totally connected, then G is totally connected.*

Proof Assume that the assertion is false. Then the condition from the definition of a totally connected set is violated for some bounded subset in G. Hence, for every positive integer n it contains points x_n and y_n such that (throughout the proof, $s(\gamma)$ stands for the length of a path γ)

$$s(\gamma) > n\|x_n - y_n\| \quad \text{for every path } \gamma \text{ connecting } x_n \text{ and } y_n \text{ in } G. \tag{23}$$

Note that $x_n \neq y_n$, since otherwise the strict inequality would fail for the constant path.

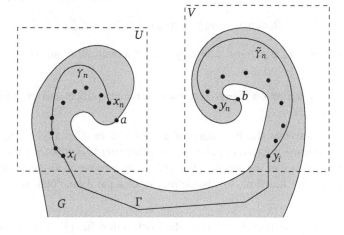

Fig. 8

Passing to subsequences if necessary, we assume that the sequences $\{x_n\}_{n \geq 1}$ and $\{y_n\}_{n \geq 1}$ are convergent: $x_n \to a$, $y_n \to b$. The points a and b have neighborhoods U and V (if $a = b$, then $U = V$) whose intersections with G are totally connected (for points of ∂G this is true by assumption, and for points of G this is obvious). Let X and Y be the sets of points x_n and y_n with $n \geq i$ that belong to the intersections $U \cap G$ and $V \cap G$, respectively, and let C and C' be the constants corresponding to the (bounded) sets X and Y according to the definition of a totally connected set. Then for every $n \geq i$ there exist paths γ_n and $\widetilde{\gamma}_n$ connecting x_n with x_i in $U \cap G$ and y_n with y_i in $V \cap G$, respectively, whose lengths do not exceed $C\|x_n - x_i\|$ and $C'\|y_n - y_i\|$. Using these paths and a rectifiable path Γ in G connecting x_i and y_i, one can easily construct a path Γ_n in G (see Fig. 8) connecting x_n and y_n such that

$$s(\Gamma_n) = s(\gamma_n) + s(\Gamma) + s(\widetilde{\gamma}_n) \leq C\|x_n - x_i\| + s(\Gamma) + C'\|y_n - y_i\|.$$

On the other hand, by (23), the paths Γ_n satisfy the inequality $s(\Gamma_n) > n\|x_n - y_n\|$. Thus,

$$n\|x_n - y_n\| < C\|x_n - x_i\| + s(\Gamma) + C'\|y_n - y_i\|.$$

Since the right-hand side of this inequality is bounded, it follows that $\|x_n - y_n\| \to 0$ and, therefore, $a = b$.

Thus, for $n \geq i$ both x_n and y_n belong to the same neighborhood U, so the union $X \cup Y$ is contained in the totally connected set $U \cap G$. Hence, for $n \geq i$ there exists a path δ_n in $U \cap G$ connecting x_n with y_n and satisfying the condition

$$s(\delta_n) \leq \widetilde{C}\|x_n - y_n\|,$$

where $\widetilde{C}$ is the constant corresponding to $X \cup Y$ according to the definition of a totally connected set. Together with (23) this shows that

$$0 < n\|x_n - y_n\| < s(\delta_n) \le \widetilde{C}\|x_n - y_n\|$$

and, therefore, $n < \widetilde{C}$. Since n is arbitrary, we get a contradiction, which proves the lemma.   □

The following definition is borrowed (with some simplification) from [15, Chapter VI, Section 3.3].

Definition We say that the boundary of a subset E in $\mathbb{R}^m$ satisfies the *minimal smoothness condition* if for every point $x_0 \in \partial E$ there exists a closed parallelepiped Q centered at x_0 such that the intersection $Q \cap E$ is congruent to the hypograph of a positive Lipschitz function (for the definition of a parallelepiped, see the end of Section 0.13).

Since (by the properties of interior) the intersection $\text{Int}(Q) \cap \text{Int}(E)$ coincides with $\text{Int}(Q \cap E)$, i.e, with the interior of the hypograph of a Lipschitz function, this intersection (see Example 2) is totally connected. Note also that a set satisfying this definition coincides with the closure of its interior.

The following theorem summarizes the results obtained earlier for sets with minimally smooth boundaries.

Theorem (Stein[3]) *Let E be a connected closed set with minimally smooth boundary. For every r, for a continuous function on E, the natural necessary conditions are not only necessary, but also sufficient for the existence of an r-smooth extension.*

Proof By Theorem 11.5, it suffices to verify that the interior of E is totally connected. By the definition of minimal smoothness, for every point from ∂E we can find a closed parallelepiped Q such that $Q \cap E$ coincides (up to a rigid motion) with the hypograph of a Lipschitz function. In this case, as we have mentioned, the intersection $U \cap \text{Int}(E)$, where $U = \text{Int}(Q)$, coincides with the interior of $Q \cap E$ and hence is congruent to the interior of the hypograph of a positive Lipschitz function defined on a convex set. As we have observed in Example 2, such sets are totally connected. Thus, every point of the boundary of E has a neighborhood U whose intersection with the interior of E is totally connected. Hence, it immediately follows from the last lemma that $\text{Int}(E)$ is totally connected.   □

Exercises

1. a) Check that the hypograph of the function $x \mapsto 1 + \sqrt{|x|}$ ($|x| \le 1$) is not totally connected.

   b) Show that the hypograph of the function $x \mapsto 1 - \frac{1}{\log 3/|x|}$ ($|x| \le 1$) (equal to 1 at the origin) is totally connected, although it does not satisfy the Lipschitz condition of any order.

---

[3] Elias Menachem **Stein** (1931–2018) was an American mathematician.

    c) Show that the hypograph of the function $(x, y) \mapsto 1 + (x^2 + y^2)^p$ defined on the disk $x^2 + y^2 \leq 1$ is totally connected for every $p > 0$.

    d) For what values of the positive parameters p, q will the hypographs of the functions $(x, y) \mapsto 1 + (|x| + y^2)^p$ and $(x, y) \mapsto 1 + (x^4 + y^2)^q$ defined on the disk $x^2 + y^2 \leq 1$ be totally connected?

2. Give examples of two- and three-dimensional totally connected closed sets whose interiors are connected but not totally connected.

3. Show that the hypograph of a positive Lipschitz function defined on a totally connected set is totally connected.

4. Show that the closure of a totally connected set is totally connected.

5. By analogy with Lemma 11.6, show that if every boundary point of a connected closed set has a totally connected relative neighborhood, then this set is totally connected.

6. Give an example of a function satisfying the natural necessary conditions for the existence of a smooth extension that cannot be smoothly extended from the union of two externally tangent closed disks.

7. By analogy with Example 4 from Section 11.5, give an example of a function that satisfies the natural necessary conditions for the existence of an r-smooth extension for all $r \in \mathbb{N}$, but does not have even a 1-smooth extension.

8. Repeating the proof of Theorem 11.5, show that every function satisfying the natural necessary conditions for the existence of a smooth extension on the set from Example 3 or on the set $\{(x, y) \in \mathbb{R}^2 : xy \geq 0\}$ can be smoothly extended to $\mathbb{R}^2$ (although neither of these sets satisfies the minimal smoothness condition and their interiors are not totally connected).

9. Let $G \subset \mathbb{R}^2$ be an open set, $f \in C(G)$, and L be a smooth simple arc contained in G. Show that if the first-order partial derivatives of f are uniformly continuous in $G \setminus L$, then a potential violation of the smoothness of f on L does not occur, i.e., $f \in C^1(G)$.

10. Show that the function from Example 4 in Section 11.5 in which $y^2 \operatorname{sign} x$ is replaced by $y^{r+1} \operatorname{sign} x$ with $r > 1$ has a 1-smooth extension but not an r-smooth extension. How smooth can it be?

11. Let $K \subset [0, 1]$ be a compact nowhere dense set. Show that the difference $(0, 1)^2 \setminus K^2$ is totally connected. If the length of K is positive, then this difference is an example of a totally connected set in the plane whose boundary has positive area (by the length and area we mean the linear and planar Lebesgue measures).

12. Show that if an open set G is totally connected and all the highest derivatives of an r-smooth function f defined on G are uniformly continuous on every bounded subset of G, then the derivatives of lower order satisfy the Lipschitz condition on such subsets. Give an example showing that the condition of G being totally connected cannot be dropped.

Chapter II. Smooth Maps

Everywhere in this chapter, O is an open subset in $\mathbb{R}^m$.

1 The Open Mapping Theorem and the Diffeomorphism Theorem

1.1 The Open Mapping Theorem

One of the main problems dealt with in this chapter is the smoothness of the inverse map. Recall that the set of r-smooth maps from an open set O to $\mathbb{R}^n$ is denoted by $C^r(O; \mathbb{R}^n)$. An invertible smooth map $F : O \to \mathbb{R}^m$ whose inverse is also smooth is called a *diffeomorphism*.

In the definition of a diffeomorphism, we consider only smooth maps from an open subset in $\mathbb{R}^m$ again to $\mathbb{R}^m$, and not to a space of some other dimension. The reason is that a homeomorphism (and hence a diffeomorphism) preserves dimension. This is a deep topological result. However, the fact that the dimension is preserved under smooth homeomorphisms can be easily proved using the chain rule for differentiating compositions. As it turns out, this rule allows one to reduce the (nonlinear) problem under consideration to a similar problem for linear maps. Indeed, let O and O' be open subsets in $\mathbb{R}^m$ and $\mathbb{R}^n$, respectively, with $n < m$. Then maps $\Phi \in C^1(O; \mathbb{R}^n)$ and $\Psi \in C^1(O'; \mathbb{R}^m)$ cannot be mutually inverse. Indeed, assuming the contrary, the composition $\Psi \circ \Phi$ is the identity map in O. Differentiating it by Theorem I.5.3 at a point $a \in O$, we obtain

$$I = d_b \Psi \circ d_a \Phi,$$

where I is the identity map in $\mathbb{R}^m$ and $b = \Phi(a)$. However, this is impossible, since the dimension of the image of the map $d_b \Psi$, which is defined in $\mathbb{R}^n$, cannot exceed n.

In Section I.5.5, we have obtained a necessary condition for the inverse map to be differentiable. Namely, the map inverse to a differentiable map $F : O \to \mathbb{R}^m$ is differentiable at a point $y_0 = F(x_0)$ only if

© The Author(s), under exclusive license to Springer Nature Switzerland AG 2021
B. M. Makarov, A. N. Podkorytov, *Smooth Functions and Maps*, Moscow Lectures 7,
https://doi.org/10.1007/978-3-030-79438-5_2

$$\text{the matrix } F'(x_0) \text{ is invertible.} \tag{1}$$

In this case, it turns out that

$$(F^{-1})'(y_0) = (F'(x_0))^{-1}. \tag{2}$$

Since differentiability is defined only at interior points of the domain of definition, for F^{-1} to be smooth it is necessary that the set of values of F be open. That is what we will establish first: that a map F satisfying condition (1) at every point $x_0 \in O$ is open, i.e., sends open sets to open sets.

Theorem *If a map $F \in C^1(O; \mathbb{R}^m)$ satisfies condition (1) everywhere in O, then the set $O' = F(O)$ is open.*

Proof We must show that an arbitrary point $y_0 \in O'$ is an interior point of this set. Let $y_0 = F(x_0)$. By Lemma I.5.5, there exist positive numbers c and δ such that

$$\bar{B} = \bar{B}(x_0, \delta) \subset O$$

and

$$\|F(x) - y_0\| = \|F(x) - F(x_0)\| \geq c\|x - x_0\| \quad \text{as long as } \|x - x_0\| \leq \delta.$$

In particular,

$$\|F(x) - y_0\| \geq c\delta \quad \text{if } \|x - x_0\| = \delta.$$

We claim that

$$B(y_0, r) \subset O' \quad \text{for } r = \frac{1}{2}c\delta.$$

To show this, we fix an arbitrary point $y \in B(y_0, r)$ and prove that the distance $\|F(x) - y\|$ vanishes for some $\bar{x} \in B(x_0, \delta)$. Indeed, since

$$\|F(x) - y_0\| \geq c\delta = 2r \quad \text{if } \|x - x_0\| = \delta,$$

for such x we have

$$\|F(x) - y\| \geq \|F(x) - y_0\| - \|y_0 - y\| > r.$$

However,

$$\|F(x_0) - y\| = \|y_0 - y\| < r,$$

and thus the values of $\|F(x) - y\|$ are greater on the boundary of $\bar{B}$ than at its center. Hence, the continuous function

$$\bar{B} \ni x \mapsto \|F(x) - y\|$$

attains an absolute minimum at an interior point $\bar{x}$ of the ball $\bar{B}$. Obviously, the smooth function

$$\bar{B} \ni x \mapsto \varphi(x) = \|F(x) - y\|^2$$

also has this property, so grad $\varphi(\bar{x}) = 0$. To write this equality in more detail, we introduce the coordinate functions $f_1, \ldots, f_m$ of the map F and the coordinates $(y_1, \ldots, y_m)$ of the vector y. Then

$$\varphi(x) = \sum_{k=1}^{m} (f_k(x) - y_k)^2$$

and, consequently,

$$\frac{\partial \varphi}{\partial x_j}(\bar{x}) = 2 \sum_{k=1}^{m} \frac{\partial f_k}{\partial x_j}(\bar{x})(f_k(\bar{x}) - y_k) = 0 \quad \text{for } j = 1, \ldots, m.$$

The matrix of this homogeneous linear system in the variables

$$f_1(\bar{x}) - y_1, \ldots, f_m(\bar{x}) - y_m$$

is nothing else than the transposed matrix $F'(\bar{x})$. Since it is invertible, the system has only the trivial solution, which is equivalent to the equality $F(\bar{x}) = y$. Thus,

$$y \in F(\bar{B}) \subset O'.$$

Since y is an arbitrary point of $B(y_0, r)$, this means that $B(y_0, r) \subset O'$. Hence, $y_0 \in \text{Int}(O')$.   □

Note that if O is a domain (a connected open set), then, obviously, the set $O' = F(O)$ is also a domain.

1.2 The Open Mapping Theorem (Continued)

The theorem proved in the previous section can be generalized to the case of a map to a smaller-dimensional space. As in the case of equal dimensions, this result will be called the open mapping theorem. Here, the condition that the matrix F' is invertible (condition (1)) must be replaced by the condition that it has maximal rank, which, as we will see below, plays an important role in the study of other properties of smooth maps too. Points at which this condition is violated are called *critical points* of F.

Theorem *Let $F \in C^1(O; \mathbb{R}^n), n \le m$. If for every $x \in O$ the rank of the matrix $F'(x)$ is equal to n (i.e., the map F has no critical points), then the set $O' = F(O)$ is open in $\mathbb{R}^n$.*

Proof The proof is based on a simple idea: apply Theorem 1.1 to the restriction of F to a cross section of the set O obtained by fixing some coordinates.

We must show that for every point $x_0 \in O$, the value $F(x_0)$ is an interior point of the set O'. In what follows, we assume that $x_0 = 0$, since otherwise F can be replaced by the map $x \mapsto F(x_0 + x)$ defined in a neighborhood of the origin. Since the matrix $F'(0)$ has maximal rank, some $n \times n$ minor of this matrix does not vanish. We assume without loss of generality that it is formed by the first n columns, so the determinant

$$\Delta(x) = \det \left(\frac{\partial f_j}{\partial x_k} \right)_{j,k \le n} (x),$$

where $f_1, \ldots, f_n$ are the coordinate functions of F, does not vanish at the origin. By continuity, it does not vanish in a ball B centered at the origin and contained in O. Let $\widetilde{B}$ be the intersection of B with the subspace generated by the first n coordinate axes (as usual, we identify this subspace with $\mathbb{R}^n$), and let $\widetilde{F}$ be the restriction of F to $\widetilde{B}$. Obviously, for $x \in \widetilde{B}$ we have

$$\det((\widetilde{F})'(x)) = \Delta(x) \ne 0,$$

and hence the matrix $(\widetilde{F})'$ is invertible in $\widetilde{B}$. Thus, the map $\widetilde{F}$ satisfies the assumptions of Theorem 1.1, which says that the set $\widetilde{F}(\widetilde{B})$ is open. Since

$$F(0) \in \widetilde{F}(\widetilde{B}) \subset O',$$

we see that $F(0)$ is an interior point of O', as required.   □

Using the open mapping theorems allows one to simplify the technical aspects of the proofs of some important results (see, e.g., Theorems 1.3 (the diffeomorphism theorem) and III.3.2 (a necessary condition for a relative extremum)).

1.3 The Diffeomorphism Theorem

This section is devoted to the proof of the theorem on the smoothness of the inverse map, also called the diffeomorphism theorem.

First, we prove a lemma which will also be useful later.

Let $GL(m)$ be the set of invertible $m \times m$ matrices. We regard it as a subset of $\mathbb{R}^{m^2}$.

Lemma *The set $GL(m)$ is open, and the map $A \mapsto A^{-1}$ is infinitely smooth on $GL(m)$.*

Proof The first assertion of the lemma holds because the function $A \mapsto \det(A)$ is continuous and $GL(m)$ coincides with the set of $m \times m$ matrices with nonzero determinant. The infinite differentiability of the map $A \mapsto A^{-1}$ follows from the fact that its coordinate functions (i.e., the elements of the matrix A^{-1}) are the ratios of the corresponding cofactors to the determinant of A and hence are infinitely differentiable functions of its elements.   □

Theorem *Let $F \in C^1(O; \mathbb{R}^m)$. If F is invertible and the matrix of the map F' is invertible at every point from O, then F^{-1} is a smooth map.*

As we have already mentioned, the invertibility of F' is necessary for F^{-1} to be smooth.

Proof Theorem 1.1 implies that the set $O' = F(O)$ is open. By the same theorem, the image of any open set $U \subset O$ is open. Setting $S = F^{-1}$, we can rewrite the equality $V = F(U)$ in the form $V = S^{-1}(U)$. Thus, the S-preimage of any open set is open, which proves that the map S is continuous. Its differentiability at an arbitrary point $y_0 \in O'$ follows from Theorem I.5.5, whose assumptions are obviously satisfied. By the same theorem (see (2)),

$$S'(y) = \left(F'(F^{-1}(y))\right)^{-1}.$$

Now we prove the smoothness of S. It is equivalent to the continuity of S' (as observed in Section I.5.2 after the definition of Jacobian matrix). Passing from y to $S'(y)$ can be represented as the composition

$$y \mapsto F^{-1}(y) = x \mapsto F'(x) = A \mapsto A^{-1} = S'(y).$$

Each of the three maps in this chain is continuous, which implies the continuity of S', i.e., the C^1-smoothness of S. $\qquad\square$

1.4 The Smoothness of the Inverse Map

Now we generalize Theorem 1.3 to maps of arbitrary (and, in particular, fractional) smoothness.

Theorem *If an invertible map F of class $C^t(O; \mathbb{R}^m)$, $1 \le t \le +\infty$, satisfies condition (1) everywhere in O, then F^{-1} also belongs to the class C^t.*

The same is also true for the class LC^r. The proof of this fact essentially coincides with the one below, as the reader can easily check.

Proof The validity of the theorem for $t = 1$ is already proved in Theorem 1.3. Now we prove it for $t = 1 + \alpha$ with $0 < \alpha < 1$. By (2) we have

$$S'(y) = (F'(S(y)))^{-1}, \tag{3}$$

where $S = F^{-1}$. Since $F' \in C^\alpha$ and $S \in C^1$, we see that the composition $F' \circ S$ belongs to the class C^α (see Section I.6.7). Due to Lemma 1.3, the inverse map $S' = (F' \circ S)^{-1}$ also belongs to the same class. By definition, this means that $S \in C^{1+\alpha}$.

The proof for $2 \le t < \infty$ proceeds by induction. Let $t = r + \alpha$ where $r \ge 2$, $0 \le \alpha < 1$. By the induction hypothesis, S and, consequently, $F' \circ S$ are $(t-1)$-smooth maps. Since, by Lemma 1.3, the operation of taking the inverse is infinitely smooth, it follows that the map $S' = (F' \circ S)^{-1}$ belongs to the same class as $F' \circ S$, i.e., is $(t-1)$-smooth, which is equivalent to the fact that $S \in C^t$.

For infinitely smooth maps, the theorem follows from the equality $C^\infty = \bigcap_{r=1}^{\infty} C^r$. $\square$

Exercises

1. Construct a C^∞-diffeomorphism
   a) from the interval $(-1, 1)$ onto $\mathbb{R}$;
   b) from the cube $(-1, 1)^m$ onto $\mathbb{R}^m$;
   c) from the ball $B(0; 1)$ onto the entire space;
   d) from a cube onto a ball.
2. Identify $\mathbb{R}^4$ with the space $\mathbb{C}^2 = \mathbb{C} \times \mathbb{C}$ by associating with a point (x, y, s, t) the point (u, v) where $u = x + iy$, $v = s + it$. Set

$$A = \{(u, v) \in \mathbb{C}^2 : 0 < |u| < |v| < 1\}$$

(in the real case, the counterpart of the set A consists of the two triangles $\{(x, y) \in \mathbb{R}^2 : 0 < |x| < |y| < 1\}$). Consider the map

$$f(u, v) = \left(u, v\sqrt{1 - \frac{|u|^2}{|v|^2}}\right), \quad (u, v) \in A.$$

Show that f is a diffeomorphism, find its Jacobian (see Section I.5.5) and the image of the set A.
3. Assume that an invertible map F of class $LC^r(O; \mathbb{R}^m)$ $(r \in \mathbb{N})$ satisfies condition (1) everywhere in O. Show that $F^{-1} \in LC^r(F(O); \mathbb{R}^m)$.

2 Local Invertibility Theorems and Dependence of Functions

2.1 Local Invertibility

As one can easily see, the invertibility of the differential of a map F, i.e., the invertibility of the matrix F', does not imply the invertibility of F. To obtain corresponding examples, consider the infinitely smooth maps

$$F_1(x, y) = (x^2 - y^2, 2xy) \qquad \text{for } (x, y) \in \mathbb{R}^2, x^2 + y^2 > 0;$$
$$F_2(x, y) = (e^x \cos y, e^x \sin y) \quad \text{for } (x, y) \in \mathbb{R}^2.$$

Identifying points in the plane with complex numbers, we can write them in the form

$$F_1(z) = z^2 \quad (z \in \mathbb{C} \setminus \{0\}); \qquad F_2(z) = e^z \quad (z \in \mathbb{C}).$$

Each of these maps has an invertible differential at every point of its domain of definition, but neither F_1 nor F_2 is one-to-one. Every nonzero point of the plane has two preimages under F_1 and infinitely many preimages under F_2. However, it turns out that if the necessary condition for the smoothness of the inverse map (condition (1) at p. 128) is satisfied, then the invertibility can be guaranteed "in the small."

Recall that throughout this section (and the entire chapter), O denotes an open subset in $\mathbb{R}^m$.

Theorem (local invertibility) *Let $F \in C^1(O; \mathbb{R}^m)$, $a \in O$. If the matrix $F'(a)$ is invertible, then there exists a neighborhood $U \subset O$ of a such that the restriction of F to U is one-to-one and, consequently, a diffeomorphism.*

Proof We must prove that the restriction of the map F to some neighborhood of the point a is invertible and satisfies condition (1) at p. 128.

Since the matrix $A = F'(a)$ is invertible, we have $\|A(h)\| \geq c\,\|h\|$ for some $c > 0$ and all $h \in \mathbb{R}^m$ (see inequality (4) at p. 21). Fix a ball $B(a, r) \subset O$ such that

$$\det(F'(x)) \neq 0 \quad \text{and} \quad \|F'(x) - A\| < \frac{c}{2} \quad \text{for } x \in B(a, r); \tag{1}$$

we will check that the neighborhood $U = B(a, r)$ is as required.

Since $\det F'(x) \neq 0$, the matrix $F'(x)$ is invertible in U; hence, by Theorem 1.3, it suffices to check that the restriction of F to U is invertible, i.e., that F is one-to-one on U. Let $x, y \in U$, $h = y - x$. We write the difference $F(y) - F(x)$ in the form

$$F(y) - F(x) = F(x + h) - F(x) - A(h) + A(h).$$

Using Corollary 1 of Lagrange's inequality (see Section I.5.4) and inequality (1), we see that for $x \neq y$,

$$\|F(y) - F(x)\| \geq \|A(h)\| - \|F(x + h) - F(x) - A(h)\|$$

$$\geq c\|h\| - \sup_{z \in [x,y]} \|F'(z) - A\|\,\|h\| \geq c\|h\| - \frac{c}{2}\|h\| = \frac{c}{2}\|y - x\| > 0.$$

Thus, we have proved that the restriction of F to U is one-to-one, which completes the proof of the theorem. $\qquad\square$

Corollary *If a smooth map $F \colon O \to \mathbb{R}^m$ has no critical points (i.e., the matrix $F'(x)$ is invertible for all $x \in O$), then the preimage of every point $b \in F(O)$ consists of isolated points.*

Proof Let a be an arbitrary point of the preimage. The theorem says that there exists a neighborhood U of a in which F is one-to-one. Therefore, U does not contain preimages of b different from a, as required. $\qquad\square$

2.2 The Partial Inversion Theorem

The local invertibility theorem proved in the previous section can be reformulated as follows.

Let $F \in C^1(O; \mathbb{R}^m)$, $a \in O$. If the rank of the matrix $F'(a)$ is equal to m, then in some neighborhood V of the point $F(a)$ one can define a diffeomorphism $\Phi \colon V \to O$ such that $a \in \Phi(V)$ and the composition $F \circ \Phi$ is the identity map.

We will prove a generalization of this theorem in which, under weakened assumptions, "inversion" is carried out with respect to only part of the coordinates. This result will be useful in Section 2.4, as well as in Sections 4 and IV.4. As usual, we identify a smaller-dimensional space with the subspace of a larger-dimensional space generated by the first vectors of the canonical basis, i.e., for $k < m$ we identify points $(y_1, \dots, y_k) \in \mathbb{R}^k$ and $(y_1, \dots, y_k, 0, \dots, 0) \in \mathbb{R}^m$. Besides, by P_k we denote the projection to $\mathbb{R}^k$, i.e., the map defined in an ambient space of $\mathbb{R}^k$ by the formula

$$y = (y_1, \dots, y_k, y_{k+1}, \dots) \mapsto P_k(y) = (y_1, \dots, y_k, 0, \dots) = (y_1, \dots, y_k).$$

Theorem Let $F \in C^1(O; \mathbb{R}^n)$, $a \in O$. If $k = \operatorname{rank} F'(a) \geq 1$, then there exist a neighborhood $U \subset O$ of a and a diffeomorphism $\Psi \colon U \to \mathbb{R}^m$ such that the composition $F \circ \Psi^{-1}$ preserves k coordinates of the argument.

If $\operatorname{rank} F'(x) = k$ everywhere in O, then U and Ψ can be chosen so that the composition $\widetilde{F} = F \circ \Psi^{-1}$ at points $y = (y_1, \dots, y_k, \dots, y_m) \in \Psi(U)$ has the form

$$\widetilde{F}(y) = (y_1, \dots, y_k, \widetilde{f}_{k+1}(u), \dots, \widetilde{f}_n(u)), \quad \text{where } u = P_k(y) = (y_1, \dots, y_k) \quad (2)$$

(if $n \leq m$ and $k = n$, then $\widetilde{F}(y) = (y_1, \dots, y_n)$ is the projection to $\mathbb{R}^n$).

If the map F belongs to the class C^t with $t > 1$, then Ψ can be taken in the same smoothness class.

Thus, if the rank of F' is constant in some neighborhood of a, then F can be represented as a composition $F = \widetilde{F} \circ \Psi$ where Ψ is a diffeomorphism and $\widetilde{F}$ is a map of a special form: it preserves the first k coordinates of the argument, while all its other coordinate functions depend only on these coordinates.

If $m \geq n$ and the matrix of F' at the point a has maximal rank (equal to n), then it is easy to see that this rank is equal to n in some neighborhood of a. Hence, we obtain the following result (a special case of the second assertion of the theorem), sometimes called the flattening theorem.

Corollary (flattening theorem) Let $n \leq m$, $F \in C^t(O; \mathbb{R}^n)$, $t \geq 1$, and $a \in O$. If $\operatorname{rank} F'(a) = n$, then there exist a neighborhood $U \subset O$ of a and a diffeomorphism $\Psi \in C^t(U; \mathbb{R}^m)$ such that for $x \in U$ and $y = \Psi(x)$ we have

$$F(\Psi^{-1}(y)) = P_n(y), \quad \text{i.e.,} \quad F(x) = P_n(\Psi(x)).$$

The first equality shows that locally, in a neighborhood of a, the map F is, up to a diffeomorphism, a projection. The second one shows that the first n coordinate functions of Ψ coincide with the coordinate functions of F. In other words, near the point a the map F can be "completed" to a diffeomorphism from U to $\mathbb{R}^m$.

Near the point a, the level sets $f_i(x) = C_i$ of the coordinate functions of F coincide with the preimages of the planes $y_i = C_i$. The diffeomorphism Ψ "flattens" these level sets, turning them into planes perpendicular to the corresponding coordinate axes.

Proof of the theorem Denote by $x_1, \ldots, x_m$ and $f_1, \ldots, f_n$ the coordinates of a point $x \in \mathbb{R}^m$ and the coordinate functions of the map F. By assumption, the rank of the matrix $F'(a)$ is equal to k. Without loss of generality (since the coordinates can be relabelled), we assume that

$$\Delta = \det\left(\frac{\partial f_i}{\partial x_j}\right)_{1 \le i, j \le k}(a) \ne 0$$

and construct a diffeomorphism Ψ such that the composition $F \circ \Psi^{-1}$ preserves the first k coordinates (in the general case, the composition must include also a permutation of coordinates in $\mathbb{R}^m$).

Consider the following map $H \colon O \to \mathbb{R}^m$:

$$H(x) = \left(f_1(x), \ldots, f_k(x), x_{k+1}, \ldots, x_m\right) \quad \text{for } x = (x_1, \ldots, x_m) \in O.$$

Since the Jacobian of H at the point a does not vanish (it is equal to Δ), by the local invertibility theorem there exists a sufficiently small neighborhood U of a such that the restriction of H to U is a diffeomorphism between U and the neighborhood $W = H(U)$ of the point $H(a)$. Now we set $\Psi = H|_U$ and claim that this diffeomorphism is as required, i.e., that the composition $F \circ \Psi^{-1}$ preserves the first k coordinates.

Indeed, let y be an arbitrary point from W. Then $x = \Psi^{-1}(y) \in U$ and

$$y = \Psi(x) = \left(f_1(x), \ldots, f_k(x), x_{k+1}, \ldots, x_m\right).$$

Hence,

$$y_1 = f_1(x) = f_1 \circ \Psi^{-1}(y), \quad \ldots, \quad y_k = f_k(x) = f_k \circ \Psi^{-1}(y),$$

which proves the claim.

Now we turn to the proof of the second part of the theorem. Without loss of generality, we assume that the set $W = \Psi(U)$ is a cube (with edges parallel to the coordinate axes).

Identify the space $\mathbb{R}^m$ with the Cartesian product $\mathbb{R}^k \times \mathbb{R}^{m-k}$, and a point of this space, with a pair (u, v) where $u \in \mathbb{R}^k$, $v \in \mathbb{R}^{m-k}$. Then the composition $\widetilde{F} = F \circ \Psi^{-1}$ can be written in the form

$$\widetilde{F}(u, v) = (u, \Theta(u, v)), \quad \text{where } (u, v) \in W, \ \Theta(u, v) \in \mathbb{R}^{n-k}.$$

In this notation, the Jacobian matrix of the map $\widetilde{F}$ has the form

$$\widetilde{F}' = \begin{pmatrix} I & \mathbb{O} \\ \Theta'_u & \Theta'_v \end{pmatrix},$$

where $\mathbb{I}$ is the $k \times k$ identity matrix, $\mathbb{O}$ is the zero matrix, while Θ'_u and Θ'_v are the left and right parts of the matrix Θ' composed of the partial derivatives with respect to the coordinates of the vectors u and v, respectively.

So far, we have not used the assumption rank $F'(x) = k$ in its entirety and relied only on its validity at the point a. Now we specify the form of the matrix $\widetilde{F}'$. A key point of the proof is that, by assumption, the rank of F' is equal to k everywhere in U and, therefore, the rank of $\widetilde{F}'$ is also equal to k everywhere in W. Hence, Θ'_v is the zero matrix everywhere in W (a nonzero element of this matrix would make it possible to form, together with $\mathbb{I}$, a nonzero minor of $\widetilde{F}'$ of order $k + 1$). Since the neighborhood W is convex, it follows (see Remark I.1.3) that the map Θ does not depend on v, so

$$\Theta(u, v) = \Theta(u, v_0) \quad \text{for all } (u, v), (u, v_0) \in W.$$

Hence, instead of $\Theta(u, v)$ we will write $\theta(u)$, assuming that the map θ is defined in the k-dimensional cube W' that is the projection of W to $\mathbb{R}^k$. Let $\theta_{k+1}, \ldots, \theta_n$ be its coordinate functions. Then for $(u, v) \in W$ we have

$$\widetilde{F}(u, v) = (u, \Theta(u, v)) = (u, \theta(u)) = \left(u_1, \ldots, u_k, \theta_{k+1}(u), \ldots, \theta_n(u)\right),$$

which is equivalent to (2).

Finally, the last assertion of the theorem follows from the method of constructing the map H and Theorem 1.4 on the smoothness of the inverse map. $\square$

2.3 Extending a Smooth Map to a Diffeomorphism

The next theorem refines the partial inversion theorem in the case where F is a map to a larger-dimensional space. As usual, we identify a smaller-dimensional space with the subspace of a larger-dimensional space generated by the first vectors of the canonical basis.

Theorem A *Let $m \leq n$, $F \in C^1(O; \mathbb{R}^n)$, $a \in O$. If the matrix $F'(a)$ has maximal rank, then there exist a neighborhood W of the point $b = F(a)$ and a diffeomorphism $\Psi \in C^1(W; \mathbb{R}^n)$ such that on the set $U = F^{-1}(W)$ the composition $\Psi \circ F$ acts as the identity map: $\Psi(F(x)) = x$ for $x \in U$.*

Here is an equivalent statement of this theorem.

Theorem B *Let $m \leq n$, $F \in C^1(O; \mathbb{R}^n)$, $a \in O$. If the matrix $F'(a)$ has maximal rank, then in $\mathbb{R}^n$ there exists a neighborhood V of the point a such that F can be extended from the intersection $V \cap \mathbb{R}^m$ to a diffeomorphism defined on V.*

To prove Theorem B, it suffices to use Theorem A and take $V = \Psi^{-1}(W)$. Then Ψ^{-1} is a required extension. In turn, Theorem A follows from Theorem B. To see this, it suffices to set $W = \widetilde{F}(V)$, $\Psi = \widetilde{F}^{-1}$, where $\widetilde{F}$ is the diffeomorphism from Theorem B.

Now we give an independent proof of Theorem B.

Proof Since the rank of the matrix $F'(a)$ is equal to m, it has a nonzero $m \times m$ minor Δ. We assume without loss of generality that this minor is formed by the first m rows.

Define a map $H: O \times \mathbb{R}^{n-m} \to \mathbb{R}^n$ by the formula

$$H(x, y) = F(x) + (0, y), \quad \text{where } x \in O \text{ and } y \in \mathbb{R}^{n-m}.$$

Obviously, $H \in C^1(O \times \mathbb{R}^{n-m}; \mathbb{R}^n)$. Identifying, as usual, a point $x \in \mathbb{R}^m$ with the point $(x, 0) \in \mathbb{R}^n$, we see that H is a smooth extension of F to $O \times \mathbb{R}^{n-m}$, with $\det H(a) = \Delta \neq 0$. By the local invertibility theorem (Theorem 2.1), the restriction of H to a sufficiently small (n-dimensional) neighborhood V of a is a diffeomorphism. This diffeomorphism, obviously, has all the required properties. $\square$

One can see from the proof that the constructed diffeomorphism has the same smoothness as the map F, and its first m coordinate functions coincide with the coordinate functions of F.

The theorem implies the following corollary.

Corollary (smoothness of a composition) *Let $m \leq n$, O and O' be open subsets in $\mathbb{R}^m$ and $\mathbb{R}^{m'}$, respectively, and $\Phi \in C^1(O; \mathbb{R}^n)$ be a map with $\operatorname{rank} \Phi'(x) = m$ everywhere in O. Further, let $\Psi \in C^1(O'; \mathbb{R}^n)$ and $\Psi(O') \subset M = \Phi(O)$. If the map Φ is invertible, then the composition $\Phi^{-1} \circ \Psi$ is a smooth map.*

Indeed, by Theorem B, for every point $t_0 \in O'$ the map Φ^{-1} is the restriction of a diffeomorphism to an intersection $M \cap B(\Psi(t_0), r)$ with small r. Hence, we see that in a sufficiently small neighborhood of t_0 the map $\Phi^{-1} \circ \Psi$ coincides with a composition of smooth maps.

2.4 Dependence and Independence of Functions

In some problems which involve finding several unknown functions $f_1, \ldots, f_n$ defined on a set X, one is only able to establish that they satisfy relations of the form

$$\varphi_i(f_1(x), \ldots, f_n(x)) = C_i, \quad \text{where } x \in X, i = 1, \ldots, k < n, \tag{3}$$

and C_i are some constants (in dealing with systems of differential equations, such relations are called first integrals). Suppose that apart from (3) we have found another similar relation:

$$\varphi(f_1(x), \ldots, f_n(x)) = \text{const} \quad \text{for } x \in X. \tag{3'}$$

Does it provide new information? In other words, doesn't (3′) follow from (3)? Simple examples show that such a situation can occur. One of them can be obtained by setting (for $k = 2, n = 3$)

$$\varphi_1(x, y, z) = x + y + z, \quad \varphi_2(x, y, z) = xy + yz + xz,$$
$$\varphi(x, y, z) = x^2 + y^2 + z^2.$$

Obviously, $\varphi = \varphi_1^2 - 2\varphi_2$, so the values of φ are determined by the values of φ_1 and φ_2. Hence, (3′) follows from (3) and, therefore, imposes no additional restrictions on the functions f_1, f_2, f_3.

To refine the statement of the problem, we introduce the following definition.

Definition We say that functions $f_1, \ldots, f_n \in C^1(O)$ are *dependent in O* if there exists a function $\Phi \in C^1(G)$, where G is an open subset in $\mathbb{R}^n$, such that

    a) $(f_1(x), \ldots, f_n(x)) \in G$ for $x \in O$;

    b) $\Phi(f_1(x), \ldots, f_n(x)) = \text{const}$ for $x \in O$;   (4)

    c) $\text{grad}\,\Phi(y) \neq 0$ for $y \in G$.

Also, we say that a system of functions is *independent in O* if it is not dependent in any open subset of O.

It follows from the implicit function theorem (Theorem I.10.3) that if $a \in O$ and $\Phi'_{y_i}(b) \neq 0$ at the point $b = (f_1(a), \ldots, f_n(a)) \in G$, then in a sufficiently small neighborhood U of a the function f_i can be expressed explicitly in terms of the other functions: there exists a smooth function φ of $n - 1$ variables such that

$$f_i(x) = \varphi\big(f_1(x), \ldots, f_{i-1}(x), f_{i+1}(x) \ldots, f_n(x)\big) \quad \text{for } x \in U.$$

Thus, if a system of functions is dependent in O, then locally the values of one of them are determined by the values of the other ones. If in a neighborhood of a point a such a function is f_i, then we say that (in this neighborhood) f_i explicitly depends on $f_1, \ldots, f_{i-1}, f_{i+1}, \ldots, f_n$.

Differentiating (4) with respect to x_j, for $x \in O$ and $y = (f_1(x), \ldots, f_n(x))$ we obtain the equalities

$$\sum_{i=1}^{n} \frac{\partial \Phi}{\partial y_i}(y) \cdot \frac{\partial f_i}{\partial x_j}(x) = 0 \quad (j = 1, \ldots, m),$$

which mean that

$$\sum_{i=1}^{n} \frac{\partial \Phi}{\partial y_i}(y) \,\text{grad}\, f_i(x) = 0.$$

Since $\text{grad}\,\Phi \neq 0$ in G, the gradients of the functions $f_1, \ldots, f_n$ are linearly dependent. Hence, the rank of the Jacobian matrix of this system does not exceed $n - 1$ everywhere in O. Thus, we obtain a necessary condition for a system of functions to be dependent:

the rank of the Jacobian matrix of a system of
dependent functions is less than the number of
functions.

In particular, if the number of functions in a dependent system does not exceed the number of variables, then its Jacobian matrix cannot have maximal rank.

This condition is close to being sufficient. The proof of the corresponding statement is simply a reformulation of the second part of Theorem 2.2.

Theorem *Let $f_1, \ldots, f_n \in C^1(O)$. If the rank of the Jacobian matrix $\left(\frac{\partial f_i}{\partial x_j}\right)_{\substack{1 \le i \le n \\ 1 \le j \le m}}$ is equal to k everywhere in O, then for $k = n$ the functions $f_1, \ldots, f_n$ are independent in O, and for $k < n$ every point $a \in O$ has a neighborhood U in which k of the functions are independent and the other ones depend on them explicitly.*

Proof The first case is obvious, since for $k = n$ the gradients of the functions $f_1, \ldots, f_n$ are linearly independent and, consequently, the necessary condition for dependence cannot be satisfied in any part of the domain O.

Now let $k < n$, $a \in O$, and F be the map whose coordinate functions are the functions of the system under consideration. To prove the theorem, it suffices to apply Theorem 2.2 to F. According to this theorem, in some neighborhood U of a the map F can be represented as the composition $F = \widetilde{F} \circ \Psi$ of a diffeomorphism Ψ defined in U and a map $\widetilde{F}$ of the form (up to relabelling coordinates and coordinate functions)

$$\widetilde{F}(y) = \left(w, \widetilde{f}_{k+1}(w), \ldots, \widetilde{f}_n(w)\right),$$

where $y \in \Psi(U)$, $w = P_k(y)$, and P_k is the projection from $\mathbb{R}^m$ to $\mathbb{R}^k$, that is, $P_k(y) = (y_1, \ldots, y_k)$.

Denoting by $\psi_1, \ldots, \psi_m$ the coordinate functions of Ψ, for $x \in U$, $y = \Psi(x)$ we obtain

$$F(x) = \widetilde{F}(\Psi(x)) = \left(P_k(y), \widetilde{f}_{k+1}(P_k(y)), \ldots, \widetilde{f}_n(P_k(y))\right)$$
$$= \left(\psi_1(x), \ldots, \psi_k(x), \widetilde{f}_{k+1}(\psi_1(x), \ldots, \psi_k(x)), \ldots, \widetilde{f}_n(\psi_1(x), \ldots, \psi_k(x))\right).$$

Rewriting this equality in coordinates, we see that

$$f_j(x) = \psi_j(x) \quad \text{for } 1 \le j \le k$$

and

$$f_j(x) = \widetilde{f}_j(\psi_1(x), \ldots, \psi_k(x)) = \widetilde{f}_j(f_1(x), \ldots, f_k(x))$$

for $k < j \le n$. This shows that the functions $f_{k+1}, \ldots, f_n$ depend in U on the first k functions of the system under consideration. □

Exercise

1. Show that the set on which the Jacobian matrix of a smooth map has maximal rank is open, while for smaller values of the rank this is generally not the case.

3 Curvilinear Coordinates and Change of Variables

3.1 Curvilinear Coordinates

By the coordinates of a point one usually means an ordered tuple of numbers that determines its position. A point in $\mathbb{R}^m$, by the very definition of this space, is an m-tuple of numbers called its Cartesian coordinates. However, in various problems it is often helpful to describe the position of a point in terms of other parameters, or, as one says, to use other coordinate systems. For example, in linear algebra, the reduction of a quadratic form to its canonical form involves various orthogonal coordinate systems. To describe the position of a point on the celestial sphere or on the surface of a ball, it is convenient to use latitude and longitude. To describe the position of a point in the three-dimensional space, one can consider concentric spheres and specify, along with latitude and longitude, also the radius of the sphere (see Example 3 in Section 3.2). Often, it is natural to assign curvilinear coordinates not to all points of a space, but only to points of some domain. What remains unchanged is the requirement that the position of a point be uniquely determined by the tuple of coordinates. More precisely, one may say that a coordinate system is a bijective map defined in some part of the space. Introducing a formal definition, we add the smoothness requirement. The result is the following definition.

Definition Let O be a domain in $\mathbb{R}^m$. A *system of curvilinear coordinates* in O is an arbitrary diffeomorphism $\Phi \colon O \to \mathbb{R}^m$. The inverse map $\Psi = \Phi^{-1} \colon O' \to O$, where $O' = \Phi(O)$, is called a *parametrization* of the set O. If $\varphi_1, \ldots, \varphi_m$ are the coordinate functions of Φ, then their values at a point $x \in O$ are called the *curvilinear coordinates* of this point.

In dealing with curvilinear coordinates and the corresponding parametrization, it is useful to think of them as being defined in "different copies" of the space $\mathbb{R}^m$ and denote the Cartesian coordinates of points of these spaces by different letters. Thus, to a point x with curvilinear coordinates

$$y_1 = \varphi_1(x), \quad y_2 = \varphi_2(x), \quad \ldots, \quad y_m = \varphi_m(x)$$

we associate the point $y = \Phi(x)$ with (Cartesian) coordinates $y_1, \ldots, y_m$. When introducing curvilinear coordinates in particular cases, it is often more convenient to first define a parametrization of the domain O, i.e., using the coordinate functions of Ψ, express the coordinates of a point x in terms of curvilinear coordinates:

$$x_1 = \psi_1(y), \quad x_2 = \psi_2(y), \quad \ldots, \quad x_m = \psi_m(y),$$

and only then, so far as is necessary for the problem, use the coordinate functions of Φ.

Constancy sets of the curvilinear coordinates, i.e., level sets of the functions $\varphi_1, \ldots, \varphi_m$, are called coordinate surfaces (in the two-dimensional case, coordinate curves). Every point $x_0 \in O$ coincides with the intersection of m coordinate surfaces defined by the equations

$$\varphi_1(x) = \varphi_1(x_0), \quad \varphi_2(x) = \varphi_2(x_0), \quad \ldots, \quad \varphi_m(x) = \varphi_m(x_0).$$

The normals to these level surfaces at the point x_0 are linearly independent (and hence form a basis), since they are proportional to the gradients of the functions $\varphi_1, \ldots, \varphi_m$, and these gradients at the point x_0 are the rows of the invertible matrix $\Phi'(x_0)$.

3.2 Examples of Curvilinear Coordinates

Here we consider some frequently used systems of curvilinear coordinates in the plane $\mathbb{R}^2$ and in the three-dimensional space $\mathbb{R}^3$.

Example 1 (polar coordinates) Without using Cartesian coordinates, the position of a point a in the plane can be determined if we know the distance r from this point to the origin (in the Cartesian coordinate system) O and the polar angle φ, i.e., the angle made by the radius vector of a with some fixed ray starting at O. The numbers r and φ are called the *polar coordinates* of x. Introducing Cartesian coordinates so that the polar angle is measured with respect to the positive part of the x-axis in the direction of the positive part of the y-axis (i.e., "counterclockwise"), we see that the relation between the polar coordinates and the Cartesian coordinates x, y is given by the formulas

$$x = r \cos \varphi, \quad y = r \sin \varphi.$$

Formally, they define a smooth map

$$(r, \varphi) \mapsto \Psi(r, \varphi) = (r \cos \varphi, r \sin \varphi)$$

sending the (r, φ)-plane to the (x, y)-plane. However, bearing in mind the geometric meaning of the parameter r (the distance to the origin), we will assume that the map Ψ is defined in the half-plane $r \geq 0$. Obviously, Ψ is not one-to-one. To make it one-to-one, we must exclude the value $r = 0$ and restrict the range of φ to an interval of length at most 2π. The reader can easily see that the restriction Ψ_α of the map Ψ to a half-strip of the form

$$P_\alpha = (0, +\infty) \times (\alpha, \alpha + 2\pi)$$

is one-to-one, and its image O_α is the plane minus the ray

$$L_\alpha = \{(r \cos \alpha, r \sin \alpha) : r \geq 0\},$$

or, as one says, the plane cut along the ray L_α. Obviously, $L_\alpha = \Psi(\partial P_\alpha)$, so $\Psi(\bar{P}_\alpha) = \mathbb{R}^2$. Since the map Ψ is not one-to-one, when changing from Cartesian to polar coordinates, one must indicate the range of the polar angle. Usually, it is either $(0, 2\pi)$ or $(-\pi, \pi)$ (which corresponds to $\alpha = 0$ and $\alpha = -\pi$).

Passing from the parametrization to the polar coordinates proper, i.e., the coordinate functions r, φ of the map inverse to Ψ_α, we see, directly from the geometric meaning of the function r, that $r(x, y) = \sqrt{x^2 + y^2}$. The second coordinate function, the polar angle, can be expressed in terms of inverse trigonometric functions. In particular, for $\alpha = 0$ or $\alpha = -\pi$, in the upper half-plane (i.e., for $y > 0$) we have $\varphi(x, y) = \arccos \frac{x}{\sqrt{x^2 + y^2}}$; while for $\alpha = -\pi$, in the right half-plane (i.e., for $x > 0$) we have

$$\varphi(x, y) = \arctan \frac{y}{x} = \arcsin \frac{y}{\sqrt{x^2 + y^2}}.$$

We leave it to the reader to obtain formulas for the polar angle in the other cases.

Coordinate curves, i.e., curves of the form $r = \text{const}$ and $\varphi = \text{const}$, are circles (centered at the origin O) and rays (starting at O), respectively. The parametrization Ψ transforms a rectangle $[r_0, r_0 + \rho] \times [\varphi_0, \varphi_0 + \xi]$ into the curvilinear quadrangle bounded by the circles $r = r_0$, $r = r_0 + \rho$ and the rays $\varphi = \varphi_0$, $\varphi = \varphi_0 + \xi$ (see Fig. 1).

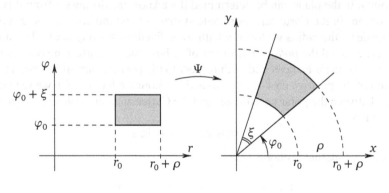

Fig. 1

Example 2 Just as the position of a point in the plane in the previous example was determined by the intersection of a circle centered at the origin with a ray starting at this center, the position of a point in the first quadrant can be described by replacing circles with hyperbolas.

More precisely, this means that in the set $\mathbb{R}_+^2 = (0, +\infty)^2$ we can introduce curvilinear coordinates u, v by setting $u = xy$, $v = \frac{y}{x}$. Coordinate curves in this case are hyperbolas $xy = \text{const}$ (constancy curves of the coordinate u) and rays $\frac{y}{x} = \text{const}$.

The reader can easily check that the map

$$(x, y) \mapsto \Phi(x, y) = \left(xy, \frac{y}{x}\right)$$

is indeed a diffeomorphism of $\mathbb{R}_+^2$ onto itself.

With the curvilinear coordinates thus introduced, the Φ-image of a curvilinear quadrangle

$$\left\{(x, y) \in \mathbb{R}_+^2 : a^2 \le xy \le b^2, \ \alpha \le \frac{y}{x} \le \beta\right\}$$

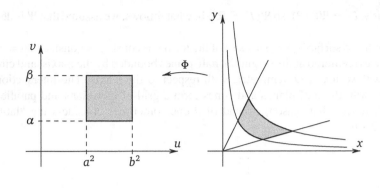

Fig. 2

is the ordinary rectangle $[a^2, b^2] \times [\alpha, \beta]$ (see Fig. 2).

Example 3 (spherical coordinates) Spherical coordinates in the three-dimensional space is an analog of polar coordinates in the plane. The position of a point (x, y, z) can be determined by three parameters as follows. First, the distance r from the point to the origin (in the Cartesian coordinate system); second, the polar angle φ in the (x, y)-plane corresponding to the projection of the point to this plane; and, finally, the angle $\theta \in [0, \pi]$ made by the radius vector of the point with the positive part of the z-axis. These numbers r, φ, θ are called the *spherical coordinates* of the point.

As is frequently the case, to establish a relation between Cartesian and spherical coordinates, it is more convenient to define not the curvilinear coordinates, but the inverse map, the parametrization. This relation is given by the formulas

$$x = r \cos \varphi \sin \theta, \quad y = r \sin \varphi \sin \theta, \quad z = r \cos \theta.$$

Formally, they define a smooth map

$$(r, \varphi, \theta) \mapsto \Psi(r, \varphi, \theta) = (r \cos \varphi \sin \theta, \, r \sin \varphi \sin \theta, \, r \cos \theta)$$

sending the (r, φ, θ)-space to the (x, y, z)-space. However, bearing in mind the geometric meaning of the parameter r (the distance to the origin), we will assume that the map Ψ is defined in the half-space $r \geq 0$.

Obviously, Ψ is not one-to-one. To make it one-to-one, we must restrict the range of r, φ, and θ. Since $\Psi(0, \varphi, \theta) = 0$ for all φ, θ, we assume, excluding the case $r = 0$, that $r > 0$. The angle φ is assumed to vary from 0 to 2π (sometimes, it is more convenient to replace these bounds by $-\pi$ and π). The reader can easily see that the restriction of Ψ to the infinite open parallelepiped

$$P = (0, +\infty) \times (0, 2\pi) \times (0, \pi)$$

is a diffeomorphism of P onto the space $\mathbb{R}^3$ minus the closed half-plane

$$L_0 = \{(r \sin \theta, 0, r \cos \theta) : r \geq 0, \, 0 \leq \theta \leq \pi\}.$$

Obviously, $L_0 = \Psi(\partial P)$, so $\Psi(\bar{P}) = \mathbb{R}^3$. In what follows, we assume that Ψ is defined on $\bar{P}$.

Coordinate surfaces, i.e., surfaces of the form $r = \text{const}$, $\varphi = \text{const}$, and $\theta = \text{const}$, are spheres (centered at the origin O), half-planes bounded by the z-axis, and circular cones with vertex at O symmetric with respect to the z-axis. The intersections of a sphere with the half-planes and cones form a grid of meridians and parallels on this sphere (for that reason, instead of θ one sometimes considers the "latitude" $\theta' = \pi/2 - \theta$).

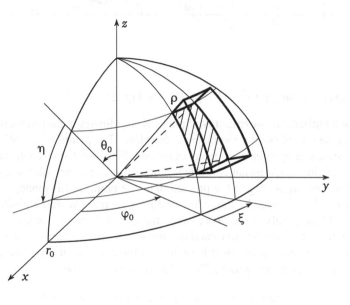

Fig. 3

The map Ψ transforms a parallelepiped $[r_0, r_0+\rho] \times [\varphi_0, \varphi_0+\xi] \times [\theta_0, \theta_0+\eta]$ into the curvilinear parallelepiped bounded by the spheres $r = r_0$, $r = r_0 + \rho$, half-planes $\varphi = \varphi_0$, $\varphi = \varphi_0 + \xi$, and conic surfaces $\theta = \theta_0$, $\theta = \theta_0 + \eta$ (see Fig. 3).

3.3 Partial Derivatives in Curvilinear Coordinates

When solving differential equations that involve partial derivatives of an unknown function U, it is often useful to make a change of variables, i.e., replace the Cartesian coordinates in the domain of definition of U by curvilinear coordinates. Then the original equation gets replaced by a new one, which involves partial derivatives of the function V defined as the composition $U \circ \Psi$, where Ψ is the parametrization corresponding to the introduced curvilinear coordinates. In this case, one says that the function V and the new equation are obtained by a change of variables, or by converting to curvilinear coordinates. We will not consider the general form of this technique, restricting ourselves to several important examples.

Example 1 (Laplacian in polar coordinates) We want to find out how the Laplacian ΔU (see Section I.6.1) of a C^2-smooth function U of two variables transforms when we pass to polar coordinates. For this, we must express its partial derivatives U''_{x^2} and U''_{y^2} in terms of the partial derivatives of the function V related to U by the formula $V(r, \varphi) = U(r \cos \varphi, r \sin \varphi)$.

Using the chain rule, we obtain (for brevity, below we omit the arguments of the functions U, V and their partial derivatives)

$$V'_r = U'_x \cos \varphi + U'_y \sin \varphi, \quad V'_\varphi = -U'_x r \sin \varphi + U'_y r \cos \varphi.$$

Repeated differentiation yields

$$V''_{r^2} = U''_{x^2} \cos^2 \varphi + U''_{y^2} \sin^2 \varphi + 2U''_{xy} \sin \varphi \cos \varphi;$$

$$V''_{\varphi^2} = (U''_{x^2} \sin^2 \varphi + U''_{y^2} \cos^2 \varphi - 2U''_{xy} \sin \varphi \cos \varphi)r^2 - (U'_x \cos \varphi + U'_y \sin \varphi)r.$$

Adding the first equality multiplied by r^2 to the second one, we obtain

$$r^2 V''_{r^2} + V''_{\varphi^2} = (U''_{x^2} + U''_{y^2})r^2 - (U'_x \cos \varphi + U'_y \sin \varphi)r = r^2 \Delta U - r V'_r.$$

Thus,

$$\Delta U(r \cos \varphi, r \sin \varphi) = V''_{r^2}(r, \varphi) + \frac{1}{r^2} V''_{\varphi^2}(r, \varphi) + \frac{1}{r} V'_r(r, \varphi). \tag{1}$$

If, as happens in some problems, it is known a priori that the unknown function U is radial, i.e., depends only on r but not on φ, that is, $U(x, y) = F(\sqrt{x^2 + y^2})$, we obtain that $V(r, \varphi) = F(r)$, and (1) takes the form

$$\Delta U(r \cos \varphi, r \sin \varphi) = F''(r) + \frac{1}{r} F'(r).$$

This is a special case ($m = 2$) of the formula obtained earlier in Example 4 from Section I.6.1.

In the next two examples we consider a similar problem for a function of three variables.

Example 2 (Laplacian in cylindrical coordinates) Every point $(x, y, z) \in \mathbb{R}^3$ is uniquely determined by its projection (x, y) to the plane $z = 0$ and the third coordinate. Hence, to specify a point (x, y, z), it suffices to know the polar coordinates (ρ, φ) of the point (x, y) and z. Accordingly, we consider the parametrization

$$(x, y, z) = \Psi(\rho, \varphi, z) = (\rho \cos \varphi, \rho \sin \varphi, z)$$

of the domain $\mathbb{R}^3 \setminus P$, where P is the half-plane $\{(x, 0, z): x \geq 0, \ z \in \mathbb{R}\}$. It is easy to see that Ψ is a diffeomorphism of the set

$$\{(\rho, \varphi, z): \rho > 0, \ \varphi \in (0, 2\pi), \ z \in \mathbb{R}\}$$

onto $\mathbb{R}^3 \setminus P$. With some abuse of language we may say that Ψ makes the change of variables to polar coordinates in the (x, y)-plane and acts identically with respect to the third variable z.

It follows from the result of the previous example that the Laplacian ΔU of a C^2-smooth function U of three variables has the following expression in terms of the partial derivatives of the composition $V = U \circ \Psi$:

$$\Delta U = V''_{\rho^2} + \frac{1}{\rho^2}V''_{\varphi^2} + \frac{1}{\rho}V'_\rho + V''_{z^2} = \Delta V + \left(\frac{1}{\rho^2} - 1\right)V''_{\varphi^2} + \frac{1}{\rho}V'_\rho$$

(for brevity, we have omitted the arguments of the functions U and V).

Example 3 (Laplacian in spherical coordinates) Now we consider the more difficult problem of expressing the Laplacian ΔU of a C^2-smooth function U of three variables in spherical coordinates (see Example 3 in Section 3.2). According to the general scheme, we must express the values of the derivatives $U''_{x^2}, U''_{y^2}, U''_{z^2}$ at a point $(x, y, z) = \Psi(r, \varphi, \theta) = (r\cos\varphi\sin\theta, r\sin\varphi\sin\theta, r\cos\theta)$ in terms of the values of the derivatives of the composition $V = U \circ \Psi$ at the point (r, φ, θ).

Instead of cumbersome direct calculations, it is helpful to introduce spherical coordinates in two stages: first introduce polar coordinates ρ, φ in the (x, y)-plane (as we essentially did in the previous example), and then introduce polar coordinates r, θ in the (ρ, z)-plane. The latter change of variables does not modify φ, as we did not modify z in passing to cylindrical coordinates (it acts identically with respect to this variable).

Consider the auxiliary function

$$W(\rho, \varphi, z) = U(\rho\cos\varphi, \rho\sin\varphi, z).$$

The result of Example 2 implies that

$$\Delta U = \Delta W + \left(\frac{1}{\rho^2} - 1\right)W''_{\varphi^2} + \frac{1}{\rho}W'_\rho. \tag{2}$$

Since $V(r, \varphi, \theta) = W(r\sin\theta, \varphi, r\cos\theta)$, we see that the function V is obtained from W by introducing polar coordinates (in the plane of the first and third variables). Hence, to find ΔW, we can again apply the result of Example 2:

$$\Delta W = \Delta V + \left(\frac{1}{r^2} - 1\right)V''_{\theta^2} + \frac{1}{r}V'_r.$$

To express the derivative W'_ρ in terms of the derivatives of V, we use the equalities

$$V'_r = W'_\rho \sin\theta + W'_z \cos\theta \quad \text{and} \quad V'_\theta = W'_\rho r \cos\theta - W'_z r \sin\theta.$$

They imply that $V'_r r \sin\theta + V'_\theta \cos\theta = rW'_\rho$, so

$$W'_\rho = V'_r \sin\theta + V'_\theta \frac{\cos\theta}{r}.$$

It remains to substitute into (2) the obtained expressions for ΔW, W'_ρ, taking into account the obvious equalities $W''_{\varphi^2} = V''_{\varphi^2}$ and $\rho = r \sin\theta$:

$$\Delta U = \Delta V + \left(\frac{1}{r^2} - 1\right)V''_{\theta^2} + \frac{1}{r}V'_r + \left(\frac{1}{r^2 \sin^2\theta} - 1\right)V''_{\varphi^2} + \frac{1}{r\sin\theta}\left(V'_r \sin\theta + V'_\theta \frac{\cos\theta}{r}\right).$$

Collecting like terms in the right-hand side, we obtain the required result:

$$\Delta U = V''_{r^2} + \frac{1}{r^2}V''_{\theta^2} + \frac{1}{r^2 \sin^2\theta}V''_{\varphi^2} + \frac{\cot\theta}{r^2}V'_\theta + \frac{2}{r}V'_r.$$

For a radial function U, this formula becomes considerably simpler: if

$$U(x, y, z) = F(r), \quad \text{where } r = \sqrt{x^2 + y^2 + z^2},$$

then (see Example 4 from Section I.6.1 with $m = 3$)

$$\Delta U(x, y, z) = F''(r) + \frac{2}{r}F'(r).$$

Example 4 (equation for the vibrating string) It is known that small vibrations of an elastic string are described by the equation

$$U''_{t^2}(x, t) = a^2 U''_{x^2}(x, t), \tag{3}$$

where $U(x, t)$ is the displacement from the rest position of the point with abscissa x on the string at time t and a is a nonzero parameter determined by the properties of the string.

Let us find the general form of a C^2-smooth solution of this equation using an appropriate (invertible) linear change of variables

$$(y, s) \mapsto (x, t) = (\alpha y + \beta s, \gamma y + \delta s).$$

Here $\alpha, \beta, \gamma, \delta$ are real parameters which can be chosen as we wish provided that the resulting change of variables is invertible.

Clearly, the function $V(y, s) = U(\alpha y + \beta s, \gamma y + \delta s)$ belongs to the class C^2, and its mixed derivative is equal (for brevity, below we omit the arguments of the functions U and V)

$$V''_{ys} = (V'_y)'_s = \left(\alpha U'_x + \gamma U'_t\right)'_s = \alpha\beta U''_{x^2} + \gamma\delta U''_{t^2} + (\alpha\delta + \beta\gamma)U''_{xt}.$$

Choose the parameters $\alpha, \beta, \gamma, \delta$ so that the coefficient of the mixed derivative in the right-hand side vanish and the ratio of the coefficients of the pure derivatives be the same as in (3). This can be achieved by taking, for example, $\alpha = \beta = a$, $\gamma = -\delta = 1$. Then we obtain $V''_{ys} = a^2 U''_{x^2} - U''_{t^2}$. Since U satisfies equation (3), we see that $V''_{ys} = 0$. It follows that V can be written as a sum of two functions, one depending only on y and the other depending only on s. Making the inverse change of variables, we see that U has the form $U(x, t) = F(x + at) + G(x - at)$, where F, G are arbitrary C^2-smooth functions.

Exercises

Below, a, b, c, p, q are positive numbers.

1. Check that the map

$$(r, \varphi) \mapsto (x, y) = (ar \cos^p \varphi, br \sin^p \varphi),$$

where $r > 0$, $0 < \varphi < \frac{\pi}{2}$, is a parametrization of the first quadrant $\mathbb{R}_+^2$ (the numbers r, φ are called the generalized polar coordinates). What are the coordinate curves? What curves correspond to the half-lines $x = $ const and $y = $ const lying in $\mathbb{R}_+^2$? How does the coordinate curve $r = $ const change as the parameter p varies in the interval $(0, +\infty)$?

2. Check that the map

$$(r, \varphi, \theta) \mapsto (x, y, z) = (ar \cos^p \varphi \sin^q \theta, br \sin^p \varphi \sin^q \theta, cr \cos^q \theta),$$

where $r \in (0, +\infty)$ and $\varphi, \theta \in \left(0, \frac{\pi}{2}\right)$, defines curvilinear (generalized spherical) coordinates in the first octant $\mathbb{R}_+^3$. What are the coordinate surfaces?

3. Express $\| \operatorname{grad} U \|^2$ in polar ($m = 2$) and spherical ($m = 3$) coordinates (U is a smooth function).

4. Consider the map $(u, v) \mapsto \Phi(u, v) = (u^2 - v^2, 2uv)$.
   a) Find the images of the lines $u = $ const and $v = $ const and the image of the upper half-plane.
   b) Show that the map Φ is one-to-one in the upper half-plane (the corresponding coordinates are called parabolic). Find the inverse map.
   c) Find the relation between the Laplacians of a C^2-smooth function f and the function $g = f \circ \Phi$ obtained from f by passing to parabolic coordinates.

5. Solve the equation $U''_{x^2}(x, y) + 2U''_{xy}(x, y) + U''_{y^2}(x, y) = 0$.

   HINT. Write the left-hand side as the repeated derivative along the vector $(1, 1)$.

6. Find all radial solutions of the equation

$$\frac{\partial^4 U}{\partial x^4}(x, y) + 2\frac{\partial^4 U}{\partial x^2 \partial y^2}(x, y) + \frac{\partial^4 U}{\partial y^4}(x, y) = 0.$$

   HINT. Write the left-hand side in the form $\Delta(\Delta U)$.

7. Generalize the result of Example 4 at p. 147 by considering the equation

$$AU''_{x^2}(x, y) + 2BU''_{xy}(x, y) + CU''_{y^2}(x, y) = 0$$

whose coefficients satisfy the inequality $B^2 > AC$.

8. Express the sum

$$x^2 U''_{x^2}(x, y) - y^2 U''_{y^2}(x, y) + x U'_x - y U'_y(x, y)$$

in the curvilinear coordinates $s = xy$, $t = \frac{x}{y}$ (here $x, y > 0$ and U is a C^2-smooth function in $\mathbb{R}^2_+$).

4 Classification of Smooth Maps

4.1 Equivalence of Smooth Maps

As is well known, in algebra rectangular matrices A and $\widetilde{A}$ are said to be equivalent if there exist invertible matrices P and Q such that $\widetilde{A} = PAQ$. The linear maps corresponding to the matrices A and $\widetilde{A}$ essentially coincide, since one of them can be expressed in terms of the other one using two linear changes of variables (one in the space of arguments, the other in the space of values). One may also say that equivalent matrices A and $\widetilde{A}$ correspond to the same linear map in different bases.

Our aim in this section is to extend the notion of equivalence to smooth maps and give a manageable condition for determining whether or not two given smooth maps are equivalent, taking into account the degree of smoothness.

In what follows, O and $\widetilde{O}$ denote open subsets in $\mathbb{R}^m$; speaking about t-smooth maps, we assume that $t \geq 1$, not excluding the case $t = +\infty$.

Definition We say that maps $F \in C^t(O; \mathbb{R}^n)$ and $\widetilde{F} \in C^t(\widetilde{O}; \mathbb{R}^n)$ are *t-equivalent* in O if they can be "intertwined" by t-smooth diffeomorphisms, i.e., there exist t-smooth diffeomorphisms $\Phi \in C^t(O; \mathbb{R}^m)$ and Ψ such that

$$\Phi(O) = \widetilde{O} \quad \text{and} \quad \Psi \circ F = \widetilde{F} \circ \Phi,$$

or, in more detail,

$$\Psi(F(x)) = \widetilde{F}(\Phi(x)) \quad \text{for all } x \in O$$

(Ψ is assumed to be defined in some neighborhood of the set $F(O)$).

The last equality can be obviously rewritten in the form

$$F(x) = \Psi^{-1}(\widetilde{F}(\Phi(x))), \quad x \in O, \tag{1}$$

or in the form

$$\widetilde{F}(y) = \Psi(F(\Phi^{-1}(y))), \quad y \in \widetilde{O}. \tag{1'}$$

In particular, the composition of a map with a t-smooth diffeomorphism yields an equivalent map.

The definition of t-equivalence shows that F and $\widetilde{F}$ coincide up to passing to curvilinear coordinates of appropriate smoothness. One may also say that the equality $\Psi \circ F = \widetilde{F} \circ \Phi$ from this definition is equivalent to the commutativity of the diagram

$$
\begin{array}{ccc}
O & \xrightarrow{\ F\ } & G \\[2pt]
{\scriptstyle\Phi}\Big\downarrow & & \Big\downarrow{\scriptstyle\Psi} \\[2pt]
\widetilde{O} = \Phi(O) & \xrightarrow{\ \widetilde{F}\ } & \mathbb{R}^n,
\end{array}
$$

in which G is a set open in $\mathbb{R}^n$ that contains $F(O)$.

We leave it to the reader to prove that the t-equivalence relation is symmetric and transitive. The latter means that if F and H are t-equivalent and H and S are t-equivalent, then F and S are also t-equivalent. In view of symmetry, transitivity means that two maps t-equivalent to a third map are t-equivalent to each other.

Before attacking our main problem, that of finding conditions for the equivalence of two smooth maps, we recall results related to classification up to equivalence of linear maps, or, which is the same, of arbitrary rectangular matrices. It is proved in linear algebra that whether or not two matrices of the same size are equivalent depends on a unique numerical characteristic, their ranks. For such matrices to be equivalent, it is necessary and sufficient that their ranks coincide. Recall that the multiplication by an invertible matrix (from the left or from the right) preserves the rank of a rectangular matrix.

In the nonlinear case, the problem of classification of smooth maps is, of course, much more complicated, but the rank of a map, understood now as the rank of its Jacobian matrix, still plays a crucial role. Our aim is to establish that the coincidence of ranks is necessary, and locally sufficient, for the equivalence of two smooth maps. With this in mind, we introduce the following definition.

Definition The *rank* of a smooth map $F\colon O \to \mathbb{R}^n$ at a point $a \in O$ is the rank of its differential at this point, i.e., the rank of the matrix $F'(a)$.

Remark Using the chain rule, from (1) we obtain

$$
F'(x) = (\Psi^{-1})'(\widetilde{F}(\Phi(x))) \cdot \widetilde{F}'(\Phi(x)) \cdot \Phi'(x).
$$

Since $(\Psi^{-1})'(\widetilde{F}(\Phi(x)))$ and $\Phi'(x)$ are invertible matrices, the ranks of equivalent maps F and $\widetilde{F}$ at the points x and $\Phi(x)$ coincide. Thus, the coincidence of the ranks of F and $\widetilde{F}$ (at the points x and $\Phi(x)$, respectively) is a necessary condition for them to be equivalent, if Φ is one of the intertwining diffeomorphisms. In particular, a necessary condition for F to be equivalent to a linear map is that its rank be constant.

It seems plausible that typically, in a small neighborhood of a point, a smooth map must "look like" its differential at this point. As we will see, this is indeed the case if near this point the rank is constant.

Of course, globally smooth maps can be extremely varied, but the example of the radial projection which maps the punctured space onto the unit sphere ($x \mapsto \frac{x}{\|x\|}$,

$x \neq 0$) makes one feel that locally a smooth map is "very much like" a projection, in this case, a projection to a plane.

4.2 The Rank Theorem

Now we can state the main result of this section, which shows that "in the small" the coincidence of ranks is not only necessary, but, as in the linear case, also sufficient for the equivalence of maps.

Recall that for $k \leq m$ (respectively, $k \leq n$) by P_k we denote the projection from $\mathbb{R}^m$ (or $\mathbb{R}^n$) to the subspace, identified with $\mathbb{R}^k$, generated by the first k vectors of the canonical basis. By $\widetilde{P}_k$, for $0 \leq k \leq \min\{m, n\}$, we denote a modified projection P_k, namely, the map from $\mathbb{R}^m$ to $\mathbb{R}^n$ differing from P_k in that for $x \in \mathbb{R}^m$ the vector $\widetilde{P}_k(x)$ is regarded as an element of $\mathbb{R}^n$. Thus, the vectors $\widetilde{P}_k(x) \in \mathbb{R}^n$ and $P_k(x) \in \mathbb{R}^m$ have the same first k coordinates, while their other coordinates are zero.

Theorem *Let $k \in \{0, 1, 2, \ldots\}$ and $F \colon O \to \mathbb{R}^n$ be a t-smooth map. If the rank of F is equal to k everywhere in O, then near every point from O it is t-equivalent to the projection $\widetilde{P}_k$.*

This theorem shows that, up to passing to curvilinear coordinates, every smooth map satisfying the assumptions of the theorem is locally a projection and, consequently, the rank is the unique invariant that locally determines a map up to introducing curvilinear coordinates. In other words, any two smooth maps of the same rank are locally indistinguishable as smooth maps.

Proof We identify the spaces $\mathbb{R}^m$ and $\mathbb{R}^n$ with the Cartesian products $\mathbb{R}^k \times \mathbb{R}^{m-k}$ and $\mathbb{R}^k \times \mathbb{R}^{n-k}$, respectively, and identify m- and n-dimensional vectors with pairs (u, v) and (u, w) where $u \in \mathbb{R}^k$, $v \in \mathbb{R}^{m-k}$, and $w \in \mathbb{R}^{n-k}$. Using this notation, we may say that $\widetilde{P}_k$ is given by the formula

$$\mathbb{R}^m \ni (u, v) \mapsto \widetilde{P}_k(u, v) = (u, 0) \in \mathbb{R}^n.$$

Let (u_0, v_0) be an arbitrary point from O. Thanks to the second claim of the partial invertibility theorem (Section 2.2) and the transitivity of the equivalence relation, we may assume, replacing F if necessary by an equivalent map, that in some m-dimensional cube $W = W' \times W''$ centered at (u_0, v_0) it has the form

$$F(u, v) = (u, \theta(u)) \quad \text{where } \theta(u) \in \mathbb{R}^{n-k}, \tag{2}$$

and thus depends only on the first element of the pair (u, v).

Let us prove the theorem in this special case. To do this, we use some auxiliary maps and rely on the local invertibility theorem. We introduce a map $H_0 \colon W' \to \mathbb{R}^n$, setting

$$H_0(u) = F(u, v_0) = (u, \theta(u)) \quad (u \in W').$$

Obviously, H_0 sends points of W' to points of the graph of θ. Now consider an extension of H_0 to the "prism" $G = W' \times \mathbb{R}^{n-k}$, defining it as

$$H(u, w) = H_0(u) + (0, w) = (u, \theta(u) + w), \quad \text{where } (u, w) \in G.$$

Clearly, $H(u, 0) = H_0(u)$ for $u \in W'$ and

$$H' = \begin{pmatrix} \mathbb{I}_k & \mathbb{O} \\ \theta' & \mathbb{I}_{n-k} \end{pmatrix},$$

where $\mathbb{I}_k, \mathbb{I}_{n-k}$ are the $k \times k$ and $(n-k) \times (n-k)$ identity matrices and $\mathbb{O}$ is the zero matrix. Since $\det(H'(u_0, 0)) \neq 0$, in some neighborhood $V \subset G$ of the point $(u_0, 0)$, the map H is invertible (by Theorem 2.1) and is a t-smooth diffeomorphism.

At the same time, H preserves the first k coordinates of the argument, and, consequently, the inverse map has the same property. Setting $U = V \cap \mathbb{R}^k$, we see that on the set $H(U)$ (which is a subset of the graph of θ), the map H^{-1} coincides with the projection from $\mathbb{R}^n$ to $\mathbb{R}^k$. In view of (2), for $\Psi = H^{-1}$ and $(u, v) \in (U \times W'') \cap F^{-1}(V)$ we can write

$$\Psi(F(u, v)) = \Psi(F(u, v_0)) = H^{-1}(H_0(u)) = H^{-1}(H(u, 0)) = (u, 0) = \widetilde{P}_k(u, v).$$

Since the diffeomorphism Ψ has the same smoothness as H, by definition this means that the maps F and $\widetilde{P}_k$ are t-equivalent. $\square$

5 *The Global Invertibility Theorem

5.1 Preliminary Remarks

Here we discuss an important special case of the invertibility problem, when the map in question is defined in the whole space $\mathbb{R}^m$. As we have seen, in general the necessary condition for the smooth invertibility of a map F, i.e., the invertibility of the matrix $F'(x)$ at every point of the domain of definition, does not guarantee the invertibility of F. For smooth maps defined in the whole space $\mathbb{R}^m$, one can easily state another condition necessary for F to be a diffeomorphism of $\mathbb{R}^m$ onto itself. Namely,

$$\|F(x)\| \xrightarrow[\|x\| \to \infty]{} \infty.$$

This follows from Weierstrass's theorem, which says that the inverse map must send compact sets to compact sets.

It turns out that these two necessary conditions together are sufficient for a smooth map to be a diffeomorphism. For the first time this result was essentially obtained by Hadamard[1] (see [8]). To prove it, we need some facts from topology.

---

[1] Jacques Salomon **Hadamard** (1865–1963) was a French mathematician.

5.2 Coverings and Their Properties

We introduce the necessary definitions restricting ourselves to a degree of generality sufficient for our purposes.

Definition Let O, O' be open connected subsets in $\mathbb{R}^m$ and F be a continuous map from O onto O'. The triple (O, O', F) is called a *covering* if every point $y \in O'$ has a neighborhood $V \subset O'$ whose preimage $F^{-1}(V)$ can be represented as the union of pairwise disjoint open sets such that the restriction of F to each of them is a homeomorphism of this set onto V. One says that a path α lying in O covers a path β lying in O' (or that α is a lifting of β) if $\beta = F \circ \alpha$.

All paths under consideration are assumed to be defined on the interval $[0, 1]$. Recall that the points $\gamma(0)$ and $\gamma(1)$ are called the beginning and the end of γ, and γ is said to be closed if $\gamma(0) = \gamma(1)$. One says that a closed path γ lying in O can be contracted to a point y_0 in O if there exists a continuous map

$$\Phi \colon [0, 1]^2 \to O$$

(called a deformation, or a homotopy) such that

$$\begin{aligned}
\Phi(t, 0) &= \gamma(t) & \text{for all } t \in [0, 1]; \\
\Phi(t, 1) &= y_0 & \text{for all } t \in [0, 1]; \\
\Phi(0, u) &= \Phi(1, u) & \text{for all } u \in [0, 1].
\end{aligned}$$

If every closed path in O can be contracted to a point in this set, then O is said to be *simply connected*. Obviously, every convex set (in particular, $\mathbb{R}^m$) is simply connected.

The proof of the following important result can be found, e.g., in the book [5, Chapter IV, Section 9.4, Lemma 2].

Proposition *Let (O, O', F) be a covering and α be a path in O that begins at a point x_0 and covers a closed path β (i.e., $\beta = F \circ \alpha$). If β can be contracted to the point $y_0 = F(x_0)$, then α is closed.*

We will use the following corollary of this result.

Corollary *Let (O, O', F) be a covering. If the set O' is simply connected, then the map F is a homeomorphism.*

Proof It follows from the definition of a covering that it suffices to prove that F is one-to-one. Assume to the contrary that a point $y_0 \in O'$ has at least two preimages, a and $b \neq a$. Consider a path α in O connecting these points, and let $\beta = F \circ \alpha$. Since $F(a) = F(b) = y_0$, we obtain $\beta(0) = \beta(1)$, i.e., β is a closed path, which by assumption can be contracted to a point. By the proposition, the path α is also closed, which is, however, impossible, because $\alpha(0) = a \neq b = \alpha(1)$. The obtained contradiction completes the proof. □

5.3 The Main Result

Now we can turn to the central result of this section.

Theorem *A smooth map $F: \mathbb{R}^m \to \mathbb{R}^m$ is a diffeomorphism from $\mathbb{R}^m$ onto $\mathbb{R}^m$ if and only if it satisfies the following conditions:*
  1) *the matrix $F'(x)$ is invertible for every $x \in \mathbb{R}^m$;*
  2) $\|F(x)\| \xrightarrow[\|x\| \to \infty]{} \infty.$

Proof We have already observed that conditions 1, 2 are necessary. Let us show that they are sufficient. First, we check that F is onto. Clearly, by condition 2, the preimage of every bounded set is bounded. In particular, the preimage of every convergent sequence is bounded. Using this fact, we will prove that the F-image of $\mathbb{R}^m$ is closed. Indeed, let

$$y_n \in O' = F(\mathbb{R}^m), \quad y_n \to y_0.$$

We claim that $y_0 \in O'$. Let $\{x_n\}_n$ be a sequence of preimages of the points y_n, i.e., a sequence such that $y_n = F(x_n)$. As we have already noted, it is bounded. Hence, it contains a subsequence $\{x_{n_k}\}_k$ converging to some point x_0. Obviously,

$$y_0 = \lim_{k \to \infty} y_{n_k} = \lim_{k \to \infty} F(x_{n_k}) = F(x_0) \in O',$$

which proves that O' is closed. On the other hand, thanks to condition 1, this set is open (by the open mapping theorem 1.1). Thus, the set O' is open and closed; since the space $\mathbb{R}^m$ is connected, it follows that O' coincides with $\mathbb{R}^m$.

Now we prove that the map F is one-to-one. For this, we show that the triple $(\mathbb{R}^m, \mathbb{R}^m, F)$ is a covering. Then the one-to-oneness of F is a special case of Corollary 5.2, since the space $\mathbb{R}^m$ is simply connected.

To begin with, note that the F-preimage of every point is finite. Indeed, it is compact and, as we have established in Corollary 2.1, consists of isolated points, which is impossible for an infinite compact set.

Now we show that such a preimage always has the same number of points. First, we check that this is the case for points close to an arbitrary point $y_0 \in O'$. Let $x_1, \ldots, x_N$ be the preimages of y_0, and let $U_1, \ldots, U_N$ be disjoint neighborhoods of these points on which F is one-to-one (and even a diffeomorphism). Such neighborhoods exist by the local invertibility theorem. Set

$$V = \bigcap_{k=1}^{N} F(U_k).$$

Clearly, every point from V has exactly N different preimages in the set

$$G = \bigcup_{k=1}^{N} U_k.$$

Hence, if some point from V has more than N preimages, then at least one of them is outside G. If arbitrarily close to y_0 there exist points having more than N preimages, then there exists a sequence of points z_n with this property converging to y_0. Let x_n be a preimage of z_n that does not belong to G ($n = 1, 2, \ldots$). We assume without loss of generality that the sequence $\{x_n\}_n$ converges. Its limit $\widetilde{x}$, obviously, does not belong to G. Meanwhile,

$$F(\widetilde{x}) = \lim_{n \to \infty} F(x_n) = \lim_{n \to \infty} z_n = y_0,$$

and thus we obtain another (different from $x_1, \ldots, x_N$) preimage of y_0, a contradiction. So, in a sufficiently small neighborhood W of y_0, all points have the same number of preimages as y_0. Clearly, we may assume that $W \subset V$. This means that the function sending a point of the set O' to the number of its preimages is locally constant, and hence constant, because O' is connected (see Proposition 0.11). It follows that the full preimage $F^{-1}(V)$ of the set V is contained in G and, consequently, splits into pairwise disjoint parts

$$U'_k = U_k \cap F^{-1}(V),$$

each homeomorphic to V (a homeomorphism is given by the restriction of F to U'_k). Thus, the triple $(\mathbb{R}^m, \mathbb{R}^m, F)$ is a covering, and, as we have already mentioned, the proof is completed by applying Corollary 5.2. $\square$

Exercise

1. Let O be a connected open subset in $\mathbb{R}^m$, $F \in C^1(O; \mathbb{R}^m)$. Show that if $\det(F'(x)) \neq 0$ for $x \in O$ and the preimage of every compact subset in $O' = F(O)$ is compact, then (O, O', F) is a covering and, consequently, F is a diffeomorphism if the set O' is simply connected.

6 *The Morse Lemma

In addition to the results of Section 4, we establish another important fact about the local structure of a sufficiently smooth function, known as the Morse[1] lemma. According to this result, near a nondegenerate critical point, the graph of such a function, after passing to appropriate curvilinear coordinates, coincides with the graph of a quadratic form.

In the first three subsections, we establish auxiliary results, and the main result will be obtained at the end of the section.

---

[1] Harold Calvin Marston **Morse** (1892–1977) was an American mathematician.

6.1 Uniform Reduction of Close Quadratic Forms to Canonical Form

Here we prove an algebraic lemma which describes the reduction of a quadratic form to a sum of squares in the way needed for our purposes. An $m \times m$ square matrix will be identified with a point of $\mathbb{R}^{m^2}$. The Cartesian coordinates of a vector $x \in \mathbb{R}^m$ will be denoted by the same letter with subscripts: $x = (x_1, \ldots, x_m)$.

Lemma *For every invertible diagonal $m \times m$ matrix A there exist a neighborhood U of A and an infinitely smooth map $\omega \colon U \to \mathbb{R}^{m^2}$ such that $\omega(A)$ is the identity matrix and*

$$\langle B(x), x \rangle = \langle A(y), y \rangle \quad \text{for } B \in U \text{ and } x = \omega(B)(y), \ y \in \mathbb{R}^m. \qquad (1)$$

Since the quadratic forms corresponding to a matrix, its transpose, and their half-sum coincide, we may assume without loss of generality that the matrix $\omega(B)$ is symmetric. Setting $\omega(B) = C$, we can write (1) in the form

$$\langle B(C(y)), C(y) \rangle = \langle A(y), y \rangle \quad \text{or} \quad \langle C(B(C(y))), y \rangle = \langle A(y), y \rangle.$$

Thus, the linear change $x = \omega(B)(y)$ transforms the quadratic form whose matrix B is sufficiently close to A to diagonal form with fixed nonzero coefficients.

Proof The proof proceeds by induction on the dimension. For $m = 1$, the assertion is obvious: the matrices A and B can be identified with numbers a (where $a \neq 0$) and b, while U can be taken to be the interval $(a - |a|, a + |a|)$. In this case, the map ω defined by the formula $\omega(B) = \sqrt{|b|/|a|}$ has the required properties. In particular, $\omega(A) = 1$.

Now we assume that the assertion is proved for $(m-1) \times (m-1)$ matrices and prove it for $m \times m$ matrices. Denote the diagonal elements of A by $a_1, \ldots, a_m$. In what follows, we assume that B is so close to A that their last diagonal elements b_{mm} and a_m satisfy the inequality $|b_{mm} - a_m| < |a_m|$ and, consequently, have the same sign.

The vector obtained from a vector x by forgetting the last coordinate and the matrix obtained from A by removing the last row and the last column will be denoted by $\tilde{x}$ and $\tilde{A}$, respectively. The neighborhood corresponding to the matrix $\tilde{A}$ by the induction hypothesis and the corresponding map will be denoted by $\tilde{U}$ and $\tilde{\omega}$.

Let $Q(x) = \langle B(x), x \rangle$ be the quadratic form generated by a matrix B with elements b_{ji} $(j, i = 1, \ldots, m)$. To simplify formulas, we assume that B is symmetric (otherwise, it should be replaced by the matrix $\frac{1}{2}(B + B^T)$); this does not affect the quadratic form. Using this convention, we complete the square in $Q(x)$ with respect to the last coordinate:

$$Q(x) = b_{mm} x_m^2 + 2 x_m \sum_{i < m} b_{mi} x_i + \sum_{j, i < m} b_{ji} x_j x_i$$

$$= b_{mm} \left(x_m + \frac{1}{b_{mm}} \sum_{i < m} b_{mi} x_i \right)^2 - \frac{1}{b_{mm}} \left(\sum_{i < m} b_{mi} x_i \right)^2 + \sum_{j, i < m} b_{ji} x_j x_i.$$

Since b_{mm} and a_m have the same sign, we can write $Q(x)$ in the form

$$Q(x) = \widetilde{Q}(\widetilde{x}) + a_m y_m^2, \tag{2}$$

where

$$\widetilde{Q}(\widetilde{x}) = \sum_{j,i<m} b_{ji} x_j x_i - \frac{1}{b_{mm}} \left(\sum_{i<m} b_{mi} x_i\right)^2, \quad y_m = \sqrt{\frac{b_{mm}}{a_m}}\left(x_m + \frac{1}{b_{mm}}\sum_{i<m} b_{mi} x_i\right).$$

Let $\widetilde{C}$ be the symmetric matrix corresponding to the quadratic form $\widetilde{Q}$. It depends continuously on the elements of B and is arbitrarily close to $\widetilde{A}$ if B is sufficiently close to A. Fix a neighborhood U of A such that for $B \in U$ the matrix $\widetilde{C}$ lies in $\widetilde{U}$ and $|a_m - b_{mm}| < |a_m|$.

By the induction hypothesis, for $\widetilde{C} \in \widetilde{U}$ we have

$$\widetilde{Q}(\widetilde{x}) = \langle \widetilde{C}(\widetilde{x}), \widetilde{x}\rangle = \langle \widetilde{A}(\widetilde{y}), \widetilde{y}\rangle, \quad \text{where } \widetilde{x} = \widetilde{\omega}(\widetilde{C})(\widetilde{y}). \tag{3}$$

It remains to set

$$\omega(B)(y) = (\widetilde{\omega}(\widetilde{C})(\widetilde{y}), y_m). \tag{4}$$

Then it follows from (2) and (3) that

$$\langle B(x), x\rangle = Q(x) = \widetilde{Q}(\widetilde{x}) + a_m y_m^2 = \langle \widetilde{A}(\widetilde{y}), \widetilde{y}\rangle + a_m y_m^2$$
$$= \sum_{i<m} a_i y_i^2 + a_m y_m^2 = \langle A(y), y\rangle.$$

One can see from the construction that the map ω is infinitely smooth (since $\widetilde{\omega}$ is infinitely smooth by the induction hypothesis). If $B = A$, then $y_m = x_m$, $\widetilde{C} = \widetilde{A}$, and, by the induction hypothesis, $\widetilde{\omega}(\widetilde{A})$ is the identity matrix. Hence, it follows from (4) that $\omega(A)$ is the identity matrix too. $\square$

Note that since the map ω is continuous, the neighborhood U can be assumed to be so small that the matrices $\omega(B)$ are invertible for $B \in U$.

6.2 Leibniz's Rule for Differentiation Under the Integral Sign

The next auxiliary result, known as Leibniz's rule, is a very special case of a more general theorem (see, e.g., [9, Section VII.1.5]). It allows one, under simple assumptions, to interchange the differentiation with respect to one variable and the integration with respect to another variable.

Points of the space $\mathbb{R}^{m+1}$ are identified with pairs (x, t) where $x \in \mathbb{R}^m, t \in \mathbb{R}$.

Lemma *Let g be an r-smooth function in a neighborhood of a cylinder $O \times [0, 1]$.*

Then the function $x \mapsto J(x) = \int_0^1 g(x, t)\, dt$ belongs to $C^r(O)$, and

$$J'_{x_j}(x) = \int\limits_0^1 g'_{x_j}(x,t)\, dt \quad (x \in O,\ j = 1,\ldots,m).$$

Proof Let e_j be the jth canonical basis vector. We claim that for every point $x \in O$, the difference

$$\Delta(h) = \frac{J(x + he_j) - J(x)}{h} - \int\limits_0^1 g'_{x_j}(x,t)\, dt$$

is infinitesimal as $h \to 0$. Since

$$\Delta(h) = \int\limits_0^1 \left(\frac{g(x + he_j, t) - g(x,t)}{h} - g'_{x_j}(x,t) \right) dt$$

$$= \int\limits_0^1 \left(\int\limits_0^1 (g'_{x_j}(x + \theta h e_j, t) - g'_{x_j}(x,t))\, d\theta \right) dt,$$

we obtain

$$|\Delta(t)| \le \int\limits_0^1 \left(\int\limits_0^1 \left| g'_{x_j}(x + \theta h e_j, t) - f'_{x_j}(x,t) \right| d\theta \right) dt.$$

Let $\bar{B}$ be a closed ball centered at x that is contained in O. Since the product $\bar{B} \times [0,1]$ is compact, the derivative g'_{x_j} is uniformly continuous in it. Taking $|h|$ less than the radius of $\bar{B}$, we obtain from the last inequality that $|\Delta(h)| \le \omega(|h|)$, where ω is the modulus of continuity of the restriction of g'_{x_j} to $\bar{B} \times [0,1]$. Hence, $\Delta(h) \to 0$ as $\omega(|h|) \to 0$. Thus, the differentiability of J with respect to the jth variable at an arbitrary point $x \in O$ and an integral representation for the derivative are established.

The smoothness of J follows from the inequalities

$$|J'_{x_j}(x) - J'_{x_j}(\widetilde{x})| \le \int\limits_0^1 |g'_{x_j}(x,t) - g'_{x_j}(\widetilde{x},t)|\, dt \ \le \omega(\|x - \widetilde{x}\|),$$

valid for all $\widetilde{x} \in \bar{B}$.

Now, to complete the proof, it suffices to check that the function J is r-smooth. This can easily be done by induction, which we leave to the reader.   □

6.3 Hadamard's Lemma

Here we obtain a convenient representation of a smooth function suggested by Hadamard.

Lemma *Let $O \subset \mathbb{R}^m$ be a convex neighborhood of the origin, $f \in C^r(O)$, $r = 1, 2, \ldots, \infty$. Then there exist functions $g_1, \ldots, g_m \in C^{r-1}(O)$ such that*

$$f(x) - f(0) = g_1(x)x_1 + \ldots + g_m(x)x_m \quad \text{for every } x \in O. \tag{5}$$

Proof Fix $x \in O$ and set $\varphi(t) = f(tx)$ for $t \in [0, 1]$. Then

$$\varphi(1) = f(x), \quad \varphi(0) = f(0), \quad \text{and} \quad \varphi'(t) = \sum_{j=1}^{m} x_j \frac{\partial f}{\partial x_j}(tx).$$

Integrating the last equality, we see that

$$f(x) - f(0) = \int_0^1 \varphi'(t) \, dt = \sum_{j=1}^{m} x_j \int_0^1 \frac{\partial f}{\partial x_j}(tx) \, dt.$$

Obviously, the functions $x \mapsto g_j(x) = \int_0^1 \frac{\partial f}{\partial x_j}(tx) \, dt$ satisfy condition (5) and, as follows from Lemma 6.2, belong to $C^{r-1}(O)$. $\qquad\square$

6.4 The Main Result

As we have already mentioned, our aim here is to prove a beautiful result due to Morse: near a nondegenerate critical point, a sufficiently smooth function can be represented as a linear combination of the squares of appropriate curvilinear coordinates. In other words, in a wide class of smooth functions, we have a local analog of the classical result on the reduction of a quadratic form to canonical form.

Recall (see Section I.9.3) that a critical point is said to be nondegenerate if the Gram matrix at this point is invertible.

Theorem (Morse) *Let $O \subset \mathbb{R}^m$ and $f \in C^r(O)$, $r = 3, 4, \ldots, \infty$. If $x_0 \in O$ is a nondegenerate critical point of f, then there exist a neighborhood $V \subset O$ of x and a diffeomorphism Φ of class C^{r-2} defined in V such that $\det \Phi'(x_0) = 1$ and for $x \in V$, $y = \Phi(x)$ we have*

$$f(x) - f(x_0) = \sum_{j=1}^{m} a_j y_j^2,$$

where $2a_j$ are the eigenvalues of the Hessian matrix $H_f(x_0)$.

Dropping the condition $\det \Phi'(x_0) = 1$ and setting $z_j = \sqrt{|a_j|}\, y_j$ $(j = 1, \ldots, m)$, we can write the increment of f in a neighborhood of x_0 in the form

$$f(x) - f(x_0) = (z_1^2 + \ldots + z_p^2) - (z_{p+1}^2 + \ldots + z_m^2),$$

where p is the number of positive eigenvalues of the Hessian matrix $H_f(x_0)$.

Proof We assume that the set O is convex and x_0 is the origin. Besides, we assume that the Hessian matrix $H_f(0)$ is diagonal (this can be achieved by an orthogonal change of variables).

By Hadamard's lemma, there exist functions $g_1, \ldots, g_m \in C^{r-1}(O)$ such that in O we have
$$f(x) - f(0) = x_1 g_1(x) + \ldots + x_m g_m(x).$$

Clearly, $g_k(0) = f'_{x_k}(0) = 0$. Applying Hadamard's lemma again, we see that for $x \in O$ and every $j = 1, \ldots, m$ we have

$$g_j(x) = x_1 h_{j1}(x) + \ldots + x_m h_{jm}(x)$$

where $h_{jk} \in C^{r-2}(O)$. Substituting the obtained expansions of g_j into the formula for the increment of f, we obtain

$$f(x) - f(0) = \sum_{j,k=1}^{m} h_{jk}(x)x_j x_k = \sum_{j,k=1}^{m} h_{jk}(0)x_j x_k + o(\|x\|^2). \qquad (6)$$

The matrix $A_x = (h_{jk}(x))_{j,k=1}^m$ can be assumed to be symmetric (otherwise, we replace it by $\frac{1}{2}(A_x + A_x^T)$, which does not affect the quadratic form). By Taylor's formula, the quadratic form in the right-hand side of (6) is equal to half the second differential. Hence, the matrix $A = (h_{jk}(0))_{j,k=1}^m$ multiplied by 2 is nothing else than the Hessian matrix $H_f(0)$.

By our assumption, the matrix $A = \frac{1}{2}H_f(0)$ is diagonal, hence the quadratic form $\langle A(x), x \rangle$ has the form $a_1 x_1^2 + \ldots + a_m x_m^2$. Obviously, the first equality in (6) can be rewritten in the form
$$f(x) - f(0) = \langle A_x(x), x \rangle, \qquad (7)$$

where $A_x \xrightarrow[x \to 0]{} A_0 = A$. Then for x sufficiently close to 0 (say, for x from a neighborhood V_0 of 0), the matrix A_x belongs to the neighborhood U from Lemma 6.1. Therefore, using (1) and the map ω constructed in Lemma 6.1, for such x we obtain

$$\langle A_x(t), t \rangle = \langle A(s), s \rangle \quad \text{where} \quad s \in \mathbb{R}^m, \ t = \omega(A_x)(s).$$

For $t = x$, this allows us to rewrite (7) in the form

$$f(x) - f(0) = \langle A(y), y \rangle,$$

where $y = (\omega(A_x))^{-1}(x)$ (as we have observed after Lemma 6.1, the matrix $\omega(B)$ is invertible for $B \in U$). The map $x \mapsto \Phi(x) = (\omega(A_x))^{-1}(x)$ is the composition of the following maps: $x \mapsto A_x$, ω, the inversion of matrices, and a bilinear map (that associates with a vector and a linear map the value of the map on the vector), of which the first one is of class C^{r-2} and the others are of class C^∞. Hence, $\Phi \in C^{r-2}(V_0, \mathbb{R}^m)$.

We claim that $\det \Phi'(0) = 1$. Indeed, since $\omega(A_0) = \omega(A) = \mathbb{I}$ is the identity matrix, we obtain $\omega(A_x) = \mathbb{I} + \Theta(x)$ where $\Theta(x) \underset{x \to 0}{\longrightarrow} 0$. Therefore,

$$\Phi(x) = (\mathbb{I} + \Theta(x))^{-1}(x) \underset{x \to 0}{=} x + o(x),$$

i.e., $\Phi'(0) = \mathbb{I}$. By the local invertibility theorem, Φ is a diffeomorphism in some neighborhood of the origin, which is precisely a required neighborhood V. $\square$

Note that since $\omega(A_x) \underset{x \to 0}{=} \mathbb{I} + o(1)$, in a sufficiently small neighborhood of the origin the diffeomorphism Φ is arbitrarily close to the identity map. Taking into account that the case considered in the proof is obtained by transforming the coordinates by a rigid motion, one may say that the diffeomorphism that brings a function to the required form is arbitrarily close to a rigid motion.

Chapter III. The Implicit Map Theorem and Its Applications

1 Implicit Maps

1.1 Statement of the Problem

This section is devoted to a generalization of the implicit function theorem proved in Section I.10. We have established that the level set of a smooth function defined by the equation $F(z) = F(z_0)$ near a noncritical point z_0 coincides with the graph of a smooth function (up to relabelling the coordinates). Now we consider a more general problem, replacing a scalar equation $F(x, y) = 0$ by a system of n equations involving functions of $m + n$ variables:

$$\begin{cases} F_1(z_1, \ldots, z_{m+n}) = C_1, \\ \cdots\cdots\cdots\cdots\cdots\cdots \\ F_n(z_1, \ldots, z_{m+n}) = C_n. \end{cases}$$

It is not difficult to formally restate this problem in the same form as in Section I.10, where we studied one equation in a scalar unknown. For this, we regard $F_1, \ldots, F_n$ as the coordinate functions of a map $F \colon E \longrightarrow \mathbb{R}^n$ defined on a subset E in the space $\mathbb{R}^{m+n}$. This space will be identified with the Cartesian product $\mathbb{R}^m \times \mathbb{R}^n$, and its points $z = (z_1, \ldots, z_{m+n})$, with pairs (x, y) where $x = (x_1, \ldots, x_m) \in \mathbb{R}^m$, $y = (y_1, \ldots, y_n) \in \mathbb{R}^n$. With this notation, the original system of equations can be written as an equation

$$F(x, y) = C, \quad \text{where } C = (C_1, \ldots, C_n) \in \mathbb{R}^n.$$

Replacing the functions F_k and the map F by the differences $F_k - C_k$ and $F - C$, we can rewrite the last equation in the equivalent form

$$F(x, y) = 0, \tag{1}$$

© The Author(s), under exclusive license to Springer Nature Switzerland AG 2021
B. M. Makarov, A. N. Podkorytov, *Smooth Functions and Maps*, Moscow Lectures 7,
https://doi.org/10.1007/978-3-030-79438-5_3

which coincides with the form of the equation considered in Section I.10 when defining an implicit function.

To make the problem more precise, we generalize Definition I.10.1.

Definition Let $E \subset \mathbb{R}^{m+n}$, X be a subset in $\mathbb{R}^m$, and $F: E \to \mathbb{R}^n$. A map $f: X \to \mathbb{R}^n$ is called an *implicit map defined by equation* (1) if

$$(x, f(x)) \in E \quad \text{and} \quad F(x, f(x)) = 0 \quad \text{for every } x \in X.$$

Besides, if $f(X) \subset Y \subset \mathbb{R}^n$ and $X \times Y \subset E$, we say that f is an *implicit map defined by equation* (1) *in the product* $X \times Y$.

Note that equation (1) defines a unique implicit map in $X \times Y$ if and only if for every $x \in X$ there is only one point $y \in Y$ such that $F(x, y) = 0$.

As in the scalar case, it is helpful to interpret the existence and uniqueness of an implicit map in geometric terms. For this, it is convenient to extend the notion of a level set to maps and to use the notion of the graph of a map.

Recall that the graph of a map $f: X \to Y$ is the subset Γ_f of the Cartesian product $X \times Y$ that consists of all points $(x, f(x))$ where $x \in X$. Further, by analogy with a level set of a function, by a *level set of a map* $F: E \to \mathbb{R}^n$ we mean the preimage of a point, i.e., a set of the form

$$M_C = \{(x, y) \in E: F(x, y) = C\},$$

where C is some vector from $\mathbb{R}^n$. Clearly, a level set of a map is the intersection of level sets of its coordinate functions.

Using these notions, one may say that an implicit map is a map whose graph is contained in a level set, which, without loss of generality, can be assumed to be the zero level set. So, in what follows we consider the set M_0 defined by equation (1), while an implicit map f is defined by this equation in the product $X \times Y$ if $f(X) \subset Y$. As in the scalar case (see Section I.10.1), equation (1) defines in $X \times Y$ a unique implicit map if the graph of f coincides with the intersection of M_0 and $X \times Y$.

As we have already mentioned, our new problem, that of solving a system of n equations, starts to look like the original scalar problem: in both cases, we must solve equation (1) for a given value of the parameter x. However, this formal similarity conceals an important difference. For $n > 1$, we lack the techniques which played a key role in solving one equation in a scalar unknown. Dealing with maps, we can neither apply the intermediate value theorem to prove the existence of a solution, nor use the monotonicity of F in y to prove its uniqueness. Now we restrict ourselves to the smooth case, assuming that the domain of definition of F is open and $F \in C^1$ and using the properties of smooth maps established in Chapter II. The result we are going to prove will be a generalization of Theorem I.10.3.

To understand how the crucial assumption of this theorem, which is the inequality $\frac{\partial F}{\partial y}(x_0, y_0) \neq 0$, must be stated in the multidimensional case, it is useful to consider a "linearization" of equation (1). Clearly, the matrix F' has the form

$$
\begin{pmatrix}
\dfrac{\partial F_1}{\partial x_1} & \cdots & \dfrac{\partial F_1}{\partial x_m} & \dfrac{\partial F_1}{\partial y_1} & \cdots & \dfrac{\partial F_1}{\partial y_n} \\
\cdots\cdots\cdots\cdots\cdots\cdots\cdots\cdots\cdots \\
\dfrac{\partial F_n}{\partial x_1} & \cdots & \dfrac{\partial F_n}{\partial x_m} & \dfrac{\partial F_n}{\partial y_1} & \cdots & \dfrac{\partial F_n}{\partial y_n}
\end{pmatrix}.
$$

Denote the left and right parts of this matrix,

$$
\begin{pmatrix}
\dfrac{\partial F_1}{\partial x_1} & \cdots & \dfrac{\partial F_1}{\partial x_m} \\
\cdots\cdots\cdots\cdots \\
\dfrac{\partial F_n}{\partial x_1} & \cdots & \dfrac{\partial F_n}{\partial x_m}
\end{pmatrix}
\quad \text{and} \quad
\begin{pmatrix}
\dfrac{\partial F_1}{\partial y_1} & \cdots & \dfrac{\partial F_1}{\partial y_n} \\
\cdots\cdots\cdots\cdots \\
\dfrac{\partial F_n}{\partial y_1} & \cdots & \dfrac{\partial F_n}{\partial y_n}
\end{pmatrix},
$$

by F'_x and F'_y, respectively. Note that F'_y is a square $n \times n$ matrix.

Near a point $z_0 = (x_0, y_0)$ satisfying equation (1), it is natural to consider also an approximate equation, replacing the vector $F(x, y)$, which coincides with the difference $F(x, y) - F(x_0, y_0)$, by the differential of F at z_0. Then instead of (1) we obtain the linear equation

$$
F'_x(z_0)(x - x_0) + F'_y(z_0)(y - y_0) = 0,
$$

which is (uniquely) solvable with respect to y if the matrix $F'_y(z_0)$ is invertible. It is this condition that is a counterpart of condition (2) in Theorem I.10.3. As we will see, it is sufficient for equation (1) to be solvable "in the small."

1.2 The Inverse Mapping Theorem

The following result is a natural generalization of Theorem I.10.3.

Theorem *Let* $r = 1, 2, \ldots, +\infty$, *G be an open subset in* $\mathbb{R}^{m+n}$, $F \in C^r(G; \mathbb{R}^n)$, $(x_0, y_0) \in G$. *If* $F(x_0, y_0) = 0$ *and*

$$
\text{the matrix } F'_y(x_0, y_0) \text{ is invertible,} \tag{2}
$$

then there exist open cubes $P \subset \mathbb{R}^m$ *and* $Q \subset \mathbb{R}^n$, $P \times Q \subset G$, *centered at the points* x_0 *and* y_0, *respectively, such that equation (1) defines a unique implicit map* f *in* $P \times Q$. *It is* C^r*-smooth, and for all* $x \in P$ *we have*

$$
f'(x) = -(F'_y(x, f(x)))^{-1} \cdot F'_x(x, f(x)). \tag{3}
$$

The theorem implies that under condition (2), the level set

$$
\{z \in G : F(z) = F(z_0)\}
$$

in a neighborhood of the point $z_0 = (x_0, y_0)$ coincides with the graph of an r-smooth map f. In particular, $y_0 = f(x_0)$.

Proof The proof relies on a reduction of the problem at hand to the local invertibility problem which we have already solved. This can be achieved by increasing the dimension of the target space of the map.

We define an auxiliary map $\Phi \colon G \to \mathbb{R}^{m+n}$ by the formula

$$\Phi(x, y) = (x, F(x, y)) \quad \text{for } (x, y) \in G.$$

Obviously, $\Phi \in C^r(G; \mathbb{R}^{m+n})$, $\Phi(x_0, y_0) = (x_0, 0)$, and $\Phi' = \begin{pmatrix} \mathbb{I} & \mathbb{O} \\ F'_x & F'_y \end{pmatrix}$, where $\mathbb{I}$ stands for the $m \times m$ identity matrix and $\mathbb{O}$ stands for the zero matrix. Since

$$\det(\Phi'(x_0, y_0)) = \det(F'_y(x_0, y_0)) \neq 0,$$

the local invertibility theorem (see Section II.2.1) implies that there is a neighborhood of (x_0, y_0) such that the restriction of Φ to this neighborhood is a diffeomorphism. The Φ-image of this neighborhood is open, hence it contains, together with the point $\Phi(x_0, y_0) = (x_0, 0)$, all close points of the form $(x, 0)$. In other words, for a point x sufficiently close to x_0, the vector $(x, 0)$ is a value taken by Φ in the neighborhood under consideration; moreover, this value is taken only once, since the restriction of Φ to the neighborhood is a bijection. Consequently, there exist open cubes P, Q centered at x_0 and y_0, respectively, such that $P \times Q \subset G$ and for every vector $x \in P$ there exists a unique vector $y \in Q$ for which $\Phi(x, y) = (x, 0)$, i.e., $(x, F(x, y)) = (x, 0)$ and, therefore, $F(x, y) = 0$. Thus, for every $x \in P$ the equation $F(x, y) = 0$ has in Q a unique solution, which we denote by $f(x)$. So, this equation defines a unique implicit map f in $P \times Q$. As follows from the definition of f, for $x \in P$ we have $F(x, f(x)) = 0$, i.e., the graph of f coincides with the part of the set of solutions of equation (1) contained in $P \times Q$ (see Fig. 1).

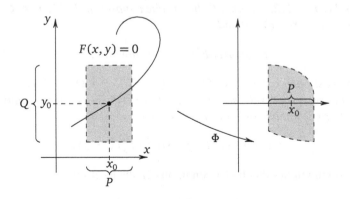

Fig. 1

Let us show that the map f is smooth. Assume without loss of generality that $G = P \times Q$ and, consequently, Φ is a diffeomorphism of class C^r. By Theorem II.1.4, the inverse map has the same smoothness, so Φ^{-1} is an r-smooth map. Given the definition of Φ, we can rewrite the equality $F(x, f(x)) = 0$ in the form $(x, f(x)) = \Phi^{-1}(x, 0)$, which, in view of the r-smoothness of Φ^{-1}, implies that f is also r-smooth.

To complete the proof, it remains to prove formula (3). Differentiating the identity $F(x, f(x)) = 0$, we obtain

$$F_x'(x, f(x)) + F_y'(x, f(x))f'(x) = 0.$$

Multiplying this equality by $(F_y'(x, f(x)))^{-1}$ from the left yields (3).   □

Remark The theorem deals with the most common class of smooth maps, namely, with maps of integer smoothness. However, it is easy to see that, repeating the above arguments. one can extend it to maps of fractional smoothness (see Exercise 1).

1.3 Concluding Remarks

As we saw in the discussion of the implicit function theorem in Section I.10.4, the structure of a level set in a neighborhood of a given point x largely depends on whether or not the gradient of the function under study vanishes at x. Now we consider a similar result, which does not involve an a priori partition of the coordinates into two groups, for maps.

Let $F_1, \ldots, F_n$ be r-smooth functions defined in an open subset G in $\mathbb{R}^{m+n}$. Consider the system of equations

$$\begin{cases} F_1(z_1, \ldots, z_{m+n}) = C_1, \\ \cdots\cdots\cdots\cdots\cdots\cdots \\ F_n(z_1, \ldots, z_{m+n}) = C_n. \end{cases} \tag{1'}$$

It can be regarded as the coordinate form of equation (1) for the map $\widetilde{F}$ with coordinate functions $F_j - C_j$ ($j = 1, \ldots, n$), whose gradients, obviously, coincide with the gradients of the functions F_j. Solutions of this system are intersection points of the level sets $F_j(z) = C_j$ of the functions F_j. If (under an appropriate identification of $\mathbb{R}^{m+n}$ with $\mathbb{R}^m \times \mathbb{R}^n$) condition (2) is fulfilled at a point z_0 satisfying (1'), this means that

the gradients $\operatorname{grad} F_1(z_0), \ldots, \operatorname{grad} F_n(z_0)$ are linearly independent   (2')

(in geometry, one says that the level sets of $F_1, \ldots, F_n$ passing through z_0 are "in general position," or satisfy the transversality condition). Relabelling the coordinates if necessary, we can reduce the general case under consideration to the special case analyzed in the previous section (when the last n columns of the matrix $F'(z_0)$ are linearly independent). Hence, the following result holds.

Corollary *Let M be the set of solutions of (1′) and $z_0 \in M$. If condition (2′) is satisfied at z_0, then in a sufficiently small neighborhood U of z_0, the set $M \cap U$ coincides with a (wide-sense) graph of an r-smooth map.*

The local invertibility theorem underlying the above proof can be, in turn, deduced from the implicit function theorem. To to this, for a given smooth map $T: O \to \mathbb{R}^m$ satisfying the condition $\det(T'(w)) \neq 0$ at a point $w \in O$, one must consider the smooth map

$$F(x, y) = x - T(y) \quad ((x, y) \in G)$$

in the open set $G = \mathbb{R}^m \times O$ and apply the implicit mapping theorem at the point (x_0, y_0) where $x_0 = T(w)$, $y_0 = w$.

Exercise

1. Prove the following claim from Remark 1.2: in the implicit function theorem, integer smoothness can be replaced by arbitrary t-smoothness with $1 < t < +\infty$.

2 Smooth Manifolds

2.1 Definitions and Notation

This section is devoted to an important geometric notion, that of a smooth manifold. It generalizes the (at least visually) familiar notions of a curve and a surface. Without going too deep into the theory of these manifolds, which is the subject of such areas of mathematics as differential geometry and differential topology, we confine ourselves to various definitions of this notion and some examples, using the properties of smooth maps established in Sections 1, 2 of Chapter II and the implicit mapping theorem.

First, we recall some notation concerning smooth maps. By $C^r(O; \mathbb{R}^m)$ (where $r = 1, 2, \ldots, \infty$) we denote the class of r times continuously differentiable maps to $\mathbb{R}^m$ from a set $O \subset \mathbb{R}^k$, which is always assumed to be open. Maps of class C^1 are said to be smooth. Recall also that the differential of a map $\Phi \in C^1(O; \mathbb{R}^m)$ at a point $t \in O$ is denoted by $d_t\Phi$ (see Section I.5), and the corresponding matrix (in the canonical bases of $\mathbb{R}^k$ and $\mathbb{R}^m$), i.e., the Jacobian matrix of Φ, is denoted by $\Phi'(t)$. As is well known, this $m \times k$ matrix (with m rows and k columns) consists of the partial derivatives $\frac{\partial \varphi_j}{\partial t_i}$ ($1 \leq j \leq m, 1 \leq i \leq k$) of the coordinate functions $\varphi_1, \ldots, \varphi_m$ of Φ. If $m = k$, then the Jacobian matrix is square. Its determinant $\det\left(\frac{\partial \varphi_j}{\partial t_i}\right)$ (the Jacobian of Φ) is also denoted by $\frac{D(\varphi_1,\ldots,\varphi_k)}{D(t_1,\ldots,t_k)}$.

Definition A set $M \subset \mathbb{R}^m$ is called a *simple k-dimensional manifold* if it is homeomorphic to an open subset O in $\mathbb{R}^k$ ($k \leq m$). A homeomorphism $\Phi \colon O \to M$ is called a *parametrization* of M. If $\Phi \in C^r(O; \mathbb{R}^m)$ for some $r = 1, 2, \ldots, +\infty$ and the matrix $\Phi'(t)$ has maximal rank everywhere in O (i.e., rank $d_t\Phi = \dim d_t\Phi(\mathbb{R}^k) = k$ at every point $t \in \mathbb{R}^k$), then the parametrization Φ is said to be *smooth of class C^r*, or *r-smooth*. A simple manifold that has such a parametrization is also said to be *smooth of class C^r*, or *r-smooth*.

This terminology is also used in cases where the statement of the problem suggests considering parametrizations and manifolds of fractional smoothness (see Section IV.1).

We draw the reader's attention to the fact that, according to our definition, the term "smooth parametrization" (as opposed to "smooth map") assumes not only that the corresponding map belongs to a certain smoothness class, but also that the derivative has maximal rank. Note also that the domain of definition of a parametrization is open.

Corollary II.2.3 implies, in particular, that a smooth manifold cannot have parametrizations defined in open subsets of spaces of different dimensions, since otherwise these sets would be diffeomorphic, which is impossible, as we know from Section II.1.1. This shows that the dimension of a smooth manifold does not depend on the choice of parametrization, i.e., is well defined.

Since the position of a point $p = \Phi(t)$ on a manifold is uniquely determined by the parameter t, the coordinates $t_1, \ldots, t_k$ are often called the curvilinear coordinates of p (cf. the terminology introduced in Section II.3). In specific examples, they often have a simple geometric meaning, which makes it possible to simplify the solution of the problem.

In what follows, we again identify the space $\mathbb{R}^m$ with the Cartesian product $\mathbb{R}^k \times \mathbb{R}^{m-k}$, and the space $\mathbb{R}^k$, with the subspace $\mathbb{R}^k \times \{0\}$, thus assuming that $\mathbb{R}^k$ is embedded into $\mathbb{R}^m$ (in other words, we identify vectors $(t_1, \ldots, t_k) \in \mathbb{R}^k$ and $(t_1, \ldots, t_k, 0, \ldots, 0) \in \mathbb{R}^m$).

An important example of a simple manifold in $\mathbb{R}^m$ is the graph Γ_f of a map $f \colon O \to \mathbb{R}^{m-k}$, where $O \subset \mathbb{R}^k$. It has a natural parametrization Φ_f, defined by the formula $\Phi_f(t) = (t, f(t))$ ($t \in O$), which will be called the canonical parametrization of Γ_f. If f is r-smooth, then so is Φ_f. Since rank $\Phi'_f(t) = k$ for all $t \in O$, the graph of an r-smooth map is a simple k-dimensional manifold in $\mathbb{R}^m$ of class C^r. Generalizing the notion of the graph of a map, we say that a subset in $\mathbb{R}^m$ is a wide-sense graph if it coincides with the graph of some map after relabelling the coordinates.

Another example of an (infinitely smooth) simple k-dimensional manifold is a k-dimensional vector subspace. A parametrization can be constructed, for instance, as follows. Fix an arbitrary basis $\tau_1, \ldots, \tau_k$ in this subspace and set

$$\Phi(t) = t_1\tau_1 + \ldots + t_k\tau_k \quad \text{for } t = (t_1, \ldots, t_k) \in \mathbb{R}^k.$$

Obviously, the map Φ satisfies all the required properties. The "curvilinear" coordinates of a vector from the subspace are its coordinates in the basis $\tau_1, \ldots, \tau_k$.

Clearly, every k-dimensional affine subspace obtained from the corresponding vector subspace by a translation is also a k-dimensional manifold of class C^∞.

The class of simple manifolds defined above does not include geometric objects such that circles, spheres, tori, and many others, which one would naturally regard as manifolds. To cure this defect, we introduce a wider notion.

Definition A set $M \subset \mathbb{R}^m$ is called a *k-dimensional manifold of class C^r* if every point $p \in M$ has a neighborhood U such that the intersection $U \cap M$ is a simple k-dimensional manifold of class C^r. The number k is called the *dimension* of M. A parametrization of the intersection $U \cap M$ is called a *local parametrization* of the manifold M near the point p.

In a similar way one defines a k-dimensional manifold of class C^0: it is a set M that is locally a simple k-dimensional manifold (without any smoothness conditions), i.e., every point of M has a neighborhood U such that the intersection $U \cap M$ is a simple k-dimensional manifold. Note that neither a manifold nor a simple manifold is assumed to be connected.

The dimension of a manifold M is denoted by $\dim M$. The difference $m - \dim M$ is called the *codimension* of M. A manifold of codimension 1 is called a *surface*. In the case $k = 1$, the term "curve" is used instead of "manifold." A connected simple curve is also called an arc. By definition, a one-point set is a smooth manifold of dimension 0.

In contrast to a local parametrization, a parametrization of a simple manifold is also called *global*. Note that a compact manifold has no global parametrization, since such a parametrization, being a homeomorphism, cannot map the open set on which it is defined onto a compact set.

If the value of $r \geq 1$ is irrelevant (in most cases, it suffices to assume that $r = 1$), we call M a smooth k-dimensional manifold, or a smooth manifold, or sometimes just a manifold, because if necessary its smoothness is stated explicitly.

If p is a point of a set $M \subset \mathbb{R}^m$, then by an *M-neighborhood* of p we mean the intersection of M with a neighborhood of p in $\mathbb{R}^m$. If M is a manifold, then every its point has a basis of M-neighborhoods whose closures are contained in M.

An M-neighborhood that is a simple manifold, i.e., admits a parametrization and, consequently, curvilinear coordinates, will be called a *coordinate neighborhood*.

Remark 1 A C^r-manifold can be represented as the union of at most countably many simple manifolds of the same smoothness.

This follows from Lindelöf's theorem (see Section 0.4).

Since the range of curvilinear coordinates is a countable union of compact sets, we obtain the following.

Remark 2 A smooth manifold is an at most countable union of compact sets, each contained in some simple manifold.

It is not uncommon in simple and important examples that one has to deal with "almost smooth" manifolds (for instance, polygonal lines, the boundaries of a cube, cone, etc.). That is why, we extend the definition of a smooth manifold as follows:

a *piecewise smooth* k-dimensional manifold is the union of a (possibly disconnected) smooth k-dimensional manifold and finitely many smooth manifolds of dimensions less than k. Obviously, with this definition, the boundaries of polyhedra are piecewise smooth surfaces. In particular, for $m = 3$ such a boundary consists of the two-dimensional faces (without "boundaries"), edges (without boundary points), and vertices of the polyhedron.

Formally, considering a smooth manifold M contained in $\mathbb{R}^m$, we do not exclude the possibility that $\dim M = m$. Then, as follows from the definition, M is just an open subset in $\mathbb{R}^m$. The definition of a parametrization in this case agrees with the terminology introduced in Section II.3 in connection with curvilinear coordinates.

2.2 Equivalent Descriptions of a Smooth Manifold

Due to its geometric nature, the notion of manifold admits some equivalent definitions, which may be preferable in specific problems. The following theorem, whose assertions II–IV lead to alternative definitions of an r-smooth manifold, provides enough freedom in this regard.

In the statement and proof of this theorem, we adhere to the convention adopted above that $\mathbb{R}^k$ is identified with a subspace in $\mathbb{R}^m$.

Theorem *Let $M \subset \mathbb{R}^m$, $1 \leq k < m$, $1 \leq r \leq +\infty$. Then for every point $p \in M$, the following four conditions are equivalent.*

I. *There exists a neighborhood $U \subset \mathbb{R}^m$ of p such that the intersection $M \cap U$ is a simple k-dimensional manifold of class C^r.*

II. *There exists a diffeomorphism $\Theta \in C^r(G; \mathbb{R}^m)$, where $G \subset \mathbb{R}^m$, such that $p \in M \cap \Theta(G)$ and $M \cap \Theta(G) = \Theta(\mathbb{R}^k \cap G)$.*

III. *The set M locally is the intersection of $m - k$ level sets of r-smooth functions "in general position." In more detail: there exist a neighborhood $V \subset \mathbb{R}^m$ of p and functions $F_1, \ldots, F_{m-k}$ of class C^r defined in this neighborhood such that for every $x \in V$,*

$$x \in M \iff F_1(x) = \ldots = F_{m-k}(x) = 0 \tag{1}$$

and the vectors

$$\operatorname{grad} F_1(p), \ldots, \operatorname{grad} F_{m-k}(p) \tag{2}$$

are linearly independent.

IV. *There exists a neighborhood $W \subset \mathbb{R}^m$ of p such that the intersection $M \cap W$ is a (wide-sense) graph of a map of class C^r defined in a k-dimensional domain.*

Thus, the definition of a smooth manifold (that is, the fact that condition I is satisfied at every point $p \in M$) can be equivalently restated in three ways according to conditions II, III, and IV. In particular, conditions III and IV imply a result already mentioned in connection with the implicit function theorem: a level set of a smooth function whose gradient does not vanish on this set is locally a (wide-sense) graph and, consequently, a smooth manifold.

Note also that the theorem remains valid also for manifolds of fractional smoothness; we leave it to the reader to verify this.

Proof The proof follows the scheme I $\Rightarrow$ II $\Rightarrow$ III $\Rightarrow$ IV $\Rightarrow$ I.

I $\Rightarrow$ II. This implication is a special case of Theorem II.2.3 B applied (with an appropriate change of notation) to a parametrization of the M-neighborhood of p.

II $\Rightarrow$ III. Set $V = \Theta(G)$, and for $F_1, \ldots, F_{m-k}$ take the last coordinate functions of the map Θ^{-1} (which is defined on the open set V). Since

$$M \cap V = M \cap \Theta(G) = \Theta(\mathbb{R}^k \cap G),$$

we obtain

$$\Theta^{-1}(M \cap V) = \mathbb{R}^k \cap G = \{(y_1, \ldots, y_k, 0, \ldots, 0) \in G\},$$

and hence for every $x \in V$ the inclusion $x \in M \cap V$ is equivalent to the equalities

$$F_1(x) = \ldots = F_{m-k}(x) = 0.$$

Since $\Theta \in C^r$, it follows from Theorem II.1.4 that the map Θ^{-1} is also r-smooth, and hence all its coordinate functions are r-smooth. Finally, the Jacobian matrix of the diffeomorphism Θ^{-1} is invertible, which implies that all its rows and, consequently, the gradients of the coordinate functions, are linearly independent. In particular, so are grad $F_1(p), \ldots,$ grad $F_{m-k}(p)$.

III $\Rightarrow$ IV. This implication coincides with Corollary I.1.3 (with $z_0 = p$ and $n = m - k$).

IV $\Rightarrow$ I. This implication is trivial, since, as we mentioned in Section 2.1, the graph of a smooth map is an example of a simple smooth manifold. $\square$

Interpreting graphs in a wide sense, one may say that every smooth manifold is an at most countable union of graphs of smooth maps.

Remark As we have mentioned in the proof of the implication I $\Rightarrow$ II, Theorem II.2.3 B implies that every parametrization Φ locally is a restriction of a diffeomorphism. Obviously, the same property is possessed by the inverse map Φ^{-1}. Hence, in particular, Φ^{-1} satisfies the local Lipschitz condition.

2.3 The Tangent Subspace

There are important notions related to a smooth manifold: a tangent vector and a tangent subspace. Recall that a *path* (in $\mathbb{R}^m$) is any continuous map from a compact interval to $\mathbb{R}^m$. A path is said to be *smooth* if its coordinate functions are smooth. We will say that a path $\gamma : \Delta \to \mathbb{R}^m$ *lies in* M if $\gamma(t) \in M$ for all $t \in \Delta$.

Definition Let M be a smooth manifold in $\mathbb{R}^m$ and $p \in M$. A vector $\tau \in \mathbb{R}^m$ is called a *tangent vector* to M at the point p if τ is a tangent vector at this point to some path[1] lying in M, i.e., there exists a smooth path $\gamma : [a, b] \mapsto \mathbb{R}^m$ lying in M such that $\gamma(t_0) = p$ and $\gamma'(t_0) = \tau$ for some $t_0 \in [a, b]$.

In what follows, unless otherwise stated, we assume without loss of generality that $t_0 = 0$.

Let Φ be a local parametrization of a k-dimensional manifold M near a point $p = \Phi(a)$. "Freezing" all coordinates of the point $a = (a_1, \ldots, a_k)$ except the jth one and varying the latter near a_j, we obtain a path passing through p. It parametrizes a curve, which is called a *coordinate curve*. The tangent vector to this path at the point p is nothing else than the jth column of the matrix $\Phi'(a)$, which will be denoted by $D_j\Phi(a)$ or $\tau_j = \tau_j(a)$. Obviously, $\tau_j(a) = d_a\Phi(e_j)$ (as usual, $e_1, \ldots, e_k$ are the canonical basis vectors in $\mathbb{R}^k$). Since rank $d_a\Phi = k$, the vectors $\tau_1, \ldots, \tau_k$ are linearly independent. They will be called the canonical tangent vectors corresponding to the parametrization Φ.

The set of all tangent vectors to a manifold M at a point p is called the *tangent subspace* to M and denoted by $T_p(M)$ or, in short, T_p. For $k = m - 1$, this subspace is also called the *tangent plane*. The validity of using these terms is guaranteed by the following lemma.

Lemma *The set $T_p(M)$ is a k-dimensional vector subspace in $\mathbb{R}^m$.*

Proof Since the set $T_p(M)$ is determined by the properties of the manifold M in an arbitrarily small neighborhood of p, in what follows we assume that M is simple.

We will show that along with any two vectors τ, τ_* the set T_p contains also any linear combination of these vectors. By the definition of T_p, the vectors τ and τ_* are the derivatives of some smooth paths γ and γ_* lying in M and passing through p:

$$\gamma(0) = \gamma_*(0) = p \quad \text{and} \quad \tau = \gamma'(0), \ \tau_* = \gamma'_*(0).$$

We will prove that the linear combination $c\tau + c_*\tau_*$ ($c, c_* \in \mathbb{R}$) also belongs to T_p. Without loss of generality, we assume that the paths γ and γ_* are defined on the same interval containing 0.

Let Φ be a smooth parametrization of the manifold M defined in a k-dimensional neighborhood U of the origin, $\Phi(0) = p$. Consider the paths

$$\widetilde{\gamma} = \Phi^{-1}(\gamma) \quad \text{and} \quad \widetilde{\gamma}_* = \Phi^{-1}(\gamma_*)$$

in U. Since Φ^{-1} locally is a restriction of a diffeomorphism (see the remark at the end of the previous subsection), they are smooth. Obviously, $\widetilde{\gamma}(0) = \widetilde{\gamma}_*(0) = \Phi^{-1}(p) = 0$ and

$$\tau = \gamma'(0) = (\Phi \circ \widetilde{\gamma})'(0) = \Phi'(0) \cdot \widetilde{\gamma}'(0),$$
$$\tau_* = \gamma'_*(0) = (\Phi \circ \widetilde{\gamma}_*)'(0) = \Phi'(0) \cdot \widetilde{\gamma}'_*(0). \tag{3}$$

Set $\widetilde{w} = c\widetilde{\gamma} + c_*\widetilde{\gamma}_*$. Clearly, $\widetilde{w}$ is a smooth path in $\mathbb{R}^k$ and $\widetilde{w}(0) = 0$. Besides, $\widetilde{w}(t) \in U$ for t sufficiently small in absolute value. Hence, the composition

---

[1] The definition of a tangent vector to a path is given at the end of Section I.5.2.

$$w = \Phi \circ \widetilde{w} = \Phi(c\,\widetilde{\gamma} + c_*\,\widetilde{\gamma}_*)$$

makes sense near the origin. This is a smooth path lying in M, with $w(0) = p$. We can find its derivative at the origin using (3):

$$w'(0) = \Phi'(0) \cdot (c\,\widetilde{\gamma}'(0) + c_*\widetilde{\gamma}_*'(0)) = c\Phi'(0) \cdot \widetilde{\gamma}'(0) + c_*\,\Phi'(0) \cdot \widetilde{\gamma}_*'(0) = c\tau + c_*\tau_*.$$

Thus, a linear combination of any two tangent vectors (at p) is again a tangent vector and, consequently, belongs to T_p.

Finally, since τ is arbitrary, it follows from (3) that $T_p \subset d_0\Phi(\mathbb{R}^k)$ and, consequently, $\dim T_p \le k$. Therefore, $\dim T_p = k$, since, as we have established after the definition, T_p contains k linearly independent canonical tangent vectors. $\square$

Assume that a smooth path γ lies in a level set M of a smooth function F, $p \in M$, $p = \gamma(t)$. Then $\langle \operatorname{grad} F(\gamma), \gamma' \rangle = (F(\gamma))' = 0$, since the composition $F(\gamma)$ is constant. Therefore, the tangent vector $\gamma'(t)$ to the manifold M at the point p is orthogonal to $\operatorname{grad} F(p)$. Hence (since γ is an arbitrary path), all tangent vectors to M at p are orthogonal to $\operatorname{grad} F(p)$. Thus, if $\operatorname{grad} F(p) \ne 0$, then the tangent subspace to M at p is the plane consisting of the vectors orthogonal to $\operatorname{grad} F(p)$, i.e., the plane defined by the equation $\langle x, \operatorname{grad} F(p) \rangle = 0$.

In the more complicated case where M is the intersection of level surfaces (1), the subspace $T_p(M)$ is the intersection of tangent planes to these surfaces. Since these planes are orthogonal to the gradients, T_p consists of the vectors $h = (h_1, \dots, h_m)$ belonging to the intersection of the tangent planes to the zero level surfaces of the functions F_j $(j = 1, \dots, m - k)$, i.e., from the vectors h satisfying the condition

$$h \perp \operatorname{grad} F_j(p) \quad (j = 1, \dots, m - k),$$

or, in more detail,

$$\begin{cases} \dfrac{\partial F_1}{\partial x_1}(p)h_1 + \dots + \dfrac{\partial F_1}{\partial x_m}(p)h_m = 0, \\ \dots\dots\dots\dots\dots\dots\dots\dots\dots\dots\dots\dots\dots\dots \\ \dfrac{\partial F_{m-k}}{\partial x_1}(p)h_1 + \dots + \dfrac{\partial F_{m-k}}{\partial x_m}(p)h_m = 0. \end{cases}$$

Note that since the canonical tangent vectors $\tau_1, \dots, \tau_k$ corresponding to the parametrization Φ of the manifold M near a point $p \in M$ are linearly independent, they form a basis in the tangent subspace. By the linearity of the map $d_a\Phi \colon \mathbb{R}^k \to \mathbb{R}^m$, for every vector $t = (t_1, \dots, t_k) \in \mathbb{R}^k$ we have

$$d_a\Phi(t) = \sum_{j=1}^{k} t_j\tau_j.$$

Thus, denoting $p = \Phi(a)$, the differential of the parametrization is an isomorphism of $\mathbb{R}^k$ onto the tangent space T_p.

It is sometimes more illustrative to consider, along with the tangent subspace T_p, the *affine tangent subspace* L_p, which is obtained by translating T_p by the vector p: $L_p = p + T_p$. Since $p + d_a\Phi(t - a) \in L_p$ for $p = \Phi(a)$ and

$$\Phi(t) \underset{t \to a}{=} p + d_a\Phi(t - a) + o(\|t - a\|),$$

for the point $x = \Phi(t)$ we have

$$\text{dist}(x, L_p) = \text{dist}(\Phi(t), L_p) \leq \left\|\Phi(t) - (p + d_a\Phi(t - a))\right\| \underset{t \to a}{=} o(\|t - a\|).$$

By the remark at the end of Section 2.2, near p the map Φ^{-1} satisfies the Lipschitz condition:

$$\|t - a\| = \|\Phi^{-1}(x) - \Phi^{-1}(p)\| \leq C\|x - p\|,$$

and hence

$$\text{dist}(x, L_p) = o(\|x - p\|) \quad \text{as } x \to p, \ x \in M. \tag{4}$$

Thus, if points of the manifold are replaced by points of the subspace L_p, the relative error tends to 0, i.e., "in the small" the manifold M is "almost plane." This is a formalization of our intuitive notion of the tangent subspace as a subspace that "closely approaches" the manifold. One can prove that the established property of the affine tangent subspace determines it uniquely (see Exercise 4).

2.4 Examples

We illustrate the notions introduced above by several examples.

Example 1 Consider in $\mathbb{R}^m$ the sphere

$$S^{m-1}(R) = \{(x_1, \ldots, x_m) \in \mathbb{R}^m : x_1^2 + \ldots + x_m^2 = R^2\}.$$

We claim that it is a surface. Every point $p = (p_1, \ldots, p_m) \in S^{m-1}(R)$ has at least one nonzero coordinate. For definiteness, let $p_m > 0$. Then p belongs to the upper half-sphere

$$S_+^{m-1}(R) = \{x \in S^{m-1}(R) : x_m > 0\},$$

which is nothing else than the graph of the function

$$f(x_1, \ldots, x_{m-1}) = \sqrt{R^2 - x_1^2 - \ldots - x_{m-1}^2},$$

defined in a ball in $\mathbb{R}^{m-1}$. It is a function of class C^∞. Therefore, the half-sphere $S_+^{m-1}(R)$, and hence the entire sphere $S^{m-1}(R)$, are surfaces of class C^∞.

Being a compact set, the sphere has no global parametrization. Still, it is not difficult to find a map that parametrizes almost the entire sphere. We restrict ourselves to the most illustrative special case of a two-dimensional sphere in $\mathbb{R}^3$ (for the case

of arbitrary dimension, see Exercise 3). By analogy with the tradition to specify
the position of a point on the celestial sphere by two values, the deviation of the
radius vector from the polar axis and the time when the Sun crosses the meridian,
we will specify the position of a point on the sphere $S^2(R)$, defined by the equation
$x^2 + y^2 + z^2 = R^2$, by the angle $\theta \in [0, \pi]$ between its radius vector and the positive
part of the OZ axis and the longitude φ, i.e., the angle between the projection of
the radius vector to the XOY plane and the positive part of the OX axis measured
counterclockwise and taking values in the interval $[0, 2\pi]$. The angles θ and φ are
called the *spherical coordinates* of a point on the sphere. Leaving it to the reader
to find simple formulas relating the angles θ and φ to the geographic coordinates
(latitude and longitude), we note only that the corresponding coordinate curves (i.e.,
level sets of these angles) coincide with constancy sets of the latitude and longitude,
respectively, i.e., with parallels and meridians.

The Cartesian and spherical coordinates of a point $(x, y, z) \in S^2(R)$ are related
by the formulas

$$x = R \cos \varphi \sin \theta, \quad y = R \sin \varphi \sin \theta, \quad z = R \cos \theta, \tag{5}$$

which makes it natural to consider the map Φ defined as follows:

$$\Phi(\theta, \varphi) = (R \cos \varphi \sin \theta, R \sin \varphi \sin \theta, R \cos \theta).$$

In the case under consideration, we are facing a rather typical situation: formally,
the map Φ is defined for any θ and φ, but we are interested only in its restriction
to the rectangle $[0, \pi] \times [0, 2\pi]$. Clearly, Φ is an infinitely differentiable map; it is
not bijective, since $\Phi(\theta, 0) = \Phi(\theta, 2\pi)$ for any θ and, besides, $\Phi(0, \varphi) = \Phi(0, \widetilde{\varphi})$
and $\Phi(\pi, \varphi) = \Phi(\pi, \widetilde{\varphi})$ for all φ and $\widetilde{\varphi}$. To make Φ one-to-one, we restrict it
to the open rectangle $(0, \pi) \times (0, 2\pi)$ (keeping the same notation for this restric-
tion). Its image is no longer the entire sphere $S^2(R)$, but the sphere minus the
meridian $\{(x, 0, z) : x^2 + z^2 = R^2, x \geq 0\}$. Removing this meridian (together with
the poles), we obtain a "slit sphere": a two-dimensional surface of class C^∞ with
global parametrization (5) defined on the open rectangle $(0, \pi) \times (0, 2\pi)$.

The condition rank $d\Phi = 2$ everywhere, i.e., the linear independence of the
tangent vectors

$$\tau_1 = D_1\Phi(\theta, \varphi) = (R \cos \varphi \cos \theta, R \sin \varphi \cos \theta, -R \sin \theta),$$
$$\tau_2 = D_2\Phi(\theta, \varphi) = (-R \sin \varphi \sin \theta, R \cos \varphi \sin \theta, 0),$$

follows from their orthogonality (τ_1 is tangent to a meridian, while τ_2 is tangent to
a parallel), since $\|\tau_1\| = R$ and $\|\tau_2\| = R \sin \theta \neq 0$.

Example 2 Consider a torus, that is, a surface obtained by revolving a circle in the
three-dimensional space about an axis in the plane of the circle that does not cut the
circle. We will assume that the circle is defined in the XOZ plane by the equation

$$(R - x)^2 + z^2 = r^2 \quad (0 < r < R),$$

and it is revolving about the OZ axis.

As one can easily check, the torus can be defined by the equation

$$\left(R - \sqrt{x^2 + y^2}\right)^2 + z^2 = r^2.$$

Like a sphere, the torus, being a compact set, has no global parametrization. The position of a point on this torus is determined by two angles θ and φ (similar to latitude and longitude) by the formulas

$$x = (R + r\cos\theta)\cos\varphi, \quad y = (R + r\cos\theta)\sin\varphi, \quad z = r\sin\theta.$$

The infinitely differentiable map (defined in $\mathbb{R}^2$)

$$\Phi(\theta, \varphi) = \left((R + r\cos\theta)\cos\varphi, \ (R + r\cos\theta)\sin\varphi, \ r\sin\theta\right)$$

sends the square $[-\pi, \pi]^2$ onto the torus. It is not bijective in view of the 2π-periodicity of trigonometric functions. Removing two circles corresponding to the angles $\theta = \pm\pi$ and $\varphi = \pm\pi$, we obtain a "torus with two slits": a surface with global parametrization Φ defined on the square $(-\pi, \pi)^2$. The coordinate curves passing through an arbitrary point of this surface are circles, one lying in the horizontal plane (the angle θ is fixed), and the other one lying in the vertical plane (the angle φ is fixed).

When one of the coordinates is fixed and the other varies, the point $\Phi(\theta, \varphi)$ moves along the corresponding coordinate curve with a constant (nonzero) velocity. It is clear geometrically that the tangent vectors $\tau_1 = D_1\Phi$ and $\tau_2 = D_2\Phi$ are orthogonal.

It is also clear that they are nonzero. Thus, the vectors τ_1 and τ_2 are linearly independent, and, consequently, Φ is an infinitely smooth parametrization of the torus with slits. The reader can easily make a formal calculation of the vectors τ_1 and τ_2.

To show that not only the torus with slits, but the entire torus itself is indeed a smooth surface, we must prove that every point $p = \Phi(\theta_0, \varphi_0)$ has a neighborhood on the torus admitting a local parametrization. Such a parametrization can be constructed by changing the domain of definition of Φ. We leave it to the reader to check that it suffices to consider the restriction of Φ to the square

$$(\theta_0 - \pi, \theta_0 + \pi) \times (\varphi_0 - \pi, \varphi_0 + \pi).$$

The corresponding neighborhood of p is the torus cut along the circles $\theta = \theta_0 \pm \pi$ and $\varphi = \varphi_0 \pm \pi$.

Note that in the limiting case $r = R$, the circle $(R - x)^2 + z^2 = R^2$ is revolving about the OZ axis. The resulting set M is not a smooth surface, since the origin has no M-neighborhood that is a simple surface. However, the reader can easily see that the set $M \setminus \{0\}$ is a smooth surface of class C^∞, and hence M is a piecewise smooth surface.

Example 3 Consider a one-dimensional manifold, i.e., a curve. Its parametrization in a neighborhood of an arbitrary point is a smooth vector function defined on an interval of the real axis. This is a homeomorphism with a nonzero derivative. Clearly, the graph of a smooth function of one variable defined on an arbitrary interval is a smooth plane curve (i.e., a curve contained in $\mathbb{R}^2$).

Another well-known curve is a circle. Generalizing this example, recall that, according to Theorem 2.2, a level set of a smooth function of two variables with a nonzero gradient is a smooth curve. Accordingly, the Cassini ovals (see Example 4 in Section I.4.1), defined by the equation

$$(x^2 + y^2)^2 - 2a^2(x^2 - y^2) = C,$$

are smooth curves if $C > 0$, $C \neq a^4$. If $C = a^4$, we obtain the lemniscate of Bernoulli passing through the origin. This example shows that the assumption on the gradient is essential: the point $(0,0)$, at which the gradient of the function $(x^2 + y^2)^2 - 2a^2(x^2 - y^2)$ vanishes, has no relative neighborhood homeomorphic to an interval (see the illustration to Example 4 in Section I.4.1). Near the origin, the lemniscate can be regarded as the union of two graphs (a "self-intersecting curve"). Removing the origin, we obtain a (disconnected) smooth curve. Thus, the lemniscate is a piecewise smooth curve.

Example 4 Let $O(n)$ be the group of orthogonal $n \times n$ matrices. We will regard it as a subset in the n^2-dimensional Euclidean space, which is identified with the set of all $n \times n$ matrices $U = \{u_{ij}\}_{i,j=1}^n$ with elements u_{ij}. This subset is defined by the system of equations

$$u_{i1}^2 + \ldots + u_{in}^2 = 1, \qquad 1 \leq i \leq n,$$

$$u_{i1}u_{k1} + \ldots + u_{in}u_{kn} = 0, \quad 1 \leq i < k \leq n.$$

The gradients of the functions

$$F_i(U) = u_{i1}^2 + \ldots + u_{in}^2 \quad \text{and} \quad F_{ik}(U) = u_{i1}u_{k1} + \ldots + u_{in}u_{kn}$$

at the points of $O(n)$ are linearly independent. To see this, imagine these gradients as matrices in which the element at the intersection of the ith row and jth column is the result of differentiating with respect to u_{ij}. Then the gradient of F_i at a point U has a unique nonzero row, which coincides with twice the ith row of the matrix U; while the gradient of F_{ik} has two nonzero rows: its ith row coincides with the kth row of U, and its kth row coincides with the ith row of U. Hence, every row of a matrix that is a linear combination of the gradients contains only linear combinations of at most n pairwise distinct (and, consequently, pairwise orthogonal) rows of U, which easily implies the required property. Thus, $O(n)$ is a smooth manifold of dimension

$$n^2 - \left(n + \frac{n(n-1)}{2}\right) = \frac{n(n-1)}{2}.$$

Exercises

1. Let
$$M = \{(x, y, u, v) \in \mathbb{R}^4 : x^2 + y^2 = 1, \; u^2 + v^2 = 1\}.$$

   Show that

   a) M is a two-dimensional C^∞-smooth compact manifold homeomorphic to the Cartesian product of two circles;

   b) the map $(x, y, u, v) \mapsto ((R + ru)x, (R + ru)y, rv)$ is a homeomorphism of M onto the torus considered in Example 2 at p. 176.

2. a) Let S^2 be the two-dimensional sphere in $\mathbb{R}^3$ of radius 1 centered at the origin and P be a point on this sphere. Consider a map from the "punctured" sphere $S^2 \setminus \{P\}$ onto the plane passing through the origin and orthogonal to the radius vector of P. We define it as follows: a point $p \neq P$ on the sphere is mapped to the intersection point of this plane and the line through p and P. This map is called the stereographic projection.

   Below, in 1) and 2) we assume for simplicity that P coincides with the north pole $N = (0, 0, 1)$ of the sphere. Show that

   1) the stereographic projection is a one-to-one map from the punctured sphere $S^2 \setminus \{N\}$ onto the plane;

   2) the map inverse to the stereographic projection is a C^∞-smooth parametrization of the punctured sphere;

   3) the angle between two curves on the sphere S^2 passing through a given point $p \neq P$ coincides with the angle between their images under the stereographic projection (this property is called the conformality of the stereographic projection).

   b) Generalize the definition of the stereographic projection to the multidimensional case, assuming that P coincides with the "north pole" of the sphere.

3. Let $R > 0$. Generalizing Example 1 at p. 175, consider the map Φ that sends every point $(\theta_1, \ldots, \theta_{m-2}, \varphi)$ to a vector $x = (x_1, \ldots, x_m) \in \mathbb{R}^m$ according to the rule

$$x_1 = R \cos \theta_1,$$
$$x_2 = R \sin \theta_1 \cos \theta_2,$$
$$x_3 = R \sin \theta_1 \sin \theta_2 \cos \theta_3,$$
$$\cdots\cdots\cdots\cdots\cdots\cdots\cdots\cdots\cdots\cdots\cdots\cdots\cdots$$
$$x_{m-1} = R \sin \theta_1 \sin \theta_2 \ldots \sin \theta_{m-2} \cos \varphi,$$
$$x_m = R \sin \theta_1 \sin \theta_2 \ldots \sin \theta_{m-2} \sin \varphi.$$

   Show that Φ maps the rectangular parallelepiped $[0, \pi]^{m-2} \times [0, 2\pi]$ onto the sphere of radius R (centered at the origin), and the restriction of Φ to the open parallelepiped $(0, \pi)^{m-2} \times (0, 2\pi)$ is a one-to-one map onto a "slit sphere" (this time, the "slit" is a compact subset of the $(m - 2)$-dimensional sphere).

    The numbers $\theta_1, \ldots, \theta_{m-2}, \varphi$ are called the spherical coordinates of a point on the boundary of a ball.

4. Let p be a point of a smooth k-dimensional manifold M. Show that among all possible k-dimensional affine subspaces L passing through p, only the tangent subspace satisfies the condition

$$\operatorname{dist}(x, L) = o(\|x - p\|) \quad \text{as } x \to p, \ x \in M.$$

5. Let L_w be the (vector or affine) tangent subspace to a smooth manifold M at a point w and P be the orthogonal projection to L_w. In addition to the obvious inequality

$$\|P(x) - P(y)\| \le \|x - y\|,$$

prove that for every $\varepsilon > 0$ in a sufficiently small M-neighborhood U of w the following lower bound holds:

$$(1 - \varepsilon)\|x - y\| \le \|P(x) - P(y)\| \quad (x, y \in U).$$

6. The result of the previous exercise implies the invertibility of the restriction of P to a sufficiently small M-neighborhood U of w. Show that $(P|_U)^{-1}$ is a smooth map (meaning that $(P|_U)^{-1}$ is the restriction of a smooth map defined in a neighborhood of the set $P(U)$).

3 Constrained Extrema

3.1 Heuristic Arguments and Statement of the Problem

Prominent examples of application of differential calculus techniques are methods for finding local extrema of a function. However, in some important cases these methods are insufficient. This is seen, for example, by considering the problem of finding extreme values of a function on a compact set K. Where can they be attained? If the function f under consideration is smooth in a neighborhood of K, then an interior point of this set can be a point of extremum only if it is a critical point of f. However, extreme values can be attained at the boundary of K, and the corresponding points are by no means necessarily critical. Hence, the results obtained in Section I.9 do not allow one to completely solve the problem of finding extreme values, and they must be complemented by techniques making it possible to "fish out" boundary points at which these values can be attained. Our methods apply to the case where the boundary of K is smooth or consists of several smooth parts of different dimensions. In what follows, we assume that the function f is defined on an open subset O of $\mathbb{R}^m$. Our objective is to learn how to find points at which f takes values that are extreme not for the entire set O, but only for a part M of O that has no interior points. Keeping the notation introduced above, we give the following definition to make the problem more precise.

Definition Let $x_0 \in M \subset O$. We say that a function f defined on O has a *relative maximum* at x_0 if there exists a neighborhood U of x_0 such that $f(x) \leq f(x_0)$ for all $x \in U \cap M$.

A *relative minimum* is defined in a similar way. The term *relative* (or *constrained*) *extremum* encompasses both relative maximum and relative minimum.

How can we find points of M at which a given function can have a relative extremum? We face this problem, in particular, in many applications. Here is one of them.

A road is marked on a geographic map, and we are asked to find a point of this road that lies at the highest altitude above sea level. In this case, we study the function f that describes the ground profile (the value of f at an arbitrary point of the map is equal to the height of the corresponding point of the landscape above sea level). The role of M ("road") is played by the support of some path (of course, both the function and the path are assumed to be smooth).

To understand where a relative extremum can by no means be attained, consider the sets of constancy of our function, called horizontals (or isohypses), passing through the points of the road. We exclude from consideration the endpoints of the road and the points where horizontals are degenerate (these are points corresponding to local extrema of f, i.e., summits of hills and lowest points of depressions). Then the typical mutual position of the road and a horizontal is as shown in the left part of Fig. 2.

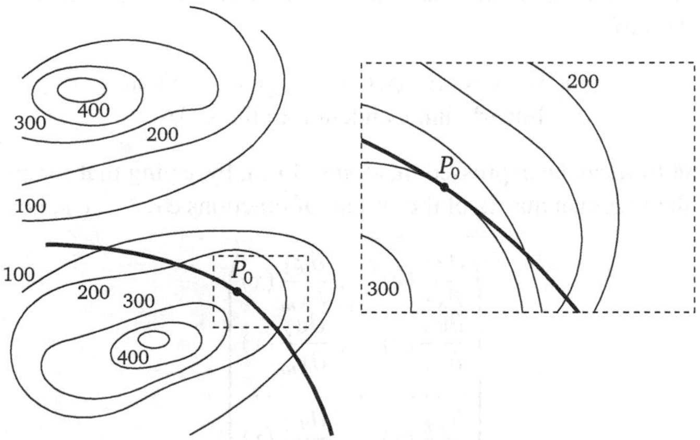

Fig. 2

Let P be an intersection point of the road and a horizontal $f = C$. It is intuitively clear that if the tangents at P to the horizontal and to the road do not coincide, then after a small perturbation of the level (that is, of the constant C) the horizontal still meets the road, and, therefore, arbitrarily close to P there are higher points of the road. Thus, a relative maximum of the height function on the road (except the points excluded from consideration) can be attained only at points where the tangents to the road and to the corresponding horizontal coincide (in Fig. 2, this happens at the point P_0). In other words, at this point the road and the horizontal must touch each other. As we will see, this condition underlies the general method of finding relative extrema which will be obtained below.

In what follows, we restrict ourselves to sets M of special form, namely, assume that M is a piecewise smooth manifold. Since the existence of a relative extremum at a point depends on the behavior of the function only in an arbitrarily small neighborhood of this point, first of all we must analyze the case where M is a smooth manifold (in the more general case, M breaks into several such manifolds, and it remains to study f in each of them separately). So, let M be a set defined by a system of k ($k < m$) equations

$$\begin{cases} \varphi_1(x_1, x_2, \ldots, x_m) = C_1, \\ \varphi_2(x_1, x_2, \ldots, x_m) = C_2, \\ \cdots\cdots\cdots\cdots\cdots\cdots\cdots \\ \varphi_k(x_1, x_2, \ldots, x_m) = C_k, \end{cases} \tag{1}$$

i.e., the intersection of the level sets

$$\{x \in O: \varphi_j(x) = C_j\}$$

of the functions φ_j. Equations (1) are called the *constraint equations*.

To guarantee that M is a smooth manifold, we must assume that $\varphi_1, \ldots, \varphi_k$ are smooth functions in O and that (see Theorem 2.2) their gradients are linearly independent on M:

$$\text{the vectors grad } \varphi_1(x), \ldots, \text{grad } \varphi_k(x) \text{ are} \tag{2}$$
$$\text{linearly independent for all } x \in M.$$

This condition can be expressed in another form, by saying that for every $x \in M$, the rank of the Jacobian matrix of the system of functions $\varphi_1, \ldots, \varphi_k$, i.e., the matrix

$$\begin{pmatrix} \dfrac{\partial \varphi_1}{\partial x_1}(x) & \cdots & \dfrac{\partial \varphi_1}{\partial x_m}(x) \\ \dfrac{\partial \varphi_2}{\partial x_1}(x) & \cdots & \dfrac{\partial \varphi_2}{\partial x_m}(x) \\ \cdots\cdots\cdots\cdots\cdots\cdots \\ \dfrac{\partial \varphi_k}{\partial x_1}(x) & \cdots & \dfrac{\partial \varphi_k}{\partial x_m}(x) \end{pmatrix},$$

is equal to k. If the system (1) reduces to one equation ($k = 1$), then condition (2) means that M contains no critical points of φ_1.

Since the functions φ_j can be replaced by the differences $\varphi_j - C_j$ without changing the gradients, we may assume without loss of generality that M is the intersection of zero level sets, i.e., has the form

$$M = \{x \in O: \varphi_1(x) = \ldots = \varphi_k(x) = 0\},$$

or, in other words, that M is defined by the system of equations

$$\begin{cases} \varphi_1(x_1, x_2, \ldots, x_m) = 0, \\ \varphi_2(x_1, x_2, \ldots, x_m) = 0, \\ \ldots\ldots\ldots\ldots\ldots\ldots\ldots \\ \varphi_k(x_1, x_2, \ldots, x_m) = 0. \end{cases} \tag{1'}$$

In the search for relative extrema, it is often useful to consider the level sets of the function under study and use geometric considerations. We illustrate this with two examples involving functions of two variables. In both cases, the system (1) consists of the single equation $\varphi(x, y) = x^2 + y^2 = 1$ and thus M is the unit circle.

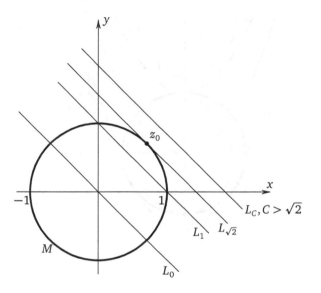

Fig. 3

Example 1 Let $f(x, y) = x + y$. The level sets of f are the lines L_C defined by the equations $x + y = C$. If C is large, then the line L_C does not meet M (see Fig. 3).

As C decreases, the line L_C approaches the circle M and eventually, for some $C = C_0$, touches it at a point $z_0 = (x_0, y_0)$. Since for $C > C_0$ the lines L_C do not intersect M, the values of f on M cannot exceed C_0 and, consequently, f has a relative maximum at z_0. Obviously, at the point symmetric to z_0 with respect to the origin, f has a relative minimum. Since it is clear from symmetry that all coordinates of z_0 are equal, we can easily find them using the constraint equation ($x_0 = y_0 = \frac{1}{\sqrt{2}}$), and then determine the value of the relative maximum: $f(x_0, y_0) = \sqrt{2}$ ($= C_0$).

Example 2 Now we consider a little more complicated function: $f(x, y) = y - x^2$. The level sets of f are the parabolas defined by the equations $y = x^2 + C$. If C is large, then the corresponding parabola L_C does not meet M.

As C decreases, the parabola L_C descends toward the circle M until eventually (for $C = 1$) its vertex coincides with the "north pole" $N = (0, 1)$ of the circle. Since for $C > 1$ the parabolas L_C do not intersect M, the function f has a relative maximum at N. For $C = -1$, the vertex of L_C coincides with the "south pole" $S = (0, -1)$ of the circle. Since $1 - x^2 \leq \sqrt{1 - x^2}$ ($|x| < 1$), the arc of L_{-1} in the strip $|x| < 1$ lies above the lower semicircle (except the common point $(0, -1)$).

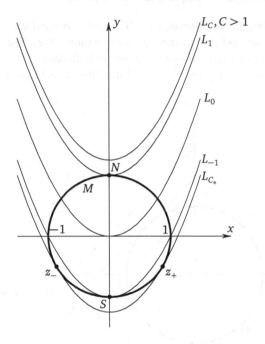

Fig. 4

As C increases, the vertex of the parabola L_C lies above the parabola L_{-1}, and hence points close to S belong to the parabolas L_C with $C < -1$. It follows that f has a relative maximum both at S and N. But where does it have a relative minimum? If C is less than -1 but close to -1, then the parabola L_C meets the circle at two points with positive x-coordinates. As C decreases, these points approach each other and eventually (say, for $C = C_*$) merge at a point $z_+ = (\bar{x}, \bar{y})$. Since for $C < C_*$ the parabolas L_C no longer intersect M, the function f has a relative minimum at z_+ and at the symmetric point $z_- = (-\bar{x}, \bar{y})$ (see Fig. 4).

At the point z_+, the level sets of f "break away" from the circle. It is intuitively clear that this can happen only if the level set passing through z_+ and the circle touch each other (i.e., have a common tangent). Although this fact allows one to determine the coordinates of z_+, we will not dwell on this here and will return to the problem of determining the coordinates of z_+ a little later.

3.2 A Necessary Condition for a Constrained Extremum

Our aim here is to give a rigorous meaning to the considerations discussed at the beginning of the section.

Theorem *Let O be an open subset in $\mathbb{R}^m$ and $f, \varphi_1, \ldots, \varphi_k \in C^1(O)$, $k < m$. Further, let M be the set defined by* (1), *and assume that the functions $\varphi_1, \ldots, \varphi_k$ satisfy condition* (2). *If f has a relative extremum at a point $a \in M$, then $\operatorname{grad} f(a)$ is a linear combination of the gradients $\operatorname{grad} \varphi_1(a), \ldots, \operatorname{grad} \varphi_k(a)$, i.e., there exist $\lambda_1, \ldots, \lambda_k$ such that*

$$\operatorname{grad} f(a) = \sum_{j=1}^{k} \lambda_j \operatorname{grad} \varphi_j(a). \tag{3}$$

If $k = 1$, i.e., there is just one constraint equation $\varphi(x) = C$, then condition (3) shows that the gradients of f and φ at the point a are proportional. This means, in turn, that at this point the level set $\{x \in O : f(x) = f(a)\}$ and the manifold M touch each other.

Proof Assume without loss of generality that the manifold M is defined by (1′) and, besides, $f(a) = 0$ (otherwise, replace f by the function $x \mapsto f(x) - f(a)$).

We will argue by contradiction and assume that the gradient of f at the point a is not a linear combination of the gradients of φ_j. Since these gradients are linearly independent (condition (2)), we see that the system of vectors

$$\operatorname{grad} \varphi_1(a), \ldots, \operatorname{grad} \varphi_k(a), \operatorname{grad} f(a)$$

is linearly independent. Obviously, this property is preserved in some neighborhood $U \subset O$ of a. Consider the auxiliary map $\Psi \colon U \mapsto \mathbb{R}^{k+1}$ defined by

$$x \mapsto \Psi(x) = \big(\varphi_1(x), \ldots, \varphi_k(x), f(x)\big) \quad (x \in U).$$

Since $a \in M$, it is clear that $\Psi(a) = (0, \ldots, 0)$. By the choice of U, the rank of the matrix $\Psi'(x)$ is equal to $k + 1$ everywhere in U. Hence, by Theorem II.1.2, the map Ψ is open. Therefore, in every ball $B \subset U$ centered at a, it takes the values $(0, \ldots, 0, \pm t)$ for all sufficiently small $t > 0$. By the definition of the coordinate functions of Ψ, this means that f takes both positive and negative values at some points of B satisfying equation (1′). In other words, in an arbitrarily small M-neighborhood of a, the function f takes both positive and negative values. Hence, it cannot have a relative extremum at this point, contradicting the assumptions of the theorem. □

In conclusion, note that the theorem proved above provides only a necessary condition for a constrained extremum, which is by no means sufficient. It is natural to say that a point satisfying this condition is a "constrained critical point of f" (subject to the constraints (1)). The reader can easily construct examples of such points at which there is no relative extrema (they are quite analogous to corresponding examples for unconstrained local extrema). The analysis of the behavior of functions near constrained critical points is the subject of the last two subsections of this section.

3.3 The Lagrange Function and Lagrange Multipliers Method

Now we restate the result obtained above in a form more convenient for practical applications. For this, we introduce an auxiliary function, called the *Lagrange function*. It is defined as

$$F(x) = f(x) - \sum_{j=1}^{k} \lambda_j \varphi_j(x) \quad (x \in O),$$

where f is the function under study, $\varphi_1, \ldots, \varphi_k$ are the functions defining the constraint equations (1), and $\lambda_1, \ldots, \lambda_k$ are as yet undetermined coefficients, called *Lagrange multipliers*.

Note that on M the Lagrange function F differs from f only by a constant. Hence, it has relative extrema on M simultaneously with f. The Lagrange multipliers are chosen so as to satisfy equation (3), i.e., so that grad $F(a) = 0$. Thus, as in the case of (unconstrained) local extrema, constrained critical points of f are again critical points, only not of f, but of the Lagrange function F.

To find constrained critical points, we must determine $k + m$ numerical parameters: k Lagrange multipliers and m coordinates of a. For this, we have k constraint equations and (writing the equation grad $F(a) = 0$ in coordinates) m equations $\frac{\partial F}{\partial x_j}(a) = 0$, $j = 1, 2, \ldots, m$. In some cases, we may restrict ourselves to determining the coordinates of constrained critical points without finding the Lagrange multipliers.

3.4 Examples of Applying the Lagrange Multipliers Method

Here we consider several problems illustrating the application of the obtained results. Condition (2) in these cases is obviously satisfied.

Example 1 First, we return to Example 2 at p. 183. Since $f(x, y) = y - x^2$ and $\varphi(x, y) = x^2 + y^2$, the Lagrange function is $F(x, y) = y - x^2 - \lambda(x^2 + y^2)$. Therefore, the system of equations for the coordinates of a constrained critical point on the unit circle M and the corresponding Lagrange multiplier λ is as follows:

$$\begin{cases} -2x - 2\lambda x = 0, \\ 1 - 2\lambda y = 0, \\ x^2 + y^2 = 1. \end{cases}$$

If $x = 0$, the last equation implies that $y = \pm 1$, and hence $\lambda = \pm\frac{1}{2}$. Thus, we obtain the points where, as we have already established, f has a relative maximum. If $x \neq 0$, the first equation implies that $\lambda = -1$, and the second one, that $y = -\frac{1}{2}$, so we obtain the points $z_\pm = \left(\pm\frac{\sqrt{3}}{2}, -\frac{1}{2}\right)$, at which f has a relative minimum.

It must be said that in order to find the absolute maximum and minimum values of f on M, there is no need to determine the type of extremum at the obtained points. It suffices to calculate the values of f at these points and compare them. Such a calculation shows that

$$\max_M f = f(0, 1) = 1, \quad \min_M f = f(z_\pm) = -\frac{5}{4}.$$

Note that determining a point of maximum does not require knowing the corresponding Lagrange multiplier.

Example 2 Consider the ellipse $\frac{x^2}{a^2} + \frac{y^2}{b^2} = 1$ and draw the normals (lines perpendicular to tangents) at all its points. The problem is to find the normal that is at the greatest distance from the origin. What is this distance?

To exclude the trivial case, below we assume that $a \neq b$.

Standard calculations show that the squared distance from the origin to the normal is equal to

$$f(x, y) = x^2 + y^2 - \frac{1}{\dfrac{x^2}{a^4} + \dfrac{y^2}{b^4}}.$$

We must find the absolute maximum value of this function subject to the condition $\frac{x^2}{a^2} + \frac{y^2}{b^2} = 1$. The Lagrange function for this problem is

$$F(x, y) = x^2 + y^2 - \frac{1}{\dfrac{x^2}{a^4} + \dfrac{y^2}{b^4}} - \lambda\left(\frac{x^2}{a^2} + \frac{y^2}{b^2}\right).$$

Setting its partial derivatives equal to zero and discarding the trivial solutions $(\pm a, 0)$, $(0, \pm b)$ (at which f vanishes), we obtain the system of equations

$$\begin{cases} \lambda = a^2 + \dfrac{1}{a^2\left(\dfrac{x^2}{a^4} + \dfrac{y^2}{b^4}\right)^2}, \\[4mm] \lambda = b^2 + \dfrac{1}{b^2\left(\dfrac{x^2}{a^4} + \dfrac{y^2}{b^4}\right)^2}. \end{cases}$$

Excluding λ yields $\frac{x^2}{a^4} + \frac{y^2}{b^4} = \frac{1}{ab}$. Taking into account the constraint equations, we obtain

$$x^2 = \frac{a^3}{a+b}, \quad y^2 = \frac{b^3}{a+b}.$$

Therefore, the maximum value of f is attained at four symmetric points and is equal to

$$f(x, y) = x^2 + y^2 - \frac{1}{\dfrac{x^2}{a^4} + \dfrac{y^2}{b^4}} = \frac{a^3 + b^3}{a + b} - ab = (a - b)^2.$$

Thus, the maximum distance in question is equal to $|a - b|$.

Example 3 Let us find the absolute maximum and minimum values of the function $f(x, y, z) = y^2(x + 2z - 1)$ on the following part of a cone:

$$E = \left\{(x, y, z) \in \mathbb{R}^3 : x^2 + y^2 \le z^2; \ x + 2z \le 2; \ z \ge 0\right\}.$$

Due to the factor $x + 2z - 1$, the function f changes sign on E, so

$$\min_E f < 0 < \max_E f.$$

Since $f(x, 0, z) = 0$, the function cannot attain an absolute maximum or minimum value at points with $y = 0$. Hence, below we assume that $y \ne 0$. In particular, since $f_x'(x, y, z) = 0$ only for $y = 0$, in what follows we do not consider interior points of E.

Let us analyze the values of f at the boundary of E. Excluding the vertex of the cone (where $f = 0$), the boundary breaks into three smooth manifolds:

1) the part of the conic surface $x^2 + y^2 = z^2$, $x + 2z < 2$, $z > 0$;
2) the elliptic base (flat "lid") of the cone $x + 2z = 2$, $x^2 + y^2 < z^2$;
3) the ellipse bounding the base $x + 2z = 2$, $x^2 + y^2 = z^2$.

Let us find the constrained critical points of f on each of these manifolds.

1. This part of the boundary is defined by one equation $x^2 + y^2 = z^2$, hence the corresponding Lagrange function is

$$F(x, y, z) = y^2(x + 2z - 1) - \lambda(x^2 + y^2 - z^2).$$

Setting its partial derivatives equal to zero, we arrive at the system

$$\begin{cases} F_x' = y^2 - 2\lambda x = 0, \\ F_y' = 2y(x + 2z - 1) - 2\lambda y = 0, \quad \text{or} \\ F_z' = 2y^2 + 2\lambda z = 0, \end{cases} \quad \begin{cases} y^2 - 2\lambda x = 0, \\ x + 2z - 1 - \lambda = 0, \\ y^2 + \lambda z = 0. \end{cases}$$

Since $y \ne 0$, we have $\lambda x \ne 0$. Excluding y yields $z = -2x$. Now the constraint equation implies the equality $y^2 = 3x^2$, which can be substituted into the first equation of the system to obtain $3x = 2\lambda$. This allows us to exclude λ from the second equation and to successively find the values of all the unknowns: $x = -\frac{2}{9}$, $y = \pm\frac{2\sqrt{3}}{9}$, $z = \frac{4}{9}$. Thus, in the case under consideration there is only one potential minimum value: $f\left(-\frac{2}{9}, \pm\frac{2\sqrt{3}}{9}, \frac{4}{9}\right) = -\frac{4}{81}$.

2. In this case, $f(x, y, z) = y^2$, and hence, obviously, $f(x, y, z) > \min_E f$. It turns out that on this part of the boundary of E (denote it by S), the function cannot

attain a maximum value too. Of course, one can see this by considering the Lagrange function (the reader is encouraged to do this), but that is not necessary. Indeed, if a point (x, y, z) satisfies the conditions $x + 2z = 2$, $x^2 + y^2 < z^2$, then so does a point with a slightly larger $|y|$ (and the same x and z). Hence, near every point in S there is a point at which f takes a greater value.

3. It remains to study the part of the boundary that is the intersection of a plane and a conic surface. Its points satisfy two constraint equations $x + 2z = 2$ and $x^2 + y^2 - z^2 = 0$, so the Lagrange function is

$$F(x, y, z) = y^2 - \lambda(x + 2z) - \mu(x^2 + y^2 - z^2).$$

Setting all its partial derivatives equal to zero, we obtain the system

$$\begin{cases} -\lambda - 2\mu x = 0, \\ 2y - 2\mu y = 0, \\ -2\lambda + 2\mu z = 0. \end{cases}$$

Since $y \neq 0$, the second equation implies that $\mu = 1$. Hence, from the first and third equations we obtain the equalities $\lambda = -2x = z$, which (together with the constraint equations) imply that $x = -\frac{2}{3}$, $y = \pm\frac{2}{\sqrt{3}}$, and $z = \frac{4}{3}$. At these points, f takes the value $f\left(-\frac{2}{3}, \pm\frac{2}{\sqrt{3}}, \frac{4}{3}\right) = \frac{4}{3}$.

Thus, we have found all constrained critical points of f at the boundary of E at which it takes nonzero values:

$$f\left(-\frac{2}{9}, \pm\frac{2\sqrt{3}}{9}, \frac{4}{9}\right) = -\frac{4}{81} \quad \text{and} \quad f\left(-\frac{2}{3}, \pm\frac{2}{\sqrt{3}}, \frac{4}{3}\right) = \frac{4}{3}.$$

Therefore, the first of them is $\min_E f$, and the second one is $\max_E f$.

Note that in some cases, specific features of the problem under consideration allow one to depart from the general scheme and reduce the amount of calculations, as we have seen when analyzing the function on the base of the cone.

Example 4 Now we give an analytic proof of an important fact known from linear algebra: every real symmetric matrix has a real eigenvalue. First, we make a simple observation. Let λ_0 be an eigenvalue of such a matrix A and x_0 be the corresponding eigenvector; we assume without loss of generality that x_0 is normalized. Taking the scalar product of the equality $A x_0 = \lambda_0 x_0$ with x_0, we see that $\langle A x_0, x_0 \rangle = \lambda_0$. Thus, an eigenvalue of a symmetric matrix is one of the values taken by the quadratic form generated by this matrix at the unit sphere. It turns out that the extreme values of the quadratic form $\langle A x, x \rangle$ at the unit sphere are attained at eigenvectors of A and coincide with its eigenvalues. Let us prove this.

Let F be the Lagrange function corresponding to our quadratic form and the constraint equation $\|x\|^2 = 1$, so

$$F(x) = \langle A x, x \rangle - \lambda\|x\|^2 = \langle C x, x \rangle,$$

where $C = A - \lambda \mathbb{I}_m$ and $\mathbb{I}_m$ is the $m \times m$ identity matrix. Let x_0 be a point of relative extremum of the quadratic form F on the unit sphere, so $d_{x_0} F = 0$. As we have established at the end of Section I.2.1, $d_{x_0} F(h) = 2\langle C x_0, h \rangle$. Hence, $\langle C x_0, h \rangle = 0$ for every h. This means that $A x_0 - \lambda x_0 = 0$. Thus, x_0 is an eigenvector of A, and the Lagrange multiplier λ is the corresponding eigenvalue.

Imposing additional constraints, one can also find all the other eigenvalues of A (see Exercise 4).

Remark The obtained result allows us to reduce the problem of finding the norm of a linear map to that of finding the greatest eigenvalue of some symmetric matrix. Indeed, let A be the matrix (not necessarily square) corresponding to the map and $\widetilde{A} = A^T A$. Since

$$\|Ax\|^2 = \langle Ax, Ax \rangle = \langle A^T Ax, x \rangle = \langle \widetilde{A}x, x \rangle,$$

we obtain

$$\|A\|^2 = \max_{\|x\|=1} \|Ax\|^2 = \max_{\|x\|=1} \langle \widetilde{A}x, x \rangle,$$

and the right-hand side is equal to the greatest eigenvalue of the symmetric matrix $\widetilde{A}$.

Example 5 Let us show that every square real matrix

$$A = \begin{pmatrix} x_{11} & \cdots & x_{1m} \\ \cdots & \cdots & \cdots \\ x_{m1} & \cdots & x_{mm} \end{pmatrix}$$

satisfies *Hadamard's inequality*:

$$|\det A| \le \sqrt{\sum_{k=1}^{m} x_{1k}^2} \cdots \sqrt{\sum_{k=1}^{m} x_{mk}^2}.$$

This inequality has a simple geometric meaning: the left-hand side is the m-dimensional volume of the parallelepiped spanned by the row vectors of A, and the right-hand side is the product of the lengths of these vectors.

To prove Hadamard's inequality, it suffices to restrict ourselves to the case where all rows of the matrix A have unit length (since the determinant of a matrix is proportional to the lengths of its rows). Besides, we may assume that the determinant of A is positive (this can be achieved by permuting its rows). Thus, the problem is reduced to the study of the function $x \mapsto f(x) = \det(x_{jk})_{1 \le j, k \le m}$, where x is an $m \times m$ matrix identified with the vector from $\mathbb{R}^{m^2}$ obtained by the concatenation of its rows.

Let us find a relative maximum of f subject to the condition that the elements $x_{11}, \ldots, x_{mm}$ satisfy m constraint equations

$$\begin{cases} \varphi_1(x) = x_{11}^2 + \ldots + x_{1m}^2 = 1, \\ \ldots\ldots\ldots\ldots\ldots\ldots\ldots\ldots \\ \varphi_m(x) = x_{m1}^2 + \ldots + x_{mm}^2 = 1. \end{cases}$$

To apply the theorem, we must check that the Jacobian matrix of the system of functions $\varphi_1, \ldots, \varphi_m$ has maximum rank (equal to m). Obviously, it has m rows and m^2 columns. Each row consists of m blocks of m elements. In the jth row, all blocks except the jth one are zero, and the jth block is equal to $2x_{j1}, \ldots, 2x_{jm}$:

$$\begin{pmatrix} 2x_{11} \ldots 2x_{1m} \ldots & 0 & \ldots & 0 & \ldots & 0 & \ldots & 0 \\ \ldots\ldots\ldots\ldots\ldots\ldots\ldots\ldots\ldots\ldots\ldots\ldots\ldots \\ 0 & \ldots & 0 & \ldots 2x_{j1} \ldots 2x_{jm} \ldots & 0 & \ldots & 0 \\ \ldots\ldots\ldots\ldots\ldots\ldots\ldots\ldots\ldots\ldots\ldots\ldots\ldots \\ 0 & \ldots & 0 & \ldots & 0 & \ldots & 0 & \ldots 2x_{m1} \ldots 2x_{mm} \end{pmatrix}.$$

Due to the constraint equations, each row of this matrix contains a nonzero element. The columns corresponding to these elements form a diagonal square submatrix of size $m \times m$ with nonzero determinant. Thus, condition (2) is satisfied.

If f attains a relative maximum at a point x, then the partial derivatives of the Lagrange function

$$F(x) = f(x) - \lambda_1 \varphi_1(x) - \ldots - \lambda_m \varphi_m(x)$$

with respect to all variables $x_{11}, \ldots, x_{mm}$ vanish at this point:

$$\frac{\partial F}{\partial x_{jk}}(x) = \frac{\partial f}{\partial x_{jk}}(x) - \sum_{i=1}^{m} \lambda_i \frac{\partial \varphi_i}{\partial x_{jk}}(x) = 0 \quad \text{for } j, k = 1, \ldots, m.$$

To differentiate the determinant $f(x)$ with respect to x_{jk}, we expand it in the jth row:

$$f(x) = x_{j1} A_{j1}(x) + \ldots + x_{jm} A_{jm}(x),$$

where $A_{ji}(x)$ is the cofactor of x_{ji} (not depending on the elements of the jth row). This equality implies that

$$\frac{\partial f}{\partial x_{jk}}(x) = A_{jk}(x) \quad (j, k = 1, \ldots, m).$$

The partial derivatives $\frac{\partial \varphi_i}{\partial x_{jk}}$ are easy to find:

$$\frac{\partial \varphi_i}{\partial x_{jk}}(x) = (x_{i1}^2 + \ldots + x_{im}^2)'_{x_{jk}} = \begin{cases} 2x_{jk} & \text{if } i = j, \\ 0 & \text{if } i \neq j. \end{cases}$$

Thus,

$$A_{jk}(x) = \frac{\partial f}{\partial x_{jk}}(x) = \sum_{i=1}^{m} \lambda_i \frac{\partial \varphi_i}{\partial x_{jk}}(x) = 2\lambda_j x_{jk}. \tag{4}$$

Therefore,

$$f(x) = \sum_{k=1}^{m} x_{jk} A_{jk}(x) = 2\lambda_j \sum_{k=1}^{m} x_{jk}^2 = 2\lambda_j \quad (j = 1, \dots, m),$$

i.e., all Lagrange multipliers are equal to $\frac{1}{2} f(x)$. Dividing both sides of (4) by $f(x)$, we obtain

$$\frac{1}{f(x)} A_{jk}(x) = x_{jk} \quad \text{for all } j, k = 1, \dots, m.$$

This means that the transposed matrix A^T coincides with the inverse matrix A^{-1}. Hence,

$$1 = \det(A \cdot A^{-1}) = \det(A \cdot A^T) = \det A \cdot \det A^T = (\det A)^2 = f^2(x).$$

So, the greatest value of a determinant with rows of unit length is equal to 1.

3.5 Sufficient Conditions for a Constrained Extremum

When seeking the absolute maximum and minimum values of a function on a compact set, there is no need to analyze its behavior at constrained critical points on its boundary. Nevertheless, for completeness, in this and the next section we present sufficient conditions that make such an analysis possible. They are stated in terms of the second differential of the Lagrange function and, as we will see, are quite analogous to the sufficient conditions for a local ("unconstrained") extremum.

Keeping the same notation as in the previous sections, we assume that M is a manifold contained in $\mathbb{R}^m$ and defined by equations (1), where the functions $\varphi_1, \dots, \varphi_k$ satisfy condition (2). As earlier, we assume that f is a function defined in a neighborhood of M and a is a constrained critical point of f on M; we also denote by $F = f - \sum_{i=1}^{k} \lambda_i \varphi_i$ the corresponding Lagrange function. Besides, we assume that the functions f and $\varphi_1, \dots, \varphi_k$ are C^2-smooth. As we have already mentioned, the sufficient conditions for the existence of a constrained extremum at a are quite analogous to the sufficient conditions for an "unconstrained" extremum, with the difference that instead of the second differential of f we must consider the second differential of F and, besides, not on the whole space, but only on the (vector) tangent subspace T_a to M at the point a. Under the assumptions described above, the following theorem holds.

Theorem *Let Q be the restriction of the second differential $d_a^2 F$ to the subspace T_a. If Q is a definite quadratic form (on T_a), then f has a relative extremum at a. It is a relative minimum or maximum depending on whether the form Q is positive or negative definite.*

In the proof we use the following simple fact: a quadratic form H satisfies the inequality

$$|H(u + v) - H(u)| \leq C_H \|u\| \, \|v\| \quad \text{for } \|v\| \leq \|u\| \tag{5}$$

with some constant C_H.

Indeed, if A is the symmetric matrix defining the form H, then

$$H(u + v) - H(u) = \langle Av, v + 2u \rangle,$$

and hence

$$|H(u + v) - H(u)| \leq \|A\| \cdot \|v\| (\|v\| + 2\|u\|) \leq 3\|A\| \cdot \|u\| \cdot \|v\|.$$

Proof We assume without loss of generality that $a = 0$. Then the affine and vector tangent subspaces coincide.

Throughout the proof, x is a point of M (close to $a = 0$) and $\tilde{x}$ is its orthogonal projection to the tangent subspace T_0. Then $\tilde{x}$ is the closest point in T_0 to x, hence

$$\omega = x - \tilde{x} \underset{x \to 0}{=} o(\|x\|)$$

(see formula (4) in Section 2.3). Since the difference $f - F$ is constant on M and $d_0 F = 0$, it follows from Taylor's formula for the Lagrange function F that

$$f(x) - f(0) = F(x) - F(0) \underset{x \to 0}{=} \frac{1}{2} d_0^2 F(x) + o(\|x\|^2).$$

For x close to 0, replacing x with $\tilde{x}$ does not essentially change the value of the second differential, since we may assume that $\|\omega\| \leq \|x\|$, which, by (5), implies that

$$\left| d_0^2 F(\tilde{x} + \omega) - d_0^2 F(\tilde{x}) \right| \leq C \|\tilde{x}\| \, \|\omega\| \leq C \|x\| \, \|\omega\| \underset{x \to 0}{=} o(\|x\|^2).$$

Hence,

$$f(x) - f(0) = \frac{1}{2} d_0^2 F(\tilde{x}) + o(\|x\|^2) \underset{x \to 0}{=} \frac{1}{2} Q(\tilde{x}) + o(\|x\|^2). \tag{6}$$

We may assume without loss of generality that the quadratic form Q is positive definite. Then, by Lemma I.9.3, for some $c > 0$ we have

$$Q(h) \geq c \|h\|^2 \quad \text{for all } h \in T_0.$$

In particular, $Q(\tilde{x}) \geq c \|\tilde{x}\|^2$. Using the relation $\omega \underset{x \to 0}{=} o(\|x\|)$, we assume that x is so small that $\|\omega\| \leq \frac{1}{2} \|x\|$. Then $\|\tilde{x}\| \geq \frac{1}{2} \|x\|$ and, consequently,

$$Q(\tilde{x}) \geq c \left(\frac{\|x\|}{2} \right)^2.$$

Substituting this inequality into (6), we see that

$$f(x) - f(0) \geq \frac{c}{8}\|x\|^2 + o(\|x\|^2) = \left(\frac{c}{8} + o(1)\right)\|x\|^2.$$

Since for x sufficiently close to 0, the sum in the brackets is positive, for such $x \neq 0$ the increment of f on M is also positive.   □

Note that the definiteness of the quadratic form Q is incompatible with the difference $f(x) - f(a)$ decreasing "too fast" as $x \to a$: indeed, the relation $f(x) - f(a) = o(\|x - a\|^2)$ along some curve, or even along some sequence of points in M, contradicts the inequality $f(x) - f(a) \geq \left(\frac{c}{8} + o(1)\right)\|x - a\|^2$ established at the end of the proof.

Now we return one last time to Example 2 at p. 183 and use the theorem proved above to determine the type of extrema at constrained critical points once again. In the case under consideration, the Lagrange function F is

$$F(x, y) = y - x^2 - \lambda(x^2 + y^2).$$

Its first differential vanishes at four points:

$$N = (0, 1), \quad S = (0, -1), \quad \text{and} \quad z_\pm = \left(\pm\frac{\sqrt{3}}{2}, -\frac{1}{2}\right).$$

The corresponding values of the Lagrange multiplier λ are $\frac{1}{2}, -\frac{1}{2}$, and -1 (in the last two cases), respectively.

Clearly,

$$d_a^2 F(h) = -2(1 + \lambda)h_1^2 - 2\lambda h_2^2 \quad \text{at every point } a \in M.$$

At the point N we have $\lambda = \frac{1}{2}$, so the differential $d_N^2 F$ is negative definite and, consequently, N is a point of relative maximum.

At the point S the situation is a bit more complicated, since

$$d_S^2 F(h) = -h_1^2 + h_2^2$$

and we must switch to the (one-dimensional) tangent subspace $T_S(M)$. Obviously, it is given by the equation $h_2 = 0$. Hence, the restriction of $d_S^2 F$ to $T_S(M)$ is negative definite, so S is also a point of relative maximum.

Finally, at the remaining two points we obviously have

$$d_{z_\pm}^2 F(h) = 2h_2^2.$$

Since the tangent subspaces $T_{z_\pm}(M)$ are not horizontal lines, the restrictions of the second differentials to $T_{z_\pm}(M)$ are positive definite. Hence, $z_\pm$ are points of relative minimum of f.

3.6 Conditions for the Absence of a Constrained Extremum

As in the previous section, we assume that Q is the restriction of $d_a^2 F$ to the tangent subspace T_a, where a is a constrained critical point of f.

Theorem *If Q is an indefinite quadratic form, then f has no relative extremum at a.*

Such points are called *saddle points* of f (on the manifold M defined by equations (1)).

Proof We will show that if f has a conditional extremum at a, then the quadratic form Q cannot change sign on T_a (the reader will undoubtedly notice that this result is similar to the refined necessary condition for an "unconstrained" extremum). This obviously implies the theorem.

For definiteness, we assume that f has a relative minimum at a. We will show that $Q(h) \geq 0$ for every vector $h \in T_a$ (it is fixed throughout the argument).

By the definition of T_a, the vector h is the tangent vector to a smooth path γ contained in M, passing through a, and satisfying the conditions $\gamma(0) = a, \gamma'(0) = h$. Therefore,

$$M \ni \gamma(t) = a + th + \omega(t) \quad \text{where } \omega(t) \underset{t \to 0}{=} o(t).$$

It follows from (5) with $u = th$ and the equality $v = \omega(t)$ that

$$d_a^2 F(th + \omega(t)) - d_a^2 F(th) = O(t\|\omega(t)\|) \underset{t \to 0}{=} o(t^2),$$

and hence as $t \to 0$ we have

$$0 \leq f(\gamma(t)) - f(a) = \frac{1}{2} d_a^2 F(th + \omega(t)) + o(\|th + \omega(t)\|^2)$$

$$= \frac{1}{2} d_a^2 F(th) + o(t^2) = \frac{t^2}{2} Q(h) + o(t^2).$$

Therefore, $0 \leq \frac{1}{2} Q(h) + \frac{o(t^2)}{t^2}$ for small $t \neq 0$, and passing to the limit yields $Q(h) \geq 0$. Since $h \in T_a$ is an arbitrary vector, this completes the proof. $\qquad \square$

In conclusion we give two examples.

Example 1 Let us find the relative extrema of the function

$$f(x, y, z) = x^2 + y^2 + z^2$$

on the surface

$$M = \left\{ (x, y, z) \in \mathbb{R}^3 : \frac{x^3}{a^3} + \frac{y^3}{b^3} + \frac{z^3}{c^3} = 1 \right\}.$$

Here a, b, c are given positive numbers.

The Lagrange function of the problem is

$$F(x, y, z) = x^2 + y^2 + z^2 - \lambda \left(\frac{x^3}{a^3} + \frac{y^3}{b^3} + \frac{z^3}{c^3} \right).$$

It is easy to see that if the first differential of F at a point (x, y, z) vanishes, then $x = 0$ or $x = \frac{2a^3}{3\lambda}$, $y = 0$ or $y = \frac{2b^3}{3\lambda}$, $z = 0$ or $z = \frac{2c^3}{3\lambda}$. Since the point $(0, 0, 0)$ does not belong to M, we must consider the cases where at least one of the coordinates is nonzero.

In further steps, the following equality is useful:

$$d^2_{(x,y,z)} F(h) = \left(2 - 6\lambda\frac{x}{a^3}\right)h_1^2 + \left(2 - 6\lambda\frac{y}{b^3}\right)h_2^2 + \left(2 - 6\lambda\frac{z}{c^3}\right)h_3^2.$$

1. Let $x, y, z \neq 0$. Then $x = \frac{2a^3}{3\lambda}$, $y = \frac{2b^3}{3\lambda}$, $z = \frac{2c^3}{3\lambda}$. Substituting these values into the equation defining the surface M, we obtain $\lambda = \frac{2}{3}(a^6 + b^6 + c^6)^{1/3}$. This allows us to find the coordinates x, y, z and arrive, after simple calculations, at the equality

$$d^2_{(x,y,z)} F(h) = -2(h_1^2 + h_2^2 + h_3^2).$$

Hence, at the point

$$(x, y, z) = \frac{(a^3, b^3, c^3)}{(a^6 + b^6 + c^6)^{1/3}}$$

the function f has a relative maximum equal to $(a^6 + b^6 + c^6)^{1/3}$.

2. Let $z = 0$, $xy \neq 0$ (the cases $x = 0$ and $y = 0$ can be considered similarly). Then $x = \frac{2a^3}{3\lambda}$, $y = \frac{2b^3}{3\lambda}$, and the equation of M implies that $\lambda = \frac{2}{3}(a^6 + b^6)^{1/3}$. As a result, we obtain the constrained critical point

$$p = (x, y, z) = \frac{(a^3, b^3, 0)}{(a^6 + b^6)^{1/3}}.$$

Here

$$d^2_p F(h) = -2(h_1^2 + h_2^2 - h_3^2).$$

This quadratic form must be analyzed on the tangent subspace $T_p(M)$ defined by the equation

$$\frac{x^2}{a^3}h_1 + \frac{y^2}{b^3}h_2 + \frac{z^2}{c^3}h_3 = 0,$$

which can be simplified to $a^3 h_1 + b^3 h_2 = 0$. Expressing h_2 in terms of h_1 and substituting the result into the formula for $d^2_p F(h)$, we see that this quadratic form changes sign on $T_p(M)$. Hence, p is a saddle point (on the surface M) of f.

3. Let $y = z = 0$ (the cases $y = x = 0$ and $x = z = 0$ can be considered similarly). Since $(x, 0, 0) \in M$, we see that $x = a$. But the first differential of the Lagrange function at the point $q = (a, 0, 0)$ vanishes only for $\lambda = \frac{2}{3}a^2$. Then

$$d^2_q F(h) = 2(-h_1^2 + h_2^2 + h_3^2).$$

As in the previous case, we must study this quadratic form on the subspace $T_q(M)$, which, as one can easily see, is given by the equation $h_1 = 0$. Hence, the restriction of $d_q^2 F$ to $T_q(M)$ is positive definite and, therefore, f has a relative minimum at $q = (a, 0, 0)$.

Thus, on the surface M the function f has one maximum, three minima, and three saddle points.

Example 2 Let us analyze the constrained critical points of the function

$$f(x) = x_1^3 + \ldots + x_m^3 \quad (m \geq 3)$$

if the constraint equations have the form

$$x_1^2 + \ldots + x_m^2 = 1, \quad x_1 + \ldots + x_m = 0.$$

Condition (2) is obviously fulfilled at all points satisfying these equations. Setting the partial derivatives of the Lagrange function

$$F(x) = x_1^3 + \ldots + x_m^3 - \lambda(x_1^2 + \ldots + x_m^2) - \mu(x_1 + \ldots + x_m)$$

equal to zero, we conclude that all coordinates $x_1, \ldots, x_m$ satisfy the quadratic equation $3t^2 - 2\lambda t - \mu = 0$, with $\mu = \frac{3}{m}$ (to see this, add up the equalities obtained for $t = x_1, \ldots, t = x_m$). Let α and β be the roots of this equation, so $\alpha\beta = -\frac{1}{m}$. In what follows, we assume that $\beta < 0 < \alpha$.

Every constrained critical point has k coordinates equal to α and $m-k$ coordinates equal to β, for some k such that $1 \leq k < m$. One can easily see that constrained critical points of this form do exist for every k. Their coordinates α, β, as well as the coefficient λ, depend on k. We leave it to the reader to check that $\lambda = \frac{3}{2} \frac{m-2k}{\sqrt{mk(m-k)}}$.

Since both the function f and the functions defining the constraint equations are invariant under permutations of coordinates, we may restrict ourselves to an analysis of a constrained critical point of the form

$$x = (x_1, \ldots, x_m) = (\alpha, \ldots, \alpha, \beta, \ldots, \beta),$$

which has k (positive) coordinates α ($k = 1, \ldots, m-1$). Clearly, all points obtained from x by a permutation of coordinates are also constrained critical points.

We consider in detail only the case where both α and β occur among the coordinates of x at least twice, i.e., $2 \leq k \leq m-2$ (this is possible only for $m \geq 4$). Let us check that the second differential of the Lagrange function corresponding to x changes sign on the tangent subspace T_x. Clearly,

$$d_x^2 F(h) = \sum_{i=1}^{m} 6x_i h_i^2 - \lambda \sum_{i=1}^{m} 2h_i^2 = 2 \sum_{i=1}^{m} (3x_i - \lambda)h_i^2 = 6 \sum_{i=1}^{m} \left(x_i - \frac{\alpha + \beta}{2}\right) h_i^2$$

($\lambda = \frac{3}{2}(\alpha + \beta)$, because α, β are the roots of the equation $t^2 - \frac{2}{3}\lambda t - \frac{\mu}{3} = 0$). Since the tangent subspace T_x at the point $x = (\alpha, \ldots, \alpha, \beta, \ldots, \beta)$ is given by the equations

$$\alpha(h_1 + \ldots + h_k) + \beta(h_{k+1} + \ldots + h_m) = 0 \quad \text{and} \quad h_1 + \ldots + h_m = 0,$$

it contains the vectors

$$h_+ = (1, -1, 0, \ldots, 0) \quad \text{and} \quad h_- = (0, \ldots, 0, 1, -1).$$

It remains to observe that

$$d_x^2 F(h_+) = 6(\alpha - \beta) > 0, \quad d_x^2 F(h_-) = 6(\beta - \alpha) < 0.$$

Therefore, all constrained critical points arising for $k = 2, \ldots, m - 2$ are not points of extrema, but saddle points.

We leave it to the reader to find α and β in the cases $k = 1$, $k = m - 1$ (including the case $m = 3$) and check that

$$f_{\max} = \frac{m - 2}{\sqrt{m(m - 1)}} \quad (k = 1), \quad f_{\min} = -\frac{m - 2}{\sqrt{m(m - 1)}} \quad (k = m - 1).$$

Exercises

1. Using the same geometric considerations as in Examples 1, 2 from Section 3.1, find the relative extrema of the function $f(x, y) = xy$ on the unit circle.

2. Let M be the torus obtained by revolving the circle

$$x^2 + (z - a)^2 = R^2 \quad (0 < R < a)$$

   about the OX axis. Find the extreme values on M of the following functions:
   a) $f(x, y, z) = z$; b) $f(x, y, z) = 3z - (x^2 + y^2)$;
   b) $f(x, y, z) = z - 2(x^2 + y^2)$.

3. Draw a line through the center of a three-dimensional cube (tetrahedron, icosahedron) so that the sum of its squared distances to the vertices is
   a) the smallest possible; b) the greatest possible.

4. Refine the result of Example 4 from Section 3.4 by proving that all eigenvalues of a symmetric matrix are real.

   HINT. The eigenvectors corresponding to different eigenvalues of such a matrix are orthogonal, hence, when seeking the next eigenvalue, we must consider the points at which the quadratic form corresponding to the matrix takes extreme values on the cross section of the sphere by the subspace orthogonal to the eigenvectors already found.

5. Let r be the greatest or smallest distance from the origin to a point of the curve

$$(x^2 + y^2 + z^2)^2 = a^2 x^2 + b^2 y^2 + c^2 z^2 > 0, \quad \alpha x + \beta y + \gamma z = 0$$

(here $a, b, c, \alpha, \beta, \gamma$ are nonzero parameters). Show that

$$\frac{\alpha^2}{r^2 - a^2} + \frac{\beta^2}{r^2 - b^2} + \frac{\gamma^2}{r^2 - c^2} = 0.$$

6. For what C the system of equations

$$x + y + z + u + v = 0, \quad x^2 + \ldots + v^2 = 1, \quad x^3 + \ldots + v^3 = C$$

in the space $\mathbb{R}^5$

a) has a solution;   b) defines a smooth manifold?

HINT. Use the result of Example 2 at p. 197.

7. (See Problem 8 in [1].) How many maxima, minima, and saddle points has the function $x^4 + y^4 + z^4 + u^4 + v^4$ on the set

$$x + \ldots + v = 0, \quad x^2 + \ldots + v^2 = 1, \quad x^3 + \ldots + v^3 = C?$$

Chapter IV. Critical Values of Smooth Maps

Unlike the previous chapters, this one is devoted to a relatively special problem of the theory of smooth maps, which, however, plays an important role in differential topology. Both the statement of the problem and the methods for solving it require the reader to be familiar with measure theory, in particular, with the notion of Hausdorff[1] measures and dimension. The background material is presented in Section V.3.

1 *Statement of the Problem and the Main Result

1.1 Critical Points and Critical Values

In many problems involving smooth maps, an important role is played by a condition imposed on the Jacobian matrix: it must have maximal rank. Thus, a natural question arises: how "massive" can the set of values of a smooth map F be on the set N where this rank is not maximal? This is important to know, for example, when studying the level sets defined by the equation $F(x) = y$, since for $y \notin F(N)$ they contain no singularities and are smooth manifolds (see Theorem III.2.2).

This chapter contains answers to the above question and some refinements thereof. We will find out how smooth a map $F \colon O \mapsto \mathbb{R}^n$ (as earlier, by O we denote an open subset in $\mathbb{R}^m$) must be in order to ensure that almost all its level sets are smooth manifolds, i.e., that the set $F(N)$ has zero (n-dimensional) Lebesgue measure. We also complement this result by characterizing the massiveness of the set $F(N)$ and some natural parts of it in terms of not only the Lebesgue measure, but also the Hausdorff measure.

Clearly, in the simplest case where F is a function and the set N where grad $F = 0$ is a smooth arc, the image $F(N)$ degenerates to a single point. However, for more complicated sets N, the situation is not that simple. In Section 5.5, we give examples showing that if a function f is "insufficiently smooth," then, although the set N has

---

[1] Felix **Hausdorff** (1868–1942) was a German mathematician.

© The Author(s), under exclusive license to Springer Nature Switzerland AG 2021
B. M. Makarov, A. N. Podkorytov, *Smooth Functions and Maps*, Moscow Lectures 7,
https://doi.org/10.1007/978-3-030-79438-5_4

zero Lebesgue measure, its image $f(\mathcal{N})$ can be a nondegenerate interval. In Section 6, we describe the famous Whitney's example of a smooth (but "not too smooth") function f that is not constant on a simple arc $\mathcal{L}$ in $\mathcal{N}$. In other words, although the rate of change of f is zero (grad $f = 0$) everywhere on $\mathcal{L}$, still f takes different values on $\mathcal{L}$, so the interval $f(\mathcal{L})$ does not degenerate to a single point.

Definition A point $x \in O$ is called a *critical point* of a smooth map $F: O \to \mathbb{R}^n$ if rank $F'(x) < \min\{m, n\}$. The image of a critical point, i.e., the point $F(x)$, is called a *critical value* of F.

The set of all critical points of F is called the *critical set* of F and denoted by $\mathcal{N}$. Clearly, $\mathcal{N}$ is closed in O. Hence, both $\mathcal{N}$ and $F(\mathcal{N})$ can be represented as unions of at most countably many compact sets. In particular, they are Borel[1] measurable.

1.2 The Main Theorem

The first and most important step towards the solution of the problems discussed in this chapter was made by A. Morse[2] [11], who established a nontrivial relation between the smoothness of a function of several variables and the massiveness of the set of its critical values. The method suggested by Morse (it will be considered in Sections 3, 4) proved to be applicable in much more general situations. Soon it was used by Sard[3], who extended Morse's result to arbitrary smooth maps. In the statement of Sard's theorem, we use the notation $C^r(O; \mathbb{R}^n)$ introduced in Section I.6.3 for the classes of r-smooth ($r \in \mathbb{N}$) maps from O to $\mathbb{R}^n$.

Theorem *The set of critical values of a map of class $C^r(O; \mathbb{R}^n)$ has zero Lebesgue measure if $r > \max\{m - n, 0\}$.*

Thus, if $r > \max\{m - n, 0\}$, then for a map from $C^r(O; \mathbb{R}^n)$ the nonempty preimages of almost all points are r-smooth manifolds.

For $m \leq n$, the conclusion of Sard's theorem is valid for all 1-smooth maps, while for $m > n$, the smoothness assumptions become more demanding (and the proof becomes substantially more involved). In particular ($n = 1$), the set of critical values of a C^m-function has zero one-dimensional Lebesgue measure, which is exactly Morse's result. The smoothness assumptions in the theorem turn out to be sharp: for a less smooth map, the conclusion of the theorem is no longer true (counterexamples are given in Section 5).

A strengthening of the above theorem obtained by Sard later involves characterizing the "smallness" of the set of critical values in terms of not only the Lebesgue measure, but also the p-dimensional Hausdorff measures μ_p. Shortly afterwards, a much more comprehensive result was published by Federer[4]. Using Hausdorff

---

[1] Émile **Borel** (1871–1956) was a French mathematician.

[2] Anthony Perry **Morse** (1911–1984) was an American mathematician.

[3] Arthur **Sard** (1909–1980) was an American mathematician.

[4] Herbert **Federer** (1920–2010) was an American mathematician.

measures allowed him not only to characterize the set of critical values as a whole more accurately, but also to estimate the massiveness of some natural parts of this set. These parts are the images of the "fibers" of the critical set of F corresponding to different values of the rank of F'. Federer's result can be stated as follows.

Theorem *Let* $F \in C^r(O; \mathbb{R}^n)$ *and* $N_i = \{x \in O: \text{rank } F'(x) = i\}$, *where* $0 \le i < \min\{m, n\}$. *Then* $\mu_p(F(N_i)) = 0$ *for* $p = i + \frac{m-i}{r}$.

This theorem implies Sard's result discussed above. Leaving aside the simple case $m < n$, where $\lambda_n(F(O)) = 0$, we see that $N = N_0 \cup \ldots \cup N_{n-1}$ for $n \le m$. For $r > m-n$ and $i = 0, \ldots, n-1$, we have $p \le n$. Hence, $\mu_n(F(N_i)) \le \mu_p(F(N_i)) = 0$. Since the measures μ_n and λ_n are proportional, this implies that $\lambda_n(F(N)) = 0$, which is precisely the conclusion of Sard's theorem.

Thus, as demonstrated by the above theorems, the massiveness of the set of critical values of a map F depends on the smoothness of F. For example, for a function of class $C^2(\mathbb{R})$, this set has not only zero Lebesgue measure, but also zero Hausdorff measure of dimension $\frac{1}{2}$; and for an infinitely smooth function (of any number of variables), this set has zero Hausdorff dimension.

Federer's formula $p = i + \frac{m-i}{r}$, which suggests using Hausdorff measures of fractional dimension, leads naturally to the following question: To what extent will this formula hold if we characterize the smoothness of F more accurately, replacing the classes C^r ($r \in \mathbb{N}$) by classes of "fractional" smoothness C^t, with $t \in [1, \infty)$ (see Section I.6.7). Our aim is to prove Federer's theorem in the form it took after the papers [3, 4, 10, 12] (see also the bibliography therein). We mean the following result, which will be called the main theorem.

Theorem *Let* $t \ge 1$, $F \in C^t(O; \mathbb{R}^n)$, $N_i = \{x \in O: \text{rank } F'(x) = i\}$, *where* $0 \le i < \min\{m, n\}$. *Then* $\mu_p(F(N_i)) = 0$ *for* $p = i + \frac{m-i}{t}$.

This gives the following estimates on the massiveness of the set of all critical values:

$$\text{if } m \le n, \quad \text{then } \mu_p(F(N)) = 0 \quad \text{for } p = m - 1 + \frac{1}{t},$$

$$\text{if } m \ge n, \quad \text{then } \mu_p(F(N)) = 0 \quad \text{for } p = n - 1 + \frac{1}{t}(m - n + 1).$$

As we will see in Section 4, if t is a positive integer, then the theorem remains valid not only for t-smooth maps, but also for $(t - 1)$-smooth maps for which all derivatives of order $t - 1$ of all coordinate functions are locally Lipschitz.

Several subsequent sections are devoted to the proof of the main theorem, while in the last two sections we discuss the sharpness of its assumptions. The proof of the theorem and refinements thereof requires much preparatory work. Meanwhile, the case $r = 1$ is quite simple. To analyze this case, it suffices to use rather elementary considerations, which will be presented in the next section.

1.3 Critical Values of C^1-Maps

For $t = 1$, the value of p in the main theorem is equal to m (for every i). If $m > n$, then the conclusion of the theorem is trivial, since the measure μ_m vanishes on all subsets in $\mathbb{R}^n$. If $t = 1$ and $m \leq n$, then the theorem can be proved without using Morse's ideas necessary for the case of smoother maps.

Theorem *Let $F \in C^1(O; \mathbb{R}^n)$, where $O \subset \mathbb{R}^m$ and $m \leq n$. Then $\mu_m(F(\mathcal{N})) = 0$.*

Proof Since the set O is open, it can be exhausted by a countable family of closed cubes entirely contained in it. Hence, it suffices to show that $\mu_m(F(\mathcal{N} \cap Q)) = 0$ for every closed cube Q in O.

First, for $x, a \in Q$ we estimate the deviation of the map $F(x)$ from the affine map $F_0(x) = F(a) + F'(a)(x - a)$. By Corollary 1 from Lagrange's inequality (see Section I.5.4),

$$\|F(x) - F_0(x)\| = \|F(x) - F(a) - F'(a)(x - a)\|$$
$$\leq \sup_{0 \leq u \leq 1} \|F'(a + u(x - a)) - F'(a)\| \, \|x - a\| \leq \omega(\|x - a\|) \, \|x - a\|, \qquad (1)$$

where (by the uniform continuity on Q of the elements of the matrix F')

$$\omega(t) = \sup_{\substack{x, a \in Q, \\ \|x - a\| \leq t}} \|F'(x) - F'(a)\| \xrightarrow[t \to 0]{} 0.$$

Now we estimate the measure of the set $F(e)$ if e contains a critical point a, $e \subset Q$, and $\operatorname{diam}(e) < \delta$. First of all, note that the F-image of e is small: for $x \in e$ we have

$$\|F(x) - F(a)\| \leq L\|x - a\| \leq L\delta$$

where L is the Lipschitz constant of F on the cube Q. On the other hand, since a is a critical point, $\operatorname{rank} F'(a) < m$; hence, the values of the affine map $F_0(x)$ lie in an $(m - 1)$-dimensional subspace E of $\mathbb{R}^n$, and $F(x)$ deviates little from this subspace: inequality (1) implies that for $x \in e$ we have

$$\operatorname{dist}(F(x), E) \leq \|F(x) - F_0(x)\| \leq \omega(\delta) \, \delta.$$

Therefore, $F(e)$ is contained in a thin "layer" near the subspace E, and for small δ the "thickness" $2\omega(\delta)\,\delta$ of this layer is considerably less than δ, because $\omega(\delta) \xrightarrow[\delta \to 0]{} 0$.

Thus, $F(e)$ is contained in a bicylinder C, which is the direct sum $B \oplus B'$ where $B \subset E$ is the $(m - 1)$-dimensional ball of radius $L\delta$ centered at $F(a)$ and B' is the $(n - m + 1)$-dimensional ball of radius $\omega(\delta)\delta$ centered at the origin that lies in the orthogonal complement to E. For all sufficiently small δ, the radius of B is considerably larger than that of B', hence for such δ, in order to refine the estimate of the measure of C, we must divide the larger ball into subsets whose diameters do not exceed the diameter of B', equal to $2\omega(\delta)\delta$. Since the ball B lies in an $(m - 1)$-dimensional subspace and its radius is equal to $L\delta$, we can achieve this by dividing B into at most $A_1/(\omega(\delta))^{m-1}$ subsets of diameter at most $2\omega(\delta)\delta$ (hereafter, $A_1, A_2, \ldots$ are coefficients depending only on F, Q, and the dimensions m, n). As a result, the bicylinder C gets divided into the same number of small cylinders of diameter at most $A_2\omega(\delta)\delta$. Hence, for the measure μ_m of the set $F(e)$ we obtain the following estimate:

$$\mu_m(F(e)) \leq \frac{A_1}{(\omega(\delta))^{m-1}} \left(A_2\omega(\delta)\,\delta\right)^m = A_3\,\omega(\delta)\,\delta^m. \tag{2}$$

Now take a positive integer N and divide the set $Q \cap \mathcal{N}$ into at most $A_4 N^m$ subsets $e_1, e_2, \ldots$ of diameter less than $\frac{1}{N}$ (obviously, this can be done for the cube Q, and hence for a subset of Q). By (2) with $\delta = \frac{1}{N}$, the measure of the set $F(e_k)$ satisfies the inequality

$$\mu_m(F(e_k)) \leq A_3\,\omega\!\left(\frac{1}{N}\right)\!N^{-m},$$

and thus

$$\mu_m(F(\mathcal{N} \cap Q)) \leq \sum_k \mu_m(F(e_k)) \leq A_4\,N^m\,A_3\,\omega\!\left(\frac{1}{N}\right)\!N^{-m} = A_3 A_4\,\omega\!\left(\frac{1}{N}\right)\!.$$

This implies that

$$\mu_m(F(\mathcal{N} \cap Q)) = 0,$$

since $\omega\!\left(\frac{1}{N}\right) \xrightarrow[N \to \infty]{} 0.$ $\qquad\qquad\square$

2 *Well-Positioned Manifolds

To estimate the measure of the set of critical values of a map, we must learn to prove that increments of its coordinate functions at critical points are sufficiently small. Due to Lagrange's inequality, this can easily be done in the case where the gradient of the coordinate function at the point in question vanishes. However, we will also need to estimate increments of coordinate functions at points where only some of the partial derivatives vanish, which makes the problem much more complicated. As we will see, it suffices to restrict attention to the special case where all critical points under consideration lie in manifolds that are sufficiently well positioned relative to the coordinate axes, which will allow us to obtain the required result using Lemma 2 from Section 2.2.

A major difficulty in the proof of the main theorem is proving it for a special kind of maps, namely, maps that leave the first few coordinates unchanged. Accordingly, we single out a class of smooth manifolds that are convenient to use in this special case.

With this in mind, from the very beginning we fix an integer i, $0 \leq i < m$, and divide the coordinates of $\mathbb{R}^m$ into two disparate groups. The first one consists of the first i coordinates, and the second one, of the remaining coordinates (if $i = 0$, the first group is empty).

2.1 The Main Definition

To begin with, we are going to describe the above-mentioned "good" manifolds. Below we assume that i is a fixed nonnegative integer $(i < m)$ and identify the space $\mathbb{R}^m$ with the Cartesian product $\mathbb{R}^i \times \mathbb{R}^{m-i}$, so that every pair (x, y), where

$$x = (x_1, \ldots, x_i) \in \mathbb{R}^i \quad \text{and} \quad y = (y_1, \ldots, y_{m-i}) \in \mathbb{R}^{m-i},$$

is identified with the point $z = (x_1, \ldots, x_i, y_1, \ldots, y_{m-i}) \in \mathbb{R}^m$. We assume (padding the sequence of coordinates with zeros as necessary) that each of the spaces $\mathbb{R}^1, \mathbb{R}^2, \ldots, \mathbb{R}^{m-1}, \mathbb{R}^m$ is a subspace of the subsequent ones. Similarly, for $i < d < m$ we identify a vector from $\mathbb{R}^d$ with a pair (x, v) where $x \in \mathbb{R}^i$, $v \in \mathbb{R}^{d-i}$, and a vector $(x, y) \in \mathbb{R}^m$, with the triple (x, y', y'') where y' and y'' are the projections of y to $\mathbb{R}^{d-i}$ and $\mathbb{R}^{m-d}$, respectively.

Let φ be a smooth map from an open subset G in $\mathbb{R}^d$ to $\mathbb{R}^{m-d}$. Recall (see Section III.2.1) that the subset in $\mathbb{R}^m = \mathbb{R}^d \times \mathbb{R}^{m-d}$ consisting of all points of the form $(u, \varphi(u)), u \in G$, is called the graph of φ, and the map $u \mapsto (u, \varphi(u))$ is called the canonical parametrization of this graph.

To single out the above-mentioned class of "good" manifolds, we introduce the following definition.

Definition Let i be a nonnegative integer, $i < m$. A manifold[1] M in $\mathbb{R}^m$ is said to be *well positioned* if $d = \dim M > i$ and M coincides with the graph of a smooth map defined in an open subset of $\mathbb{R}^d$ or can be obtained from such a graph by a permutation of the last $m - i$ coordinates.

We emphasize that the definition allows only permutations of the last $m - i$ coordinates leaving the first i ones unchanged. Thus, a well-positioned manifold can be viewed as a wide-sense graph of a map ψ that depends on the first i coordinates and some $d - i$ of the remaining $m - i$ coordinates. The subspace corresponding to the first i coordinates is always included in the subspace containing the domain of definition of ψ.

In particular, for $i = 0$ a well-positioned manifold is just a (wide-sense) graph in $\mathbb{R}^m$ of a smooth map.

For consistency, we declare that any manifold of dimension m, i.e., any open subset in $\mathbb{R}^m$, is well positioned, meaning that its canonical parametrization is the identity map.

To illustrate the notion of a well-positioned manifold, we indicate that (for $m = 3$, $d = 2, i = 1$) the parabolic cylinders $z = (x-y)^2$ and $y = (x-z)^2$ are well positioned,

---

[1] Unless otherwise stated, all manifolds under consideration are assumed to be smooth.

while the surface obtained from them by permuting x and z (in the first case) and x and y (in the second case) does not have this property.

Remark By definition, the notion of a well-positioned manifold depends on the integer i, hence it would be more correct to call it a i-well-positioned manifold. But since we do not have to simultaneously consider well-positioned manifolds with different values of i, we omit this parameter and use the shorter term.

2.2 Two Lemmas About Increments

As usual, by a smooth map on a (not necessarily open) set we mean a map that is smooth in some neighborhood of this set.

First, we deduce a simple corollary from Lagrange's inequality.

Lemma 1 Let $t \geq 1$, let M be a smooth manifold in $\mathbb{R}^m$, and let f be a smooth function on M. If $c \in M$ and

$$\| \operatorname{grad} f(z)\| = O(\|z - c\|^{t-1}) \quad as \ z \to c, z \in M, \tag{1}$$

then

$$f(z) - f(c) = O(\|z - c\|^t) \quad as \ z \to c, z \in M.$$

It is clear from the proof that the O-symbols in the lemma can be simultaneously replaced by o-symbols.

As we will see below, to establish a property of a smooth function f on a manifold, it often suffices to establish the corresponding property of an "associate" function: the composition of f and a parametrization of the manifold. Now we are going to use exactly this trick.

Proof Let Φ be a smooth parametrization of an M-neighborhood of c defined in a ball centered at the origin and $c = \Phi(0)$.

By (1), the gradient of the composition $g = f \circ \Phi$ satisfies the relation

$$\| \operatorname{grad} g(u)\| \leq \| \operatorname{grad} f(\Phi(u))\| \, \|\Phi'(u)\| = O(\|\Phi(u) - \Phi(0)\|^{t-1}) \underset{u \to 0}{=} O(\|u\|^{t-1})$$

(since $\|\Phi'(u)\| \underset{u \to 0}{=} O(1)$). Hence, the required result can be obtained by applying Lagrange's inequality to g. Indeed, for $z = \Phi(u)$ this inequality ensures that

$$|f(z) - f(c)| = |g(u) - g(0)| \leq \| \operatorname{grad} g(\theta u)\| \cdot \|u\| \underset{u \to 0}{=} O(\|u\|^t)$$

for some $\theta = \theta(u) \in (0, 1)$. It remains to observe that since the map Φ^{-1} satisfies the local Lipschitz condition (see the remark at the end of Section III.2.2), we have

$$\|u\| = \|\Phi^{-1}(z)\| = \|\Phi^{-1}(z) - \Phi^{-1}(c)\| = O(\|z - c\|) \quad as \ z \to c, z \in M$$

and, therefore,

$$|f(z) - f(c)| = O(\|u\|^t) = O(\|z - c\|^t) \quad \text{as } z \to c, z \in M. \qquad \square$$

In the next lemma, we also estimate the increment of a smooth function "along" a manifold, but with the important difference that a restriction is imposed only on some partial derivatives and not on all of them. With the assumptions weakened in this way, the required estimate can be obtained only for well-positioned manifolds, which, however, suffices for our purposes.

Given a smooth function f defined on an open subset of $\mathbb{R}^m = \mathbb{R}^i \times \mathbb{R}^{m-i}$, by $f_y'(x, y)$ we denote the projection of its gradient to the second factor in the Cartesian product, i.e., the vector $\left(\frac{\partial f}{\partial y_1}(x, y), \ldots, \frac{\partial f}{\partial y_{m-i}}(x, y)\right)$.

Lemma 2 *Let M be a well-positioned manifold. If f is a smooth function on M that at a point $(a, b) \in M$ satisfies the condition*

$$\|f_y'(a, y)\| = O(\|y - b\|^{t-1}) \quad as \ y \to b, (a, y) \in M \qquad (2)$$

for some $t \geq 1$, then

$$f(x, y) - f(a, b) = O(\|x - a\| + \|y - b\|^t) \quad as \ (x, y) \to (a, b), (x, y) \in M.$$

As in the proof of Lemma 1, we use the associate function g, which allows us to reduce the problem of estimating the increment of f on M to that of estimating the increment of g without additional conditions.

Note that the assumption that M is well positioned is indispensable in Lemma 2; without it, the lemma is no longer true (see the remark after the proof).

Proof Let dim $M = d$, and let Φ be the canonical parametrization of M defined in an open subset G of $\mathbb{R}^d = \mathbb{R}^i \times \mathbb{R}^{d-i}$. Points of the space $\mathbb{R}^d$ will be identified with pairs (x, v) where $x \in \mathbb{R}^i, v \in \mathbb{R}^{d-i}$. Thus,

$$\Phi(x, v) = (x, \Psi(x, v)), \quad \text{where } (x, v) \in G, \ \Psi \in C^1(G; \mathbb{R}^{m-i}).$$

Since Φ does not change the first i coordinates, we have $\Phi^{-1}(a, b) = (a, \widetilde{b})$ where $\widetilde{b} \in \mathbb{R}^{d-i}$.

Now we "transplant f to G" by introducing the associate function $g = f \circ \Phi$. By the chain rule,

$$g_v'(x, v) = f_y'(\Phi(x, v)) \cdot \Phi_v'(x, v).$$

Hence, by (2) we have

$$g_v'(a, v) = O(\|f_y'(\Phi(a, v))\|) = O(\|\Psi(a, v) - b\|^{t-1})$$
$$= O(\|\Psi(a, v) - \Psi(a, \widetilde{b})\|^{t-1}) = O(\|v - \widetilde{b}\|^{t-1}) \quad \text{as } v \to \widetilde{b}.$$

Applying the last estimate to the function $v \mapsto g(a, v)$ (defined in a neighborhood of $\widetilde{b}$) and using Lagrange's inequality, we obtain

$$g(a, v) - g(a, \widetilde{b}) = O(\|v - \widetilde{b}\|^t) \quad \text{as } v \to \widetilde{b}. \qquad (3)$$

Now we estimate the increment $f(x, y) - f(a, b)$, where $(x, y) \in M$. Set $\Phi^{-1}(x, y) = (x, \widetilde{y})$. Clearly, $(x, \widetilde{y}) \in G$ for every point (x, y) sufficiently close to (a, b), and

$$f(x, y) - f(a, b) = g(x, \widetilde{y}) - g(a, \widetilde{b}) = g(x, \widetilde{y}) - g(a, \widetilde{y}) + g(a, \widetilde{y}) - g(a, \widetilde{b})$$
$$= O(\|x - a\|) + g(a, \widetilde{y}) - g(a, \widetilde{b}).$$

Since Φ^{-1} satisfies the local Lipschitz condition, we use (3) to estimate the last difference and conclude that for $(x, y) \in M$, $(x, y) \to (a, b)$,

$$f(x, y) - f(a, b) = O(\|x - a\|) + O(\|\widetilde{y} - \widetilde{b}\|^t) = O(\|x - a\|) + O(\|y - b\|^t),$$

as required. $\square$

Remark Let $m = 3$, $d = 2$, $i = 1$, $t > s > 1$. Consider the cylindrical manifold

$$M = \{(x, y, z) \in \mathbb{R}^3 : x = |y - z|^t\}$$

(which is not well positioned) and the function $f(x, y, z) = |y - z|^s$. Analyzing the behavior of f on the cross section of M by the plane $z = -y$ as $(x, y, z) \to 0$, one can easily see that the conclusion of Lemma 2 does not hold, so the assumption of this lemma that the manifold is well positioned cannot be dropped.

3 *Morse's Theorem on t-Representations

3.1 Preliminaries

The proofs of the theorems stated in Section 1.2 rely on the fact that the increments of a map $F : O \to \mathbb{R}^n$ at critical points are "sufficiently small." The corresponding general results about estimation of the Hausdorff measure of the F-image of a set E taking into account the behavior of $F(x)$ as x approaches a point $x_0 \in E$ "from different directions" are given in Section V.5. Here, we show that every subset of a Euclidean space has a special structure which makes it possible to use these general results in our problem.

We begin with an informal discussion of Morse's idea underlying the proof of the main result of this section. In an effort to avoid technical complications, for the moment we restrict ourselves to the principal special case, namely, consider not arbitrary maps, but only smooth functions of several variables (it is for this case that Morse himself obtained his remarkable result).

At points of the critical set N, the differential of a smooth function f vanishes, hence the increment of f at these points tends to zero faster than the increment of x. However, if $m > 1$, this still does not guarantee that $\lambda_1(f(N)) = 0$. It is easy to obtain such an equality for an m-smooth function f if we replace the set N of all critical points by the set N_* of critical points at which not only the first differential of f vanishes, but all its differentials up to order m. At these points, the increment of f coincides with the remainder in Taylor's formula (of order m), and hence

$$f(x) - f(a) \underset{x \to a}{=} o(\|x - a\|^m) \quad \text{for } a \in \mathcal{N}_*. \tag{1}$$

It easily follows that the f-image of the set $\mathcal{N}_*$ has zero Lebesgue measure. Nevertheless, our task is obviously still very far from completed. Indeed, Morse's theorem says that the image of the entire critical set has zero measure, and not only of the set $\mathcal{N}_*$ of "strongly critical" points.

To obtain Morse's result in its entirety, we must take into account two additional considerations. First, relation (1) provides an estimate for the increment of f as x deviates from a in all directions. However, this requirement is too strong for our purposes: such an estimate is not always needed. Indeed, we are interested in the image of the critical set $\mathcal{N}$, and hence in the values of $f(x) - f(a)$ for $x \in \mathcal{N}$. For example, if $\mathcal{N}$ is contained in some coordinate subspace L (for definiteness, let it be spanned by the first k coordinates), then we can reduce the problem to the case of a smaller number of variables, since the increments of f in directions orthogonal to L are irrelevant, and hence the increments of $x_{k+1}, \dots, x_m$ can be assumed to be zero. In this case, it is important that the higher derivatives vanish with respect to the first k coordinates only, and that the corresponding increments satisfy an analog of condition (1). Obviously, this is possible even if not all higher differentials of f are identically zero on $\mathcal{N}$. We face a similar situation in the more general case where the set $\mathcal{N}$ of critical points of f is contained in a smooth manifold M of dimension less than m. If Φ is a parametrization of M, then, introducing the function $\widetilde{f} = f \circ \Phi$, which again will be called the associate function of f, we reduce the problem to the case of a smaller number of variables. Indeed, if a is a critical point of f, then $a \in M$ and $\widetilde{a} = \Phi^{-1}(a)$ is a critical point of $\widetilde{f}$, with $f(a) = \widetilde{f}(\widetilde{a})$. Hence, the set $f(\mathcal{N})$ of critical values of f coincides with the set of critical values of $\widetilde{f}$. This allows us, instead of estimating the measure of the former set, estimate the measure of the latter one. Since a parametrization is a locally bi-Lipschitz map, relation (1) for $\widetilde{f}(u) - \widetilde{f}(\widetilde{a})$ as $u \to \widetilde{a}$ is satisfied simultaneously with the weakened relation (1) for f, when it holds under the additional assumption that $x \in M$:

$$f(x) - f(a) \underset{x \to a}{=} o(\|x - a\|^m) \quad (a \in \mathcal{N}, \ x \in M). \tag{1'}$$

Of course, this assumption that condition (1) is satisfied only "along M" is much less restrictive than condition (1) in its entirety.

Thus, the first idea to guide us in what follows is to use the associate function in order to decrease the number of variables and weaken condition (1). However, there is an obvious obstacle to making this idea work: the critical set is not necessarily contained in a smooth manifold of smaller dimension. In the two-dimensional situation, this is already the case when this set consists of two intersecting segments.

To overcome this difficulty, we bring in yet another consideration, namely, that it is not at all necessary to require that condition (1') be satisfied with the same M for all points of the critical set. Since a countable union of sets of zero measure also has zero measure, it suffices to represent N as an at most countable union of parts each having its own ambient manifold. Then, using the above argument for each part, we conclude that the f-images of these parts have zero measure.

Of course, it is not at all obvious that the critical set can be represented as a countable union of parts that can be immersed into "good" ambient manifolds, which is what makes the main result of this section quite unexpected.

The above considerations are formalized in the definition below, where we consider smooth functions of class C^t for every $t > 1$, not only for $t = m$ as in (1) and (1').

This is a preliminary definition, which applies only to functions and not to maps. Its purpose is to clarify the heuristic arguments discussed above. In the next subsection, we will introduce a more general definition used in what follows.

Definition (preliminary) Let $t > 1$, and let $E \subset \mathbb{R}^m$. We say that sets $e, E_1, E_2, \ldots$ form a *t-representation* of E if $E = e \cup E_1 \cup E_2 \cup \ldots$ where[1] $\dim_H(e) = 0$ and each set E_j is contained in a smooth manifold $M_j \subset \mathbb{R}^m$ such that

for every function f that is constant on E and t-smooth in a neighborhood of E, at every point $a \in E_j$ we have

$$f(x) - f(a) = O(\|x - a\|^t) \quad \text{as } x \to a, x \in M_j.$$

The family of sets E_j may be infinite, finite, or even empty.

For $t = r + 1$ with $r \in \mathbb{N}$, the last relation must be satisfied not only for t-smooth functions, but also for functions of class LC^r.

As we will see, using a t-representation of the critical set of a sufficiently smooth function allows us to prove that the set of its critical values has zero measure. Thus, to solve our problem, we must answer a geometric question: What subsets in $\mathbb{R}^m$ have a t-representation? The answer obtained by Morse is not at all obvious: it turns out that every subset of a Euclidean space has such a representation. Together with the above considerations leading to the notion of a t-representation, this opens the way for the first nontrivial result in the problem under study, which is Morse's theorem (see Section 1.2). We will not extensively explain the argument that proves this theorem according to the scheme discussed above, because it solves only part of the problem we are interested in (for functions, but not for arbitrary maps). To solve it in full generality, we must invoke a somewhat more complicated version of this argument suggested by Federer. This will be done in the subsequent subsections.

3.2 t-Representations

For $m, n > 1$, the critical set of a smooth map $F: O \to \mathbb{R}^n$ can be naturally divided into the subsets N_i where the rank of $F'(x)$ is constant and equal to i ($i = 0, 1, \ldots < \min\{m, n\}$). In what follows, we consider the images of these

---

[1] For the symbol $\dim_H(e)$, see Section V.3.5.

"fibers" separately and estimate the "massiveness" not only of the set of critical values "as a whole," but also of its parts $F(N_i)$. For this reason, the definition below involves a fixed integer parameter $i \geq 0$, the space $\mathbb{R}^m$ is identified with the Cartesian product $\mathbb{R}^i \times \mathbb{R}^{m-i}$, and its points, with pairs (x, y) where $x \in \mathbb{R}^i$, $y \in \mathbb{R}^{m-i}$.

Definition Let i be a nonnegative integer, $m \geq i$, $t > 1$, and let $E \subset \mathbb{R}^m$. We say that sets $e, E_1, E_2, \ldots$ form a t-*representation* of E if $E = e \cup E_1 \cup E_2 \cup \ldots$ where $\dim_H(e) \leq i$ and each set E_j (the family of these sets may be infinite, finite, or even empty) is contained in a well-positioned manifold $M_j \subset \mathbb{R}^m$ having the following property:

for every function f that is constant on E and t-smooth in a neighborhood of E, at every point $(a, b) \in E_j$ ($a \in \mathbb{R}^i$, $b \in \mathbb{R}^{m-i}$) we have

$$f(x, y) - f(a, b) = O(\|x - a\| + \|y - b\|^t)$$
$$\text{as } (x, y) \to (a, b), \ (x, y) \in M_j. \quad (2)$$

For $t = r + 1$ with $r \in \mathbb{N}$, this relation must be satisfied not only for t-smooth functions, but also for functions of class LC^r.

We draw the reader's attention to the fact that whatever t may be, the ambient manifolds M_j are assumed to be only 1-smooth.

For $i = 0$, this definition coincides with the preliminary one: in the left-hand side of (2), we must replace $f(x, y)$ and $f(a, b)$ by $f(y)$ and $f(b)$, and in the right-hand side, remove the term $\|x - a\|$.

Note that every set E whose Hausdorff dimension is at most i forms a t-representation of itself ($e = E$, and there are no E_j). In particular, for $i = m$ every set in $\mathbb{R}^m$ has a t-representation, because its Hausdorff dimension is at most m.

In the definition it is assumed that $t > 1$, because for $t = 1$ relation (2) is trivial: it holds for every function that is smooth (and even locally Lipschitz) in a neighborhood of E.

As in the definition of a well-positioned manifold (see Remark 2.1), when speaking about t-representations, for brevity we omit indicating the parameter i.

Remark 1 The union of an at most countable family of subsets in $\mathbb{R}^m$ each having a t-representation also has such a representation. Indeed, since the Hausdorff dimension of the union of a sequence of sets coincides with the supremum of their Hausdorff dimensions (see formula (2) at p. 261), the proof consists in renumbering the parts forming the t-representations of the subsets.

Remark 2 Let $i < m$, let $M \subset \mathbb{R}^m$ be a well-positioned manifold of class C^t (or LC^r), and let Φ be its canonical parametrization. Every function f that is t-smooth (or LC^r-smooth) in a neighborhood of a point $c \in M$ satisfies condition (2) if and only if the associate function $\widetilde{f} = f \circ \Phi$ satisfies the analogous condition in a neighborhood of the point $\widetilde{c} = \Phi^{-1}(c)$ (since Φ and Φ^{-1} are locally Lipschitz maps that leave the first i coordinates unchanged). It follows that sets $e, E_1, E_2, \ldots$ form a t-representation of a set E in M if and only if their Φ-preimages form a t-representation of $\Phi^{-1}(E)$ (the role of the ambient manifold for all sets $\Phi^{-1}(E_j)$ is played by the domain of definition of the parametrization Φ, i.e., the set $\Phi^{-1}(M)$).

In particular, if M is a well-positioned graph of a function of $m - 1$ variables defined in a domain G and Φ is its canonical parametrization, then the preimages of points of this graph coincide with their orthogonal projections to the coordinate subspace $\mathbb{R}^{m-1}$, and sets $e, E_1, E_2, \ldots$ form a t-representation of a set E in M if and only if their orthogonal projections form a t-representation of the projection of E, i.e., the set $\Phi^{-1}(E)$.

3.3 The Existence of a t-Representation

The following theorem, which is of fundamental importance for what follows, was proved by Morse in the case $t \in \mathbb{N}$, $i = 0$. The generalization involving well-positioned manifolds is due to Federer; the generalization to the case of fractional smoothness is proved in [3, 4, 10, 12].

Theorem *Let i be a nonnegative integer. For every $m \geq i$ and $t > 1$, every set $E \subset \mathbb{R}^m$ has a t-representation.*

Proof Set $\alpha = t - r$ where r is the greatest integer less than t, so[2] $0 < \alpha \leq 1$. Recall that, according to our definition of a t-representation for integer t (i.e., for $\alpha = 1$), condition (2) must be satisfied not only for functions of class C^t, but also for $(t - 1)$-smooth functions whose higher derivatives satisfy the local Lipschitz condition (functions of class LC^{t-1}).

We proceed by induction on the parameter $n = m + r$, assuming that i and α are fixed.

I. First, we prove the base case. If $i \geq 1$, then it is the conclusion of the theorem for $n = i + 1$ (i.e., for $m = i$ and $r = 1$). This conclusion is true because, as we have already observed, a t-representation for $m = i$ does always exist (for every t). This allows us to assume below that $m > i$.

If $i = 0$, then the base case is the conclusion of the theorem for $m = r = 1$, i.e., for $m = 1$ and $1 < t \leq 2$. To prove it, we represent an arbitrary set $E \subset \mathbb{R}$ in the form $E = e \cup E_1$ where e is the set of isolated points of E, $E_1 = E \setminus e$, and the ambient manifold for E_1 coincides with $\mathbb{R}$.

The Hausdorff dimension of the set e is zero, because it is at most countable. Since the set E_1 consists of accumulation points of E and $E_1 \subset E$, the derivative of every function f of class C^t that is constant on E vanishes on E_1. Besides, f' satisfies the local Lipschitz condition of order $t - 1$. Hence, condition (2), which here takes the form

$$f(y) - f(b) = O(|y - b|^t) \quad \text{as } y \to b,$$

follows from the mean value theorem. Indeed, for y sufficiently close to b, there is a point $\bar{y}$ between them such that

---

[2] In other words, r is the entire part of t for fractional t and $r = t - 1$ for integer t.

$$f(y) - f(b) = f'(\bar{y})(y - b) = \big(f'(\bar{y}) - f'(b)\big)(y - b)$$
$$= (y - b) \cdot O(|\bar{y} - b|^{t-1}) = O(|y - b|^{t}).$$

Thus, the sets e and E_1 form a t-representation of E, which proves the base case for $i = 0$.

II. To prove the induction step, we must show that the conclusion of the theorem holds for a pair m, r if it holds (with the same i and α) for the pairs $m-1, r$ and $m, r-1$ in the case $r > 1$, and for the pair $m - 1, 1$ in the case $r = 1$.

Let us single out a subset of E to be considered first. A point of E will be said to be t-regular if it has a neighborhood U (in $\mathbb{R}^m$) such that the intersection $E \cap U$ is contained in a well-positioned graph of a t-smooth function φ of $m - 1$ variables. Denote the set of all t-regular points by P; we claim that it has a t-representation. First, we prove this "in the small," namely, we prove that every point of P has a neighborhood U such that the intersection $P \cap U$ has a t-representation. Let U, φ be as in the definition of a t-regular point and $\widetilde{P}$ be the projection of $P \cap U$ to the subspace containing the domain of definition of φ. By the induction hypothesis (more exactly, the conclusion of the theorem for the pair $m - 1, r$), the set $\widetilde{P}$ has a t-representation. Since φ is t-smooth, this means, as observed in Remark 1 at p. 212, that the set $P \cap U$ also has a t-representation. By Lindelöf's theorem, there exists an at most countable family of such neighborhoods U covering the whole of P. Hence, to obtain a t-representation of P, it suffices to refer to Remark 1 at p. 212.

Now, we show how a t-representation of P can be extended to a t-representation of the entire set E. For this, note that for every function f that is constant on E and t-smooth in a neighborhood of E, the vector

$$f'_y = \left(\frac{\partial f}{\partial y_1}, \ldots, \frac{\partial f}{\partial y_{m-i}} \right)$$

vanishes on the set $Q = E \setminus P$. Indeed, otherwise, by Proposition I.10.4, the constancy set of f near the point (a, b) coincides with a (wide-sense) graph of a function of class C^t. It follows from the same proposition that this graph is well positioned, which implies that $(a, b) \in P$, a contradiction. Hence, the vector f'_y can be nonzero only on P.

If $t \leq 2$, then $r = 1$, and in this case it suffices simply to add the set Q (assuming that its ambient manifold coincides with $\mathbb{R}^m$) to the t-representation of P we already have. Indeed, for every point $(a, b) \in Q$ we have

$$f(x, y) - f(a, b) = f(x, y) - f(a, y) + f(a, y) - f(a, b)$$
$$= O(\|x - a\|) + O(\|y - b\|^{t}) \quad \text{as } (x, y) \to (a, b)$$

(the estimate of the first difference in the middle part holds because f satisfies the local Lipschitz condition, and the estimate of the second difference follows from the fact that $f'_y(a, b) = 0$ and an estimate of the remainder in Taylor's formula for the function $y \mapsto f(a, y)$). Thus, adding Q to the t-representation of P, we obtain, for $t \leq 2$, a t-representation of the entire set E.

Now we claim that for $t > 2$ (which means that $r > 1$), a required representation of E can be obtained by adding a $(t - 1)$-representation of the difference $Q = E \setminus P$ (which exists by the induction hypothesis) to the constructed t-representation of P.

Let $e, E_1, E_2, \ldots$ be a $(t - 1)$-representation of Q; the Hausdorff dimension of e is at most i, and each set E_j is contained in a well-positioned manifold M. We claim that (2) holds for a function f that is constant on E and t-smooth in a neighborhood of E. As we have already mentioned, $\frac{\partial f}{\partial y_j} = 0$ on Q for all $j = 1, \ldots, m - i$. By the definition of a $(t - 1)$-representation, at every point $(a, b) \in E_j$ all these partial derivatives satisfy the relation

$$\frac{\partial f}{\partial y_j}(x, y) = O(\|x - a\| + \|y - b\|^{t-1}) \quad \text{as } (x, y) \to (a, b), \ (x, y) \in M_j.$$

In particular, the assumptions of Lemma 2 from Section 2.2 are satisfied, which ensures that (2) holds and thus completes the induction step.   □

The reader is encouraged to work out where this proof fails if in the definition of a t-representation we assume that all ambient manifolds are not only smooth, but a) t-smooth; b) infinitely smooth.

Remark The set e constructed for $i = 0$ in the first part of the proof (the set of isolated points) is at most countable. This property is preserved under the induction step, so for $i = 0$ our argument proves the existence of a t-representation with a restriction on the set e stronger than $\dim_H(e) = 0$, as in the definition of a t-representation.

4 *The Main Results

For the reader's convenience, we repeat the statement of what in Section 1 was called the main theorem, extending it to maps of class LC^r. Below, O is still an open subset in $\mathbb{R}^m$ and i is a nonnegative integer.

Theorem *Let $t \geq 1$, $F \in C^t(O; \mathbb{R}^n)$. For $0 \leq i < \min\{m, n\}$, set*

$$\mathcal{N}_i = \{x \in O : \operatorname{rank} F'(x) = i\}, \quad p = i + \frac{m - i}{t}.$$

Then $\mu_p(F(\mathcal{N}_i)) = 0$. This is also true for maps of class LC^r with $p = i + \frac{m-i}{r+1}$.

At the end of the section, we give a generalization of Theorem 1.3 to locally Lipschitz maps.

4.1 Proof of the Main Theorem

The case $t = 1$, which does not require using Morse's ideas, was already considered in Section 1.3. So, in what follows we assume that $t > 1$. We draw the reader's attention to the fact that in the case $t = r + 1$, the argument below does not require the $(r + 1)$-smoothness of F, it suffices that F belong to LC^r.

The proof is divided into three steps. At the first step, we obtain, as a special case, Morse's result discussed in Section 1.2 and its generalization for functions of fractional smoothness. At the second step, the theorem is proved for a special kind of maps, and at the third step, we reduce the general case to this special one.

I. The case $i = 0$. Then $p = \frac{m}{t}$. For the moment we assume that $t > 2$ and consider a $(t - 1)$-representation $e, E_1, E_2, \ldots$ of the set $\mathcal{N}_0$ with ambient manifolds $M_1, M_2, \ldots$. Since the set e has zero Hausdorff dimension, the same is true for its (smooth) image. Therefore, $\mu_p(F(e)) = 0$. We claim that $\mu_p(F(E_j)) = 0$ for an arbitrary positive integer j. Below, to simplify notation, we omit the subscript j and write E and M instead of E_j and M_j, respectively.

Since every first-order partial derivative (denote it by φ) of every coordinate function of F vanishes on $\mathcal{N}_0$, for every point $c \in E$ we have, by the definition of a $(t - 1)$-representation (with $i = 0$),

$$\varphi(z) = O(\|z - c\|^{t-1}) \quad \text{as } z \to c, z \in M$$

(it is here that we use the t-smoothness of F). Therefore,

$$\|F'(z)\| = O(\|z - c\|^{t-1}) \quad \text{as } z \to c, z \in M. \tag{1}$$

In the case $1 < t \leq 2$ (where we cannot speak of a $(t - 1)$-representation of $\mathcal{N}_0$), relation (1) (with M replaced by $\mathbb{R}^m$) holds on the entire set $\mathcal{N}_0$, because the coordinate functions of F are smooth by assumption. This allows us to consider below not only $t > 2$, but any $t > 1$.

Now we use the map "associate" with the given one. Let Φ be a parametrization of M, $d = \dim M$, and $\widetilde{O}$ be an open subset in $\mathbb{R}^d$ in which the parametrization is defined. Still assuming that $c \in E \subset M$, set

$$\widetilde{F} = F \circ \Phi, \quad \widetilde{E} = \Phi^{-1}(E), \quad \widetilde{c} = \Phi^{-1}(c).$$

Then $\widetilde{F} \in C^1(\widetilde{O}; \mathbb{R}^n)$ and $\widetilde{F}(\widetilde{E}) = F(E)$. Since the maps Φ and Φ^{-1} are Lipschitz near the points $\widetilde{c}$ and c, respectively, (1) is equivalent to

$$\|\widetilde{F}'(u)\| = O(\|u - \widetilde{c}\|^{t-1}) \quad \text{as } u \to \widetilde{c}.$$

Hence, as follows from Lagrange's inequality for maps, we have

$$\|\widetilde{F}(u) - \widetilde{F}(\widetilde{c})\| = O(\|u - \widetilde{c}\|^t) \quad \text{as } u \to \widetilde{c}.$$

Besides, due to Lemma 2 at p. 276 (with the difference that instead of g we must consider the coordinate functions of $\widetilde{F}$, which are defined in an open subset of a d-dimensional space, not an m-dimensional one), for almost all points of $\widetilde{E}$ with respect to the Lebesgue measure in $\mathbb{R}^d$, we have

$$\|\widetilde{F}(u) - \widetilde{F}(\widetilde{c})\| \underset{u \to \widetilde{c}}{=} o(\|u - \widetilde{c}\|^t).$$

Therefore, $\widetilde{E}$ can be divided into two parts: a set e_0 of zero measure and the set H at which the last relation holds.

Now we apply Corollaries 1 and 2 at p. 274: on e_0, the assumptions of Corollary 1 are satisfied, and on H, the assumptions of Corollary 2 with $i = 0$. Hence, by these corollaries, $\mu_{d/t}(\widetilde{F}(e_0)) = 0$, $\mu_{d/t}(\widetilde{F}(H)) = 0$, and hence

$$\mu_{d/t}(\widetilde{F}(\widetilde{E})) = \mu_{d/t}(\widetilde{F}(e_0)) + \mu_{d/t}(\widetilde{F}(H)) = 0.$$

Taking into account that $d \le m$ and $F(E) = \widetilde{F}(\widetilde{E})$, we obtain

$$\mu_{m/t}(F(E)) = \mu_{m/t}(\widetilde{F}(\widetilde{E})) = 0,$$

as required. So, the F-images of both all the sets E_j and the set e have zero μ_p-measure, hence $\mu_p(F(\mathcal{N}_0)) = 0$. This completes the first part of the proof.

In the case $n = 1$, i.e., for functions of several variables, the proof would end here.

II. Now we turn to the case $i \ge 1$. Here we will prove the conclusion of the theorem for maps leaving the first i coordinates unchanged. Let

$$F(x, y) = (x, \Theta(x, y)) \quad \text{for } (x, y) \in O,$$

where $\Theta \in C^t(O; \mathbb{R}^{n-i})$, $x \in \mathbb{R}^i$, $y \in \mathbb{R}^{m-i}$. Obviously,

$$F'(x, y) = \begin{pmatrix} \mathbb{I} & \mathbb{O} \\ \Theta'_x(x, y) & \Theta'_y(x, y) \end{pmatrix},$$

where $\mathbb{I}$ is the $i \times i$ identity matrix, $\mathbb{O}$ is the zero matrix, while Θ'_x and Θ'_y are the left and right parts of the matrix Θ' corresponding to differentiating with respect to the coordinates of the vectors x and y, respectively. In particular (below $\Theta_1, \ldots, \Theta_{n-i}$ are the coordinate functions of Θ),

$$\Theta'_y = \begin{pmatrix} \dfrac{\partial \Theta_1}{\partial y_1} & \cdots & \dfrac{\partial \Theta_1}{\partial y_{m-i}} \\ \cdots\cdots\cdots\cdots\cdots\cdots \\ \dfrac{\partial \Theta_{n-i}}{\partial y_1} & \cdots & \dfrac{\partial \Theta_{n-i}}{\partial y_{m-i}} \end{pmatrix}.$$

Clearly, the rank of F' is not less than i everywhere in O, and it is greater than i if the matrix Θ'_y is nonzero. But on $\mathcal{N}_i$ this rank is equal to i, hence the matrix Θ'_y is zero everywhere on $\mathcal{N}_i$.

The further reasoning essentially reproduces (with minor additions) the reasoning from the first step of the proof. Assuming for the moment that $t > 2$, consider a $(t-1)$-representation $e, E_1, E_2, \ldots$ of the set $\mathcal{N}_i$, and let $M_1, M_2, \ldots$ be the corresponding well-positioned ambient manifolds. Since being well-positioned is understood according to the value of i which we have fixed, the dimensions of M_j are greater than i. Taking into account that $\dim_H(e) \leq i$ and $p > i$, we conclude that $\mu_p(F(e)) = 0$. Thus, it remains to prove that $\mu_p(F(E_j)) = 0$ for arbitrary $j \geq 1$. Below, as at the first step of the proof, we use an abbreviated notation and omit the subscript j.

Since $(\Theta_k)'_y = 0$ on $\mathcal{N}_i$ for $k = 1, \ldots, n-i$, it follows from the definition of a $(t-1)$-representation that at every point $(a, b) \in E$ the following relation holds (cf. (1)):

$$(\Theta_k)'_y(a, y) = O(\|y - b\|^{t-1}) \quad \text{as } y \to b, \, (a, y) \in M. \tag{2}$$

This estimate (with M replaced by $\mathbb{R}^m$) is valid on the whole of $\mathcal{N}_i$ also in the case $1 < t \leq 2$ (when we cannot speak of a $(t-1)$-representation), since in this case it follows from the degree of smoothness of the functions Θ_k. This allows us to consider below not only $t > 2$, but any $t > 1$.

Let Φ be the canonical parametrization of M, $d = \dim M$, and $\widetilde{O}$ be the open subset in $\mathbb{R}^d$ in which the parametrization is defined. Identifying the space $\mathbb{R}^d$ with $\mathbb{R}^i \times \mathbb{R}^{d-i}$, we write its points in the form (x, v) where $x \in \mathbb{R}^i$, $v \in \mathbb{R}^{d-i}$. Since Φ is a canonical parametrization, both Φ^{-1} and Φ leave the first i coordinates unchanged. Hence, the preimage of a point $(x, y) \in M$ has the form $(x, \widetilde{y})$.

Still assuming that $(a, b) \in E \subset M$, set

$$\widetilde{F} = F \circ \Phi, \quad \widetilde{\Theta} = \Theta \circ \Phi, \quad \widetilde{E} = \Phi^{-1}(E), \quad (a, \widetilde{b}) = \Phi^{-1}(a, b).$$

Clearly, $F(E) = \widetilde{F}(\widetilde{E})$.

We will estimate $\widetilde{\Theta}'_v(a, v)$ as $v \to \widetilde{b}$ using the fact that $\Phi(a, v) \in M$. Since $\widetilde{\Theta} = \Theta \circ \Phi$ and the map Φ' is bounded near $(a, \widetilde{b})$, we have

$$\widetilde{\Theta}'_v(a, v) = \Theta'_y(\Phi(a, v)) \circ \Phi'_v(a, v) = O(\|\Theta'_y(\Phi(a, v))\|) \quad \text{as } v \to \widetilde{b}.$$

Let P be the operator projecting the product $\mathbb{R}^i \times \mathbb{R}^{m-i}$ to the second factor. Then it follows from (2) that

$$\Theta'_y(\Phi(a, v)) = O(\|P(\Phi(a, v)) - b\|^{t-1})$$
$$= O(\|P(\Phi(a, v)) - P(\Phi(a, \widetilde{b}))\|^{t-1}) = O(\|v - \widetilde{b}\|^{t-1}).$$

Therefore,

$$\widetilde{\Theta}'_v(a, v) = O(\|v - \widetilde{b}\|^{t-1}) \quad \text{as } v \to \widetilde{b}.$$

Hence, we can apply (after appropriate changes of notation) Lemma 3 at p. 277 to each coordinate function of $\widetilde{\Theta}$ and conclude that at almost every point $(a, \widetilde{b}) \in \widetilde{E}$, the map $\widetilde{\Theta}$ satisfies the relation

$$\widetilde{\Theta}(x, v) - \widetilde{\Theta}(a, \widetilde{b}) \underset{(x,v) \to (a,\widetilde{b})}{=} O(\|x - a\|) + o(\|v - \widetilde{b}\|^t)$$

(a formal interpretation of this relation is given in Corollary 2 at p. 274). Thus, as at the previous step of the proof, $\widetilde{E}$ breaks into two parts: a set e_0 of zero measure (with respect to the d-dimensional Lebesgue measure) and the set H on which the last relation holds. By Corollaries 1 and 2 from the lemma on the Hausdorff measure of an image (see Section V.5.2), we conclude that $\mu_q(\widetilde{F}(e_0)) = 0$ and $\mu_q(\widetilde{F}(H)) = 0$, where $q = i + \frac{d-i}{t}$. Hence, $\mu_q(\widetilde{F}(\widetilde{E})) = 0$. Since $q \leq p$ and $F(E) = \widetilde{F}(\widetilde{E})$, we obtain $\mu_p(F(E)) = 0$. So, the F-images of both the set e_0 and all the sets E_j have zero μ_p-measure, hence $\mu_p(F(\mathcal{N}_i)) = 0$. Therefore, for maps of the special kind we consider, the theorem is proved.

III. The end of the proof. Now we proceed to prove the equality $\mu_p(F(\mathcal{N}_i)) = 0$ in the general case. It suffices to find, for each point $c \in \mathcal{N}_i$, a neighborhood U such that the F-image of the intersection $\mathcal{N}_i \cap U$ has zero measure. For this, we will show that for a sufficiently small neighborhood U, this image coincides with the set $F_*(\mathcal{N}_*)$ where F_* is a C^t-map of the special kind already considered above (i.e., a map that leaves the first i coordinates unchanged) for which the matrix F_*' has rank i everywhere on $\mathcal{N}_*$. Then, as we have already proved, $\mu_p(F_*(\mathcal{N}_*)) = 0$, and hence $\mu_p(F(\mathcal{N}_i \cap U)) = 0$ in view of the equality $F(\mathcal{N}_i \cap U) = F_*(\mathcal{N}_*)$. After that, the proof is completed by using Lindelöf's theorem to cover $\mathcal{N}_i$ by a countable family of such neighborhoods U.

Let $c \in \mathcal{N}_i$, so, by the definition of $\mathcal{N}_i$, the rank of $F'(c)$ is i. Then, by the partial invertibility theorem (see Section II.2.2), there exist a neighborhood U of c and a diffeomorphism $\Phi : U \to \mathbb{R}^m$ of class C^t such that the C^t-smooth map

$$F_* = F \circ \Phi^{-1} : U_* \to \mathbb{R}^n,$$

where $U_* = \Phi(U)$, leaves the first i coordinates unchanged. Besides, the matrices $F_*'(w)$ and $F'(z) = F_*'(w) \cdot \Phi'(z)$, where $w = \Phi(z)$, have the same rank, because Φ is a diffeomorphism and, consequently, the matrix $\Phi'(z)$ is invertible.

Let $\mathcal{N}_*$ be the set of points of U_* at which the rank of F_*' is equal to i. Then $z \in \mathcal{N}_i \cap U$ if and only if $\Phi(z) \in \mathcal{N}_*$. Therefore, $F(\mathcal{N}_i \cap U) = F_*(\mathcal{N}_*)$. By the result obtained at the previous step, we conclude that $\mu_p(F_*(\mathcal{N}_*)) = 0$, which completes the proof. □

Remark In the case of integer smoothness (i.e., for $t = r \in \mathbb{N}$), the proof can be simplified by applying to the coordinate functions, instead of Lemma 3 (see Section V.5.3), Taylor's formula with Peano remainder term (see Section I.8.2).

4.2 Generalization to Lipschitz Maps

Here we refine the main theorem and extend its special case, Theorem 1.3, to locally Lipschitz maps. While the main idea of the proof remains the same, we will have to resort to more powerful technical tools. First of all, we must generalize the definition of a critical point, since a Lipschitz map is not necessarily differentiable. By definition, a point at which the map is not differentiable is considered critical. The reader will soon see that this does not lead to an excessive expansion of the critical set, thanks to Rademacher's[1] theorem. Another essential point (of a technical nature) is that instead of uniform continuity, which in the proof of Theorem 1.3 guaranteed the smallness of $\omega\left(\frac{1}{N}\right)$, we will have to rely on Egorov's[2] theorem. For the reader's convenience, we reproduce the statements of these two theorems (their proofs can be found, e.g., in [9, Sections IX.4.2 and III.3.6]).

Theorem (Rademacher) *A function satisfying the local Lipschitz condition on an open subset of a Euclidean space is differentiable almost everywhere (with respect to the Lebesgue measure).*

Egorov's theorem will be given in a somewhat simplified form sufficient for our purposes.

Theorem (Egorov) *Let $f_1, f_2, \ldots$ be functions defined and measurable on a set E of finite measure. If $f_j \to 0$ pointwise on E, then for every $\varepsilon > 0$ there exists a subset $e \subset E$ of measure less than ε such that on $E \setminus e$ the functions f_j converge to 0 uniformly.*

Throughout this section, by a measure we always mean the m-dimensional Hausdorff measure μ_m or the Lebesgue measure of the same dimension (which is proportional to μ_m).

By Rademacher's theorem, a map $F: O \to \mathbb{R}^n$ is differentiable almost everywhere on O if it satisfies the local Lipschitz condition on this set. In this case, the set O breaks into two parts: the set D at whose points F is differentiable and the set $\widetilde{e}$ of zero measure where F is not differentiable. The set of all critical points of F (which, by the above remark, contains $\widetilde{e}$) will still be denoted by N. The aim of this subsection is to prove the following strengthening of Theorem 1.3 (as in Section 1.3, we disregard the trivial case $m > n$, where $\mu_m = 0$ in $\mathbb{R}^n$).

Theorem *If $n \geq m$ and $F: O \to \mathbb{R}^n$ is a locally Lipschitz map, then $\mu_m(F(N)) = 0$.*

Proof Since O can be represented as an at most countable union of balls such that on each of them F satisfies the Lipschitz condition, we assume from the very beginning that O is bounded and F satisfies on O the Lipschitz condition with constant L.

Let D and $\widetilde{e}$ be the subsets of O introduced before the statement of the theorem. The set $\widetilde{e}$ has zero measure, hence $\mu_m(F(\widetilde{e})) = 0$ (see Property 2 at p. 258). It remains to study the image of the part of N contained in the set D where F is differentiable.

We introduce auxiliary functions that characterize the deviation of the increment of F from the differential at the points of D. For $a \in D$ and $j \in \mathbb{N}$, set

---

[1] Hans Adolph **Rademacher** (1892–1969) was a German mathematician.

[2] Dmitri Fyodorovich **Egorov** (1869–1931) was a Russian mathematician.

$$\omega_j(a) = \sup_{\substack{x \in O \\ 0 < \|x-a\| < \frac{1}{j}}} \frac{\|F(x) - F(a) - F'(a)(x-a)\|}{\|x-a\|}.$$

Estimating this value in the smooth case (in Theorem 1.3), we relied on Lagrange's inequality, but now we are forced to proceed in a more sophisticated way.

Clearly, the functions ω_j are measurable on D (since the supremum can be taken only over a countable set of points x dense in O). By the definition of differentiability,

$$\omega_j(a) \xrightarrow[j \to \infty]{} 0 \quad \text{at every point } a \in D.$$

Setting the number ε in Egorov's theorem to $1, \frac{1}{2}, \ldots$, we can, for each $k \in \mathbb{N}$, represent D in the form $D = e_k \cup D_k$ where $\mu_m(e_k) < \frac{1}{k}$ and on each set D_k the functions ω_j converge to 0 uniformly as $j \to \infty$. It follows that

$$D = e_0 \cup \bigcup_{k=1}^{\infty} D_k \quad \text{where } \mu_m(e_0) = 0,$$

and, consequently,

$$\mu_m(F(N)) = \mu_m\left(F\left(N \cap \left(\widetilde{e} \cup e_0 \cup \bigcup_{k=1}^{\infty} D_k\right)\right)\right)$$

$$\leq \mu_m(F(\widetilde{e})) + \mu_m(F(e_0)) + \sum_{k=1}^{\infty} \mu_m(F(N \cap D_k)) = \sum_{k=1}^{\infty} \mu_m(F(N \cap D_k)).$$

Hence, it suffices to prove that $\mu_m(F(N \cap D_k)) = 0$ for every k.

Having fixed k, consider the sequence of numbers

$$\varepsilon_N = \sup_{a \in D_k} \omega_N(a), \quad N \in \mathbb{N}.$$

Since on D_k the functions ω_N converge to 0 uniformly, it follows that $\varepsilon_N \xrightarrow[N \to \infty]{} 0$. By the definition of ω_N and ε_N, for $a, x \in D_k$ and $\|x - a\| < 1/N$ we have the inequality

$$\|F(x) - F(a) - F'(a)(x-a)\| \leq \omega_N(a)\|x-a\| \leq \frac{\varepsilon_N}{N},$$

which is similar to inequality (1) from Section 1.3 underlying the proof of Theorem 1.3.

Repeating the arguments from the proof of Theorem 1.3 between inequalities (1) and (2) (with δ replaced by $1/N$ and $\omega(\delta)$ replaced by ε_N), we conclude, as in that proof, that if $a \in e$ is a critical point and $\text{diam}(e) < 1/N$, then the set $F(e)$ is contained in a bicylinder which splits into $O(1/\varepsilon_N^{m-1})$ subsets of diameter at most ε_N/N. This gives an estimate similar to inequality (2) from the proof of

Theorem 1.3:

$$\mu_m(F(e)) = O(\varepsilon_N N^{-m}),$$

which is valid for every set e with $\text{diam}(e) < 1/N$ contained in $\mathcal{N} \cap D_k$. Now, to complete the proof, it suffices to divide $\mathcal{N} \cap D_k$ into $O(N^m)$ subsets of diameter less than $1/N$ and repeat the final part of the proof of Theorem 1.3 following inequality (2).   □

5 *The Sharpness of the Conditions in the Main Theorem

5.1 Preliminaries

In this section, we analyze the sharpness of the relation, established in the main theorem, between the smoothness of a map and the Hausdorff dimension of the set of its critical values. The construction of the corresponding counterexamples begins with the case where the map under consideration is a function, and the subsequent discussion of the more general problem for maps relies on this initial result.

In Sections 1 and 5, the critical set of a map was denoted by $\mathcal{N}$. Now, we will sometimes have to simultaneously consider critical sets of different maps. So, to avoid confusion, we adjust the notation. The critical set of a smooth map (in particular, function) F will be denoted by $\mathcal{N}_F$, and the set $F(\mathcal{N}_F)$ of its critical values, by $\mathcal{K}_F$.

By the main theorem (see Section 1.2), for a t-smooth function f of m variables we have $\mu_p(\mathcal{K}_f) = 0$, where $p = m/t$ and μ_p is the Hausdorff measure. Since this result is trivial for $p > 1$, below we assume that $p \leq 1$, i.e., $m \leq t$. We will show that the smoothness condition imposed on f cannot be weakened: for $s < t$, there exists a function f of class C^s such that $\mu_p(\mathcal{K}_f) > 0$. Moreover, there exists an "extremely smooth" function, belonging to all classes C^s with $s < t$, that has an equally massive set of critical values. We will also show that under the original smoothness condition, the theorem is no longer valid for the measure μ_q with $q < p$ (here $p = m/t \leq 1$).

To see this, we first consider the one-dimensional situation and construct a family of auxiliary functions of one variable that have the required smoothness and "sufficiently massive" sets of critical values. The existence of such functions shows, in particular, that the main theorem is sharp for functions of one variable. With such a function φ at hand, we construct a counterexample in the case of functions of several variables by seeking a function f in the form of a linear combination

$$f(x_1, \ldots, x_m) = c_1\varphi(x_1) + \ldots + c_m\varphi(x_m)$$

($c_1, \ldots, c_m$ are nonzero coefficients). Then f has the same smoothness as φ, and its critical set $\mathcal{N}_f$ coincides with $\mathcal{N}_\varphi^m$. Clearly, all numbers of the form

$$c_1 y_1 + \ldots + c_m y_m \quad \text{where } y_1, \ldots, y_m \in \mathcal{K}_\varphi$$

(and only them) are critical values of f. In other words, the set $\mathcal{K}_f$ of critical values of f coincides with the algebraic sum $c_1\mathcal{K}_\varphi + \ldots + c_m\mathcal{K}_\varphi$, which is often much "more massive" than the summands. In particular, the sum of sets of zero Lebesgue measure can have positive measure (for example, $C + C = [0, 2]$ for the Cantor set C). This allows us to obtain a desired counterexample for an appropriate choice of $c_1, \ldots, c_m$.

5.2 Cantor-Like Sets

Before proceeding to the construction of φ, we describe a family $\{C_a\}_{0<a<1/2}$ of subsets of the closed interval $[0, 1]$ that will be used as sets of critical points and critical values of φ. The sets C_a are similar to the classical Cantor set C, which is included in this family for $a = \frac{1}{3}$. For every $a \in \left(0, \frac{1}{2}\right)$, the set C_a, like C, is obtained by successively removing from $[0, 1]$ finite collections of open intervals of finite length. Now we describe this construction in more detail.

Let $a \in \left(0, \frac{1}{2}\right)$. At the first step, we remove from $[0, 1]$ the open middle interval $\delta = (a, 1 - a)$, so the difference $[0, 1] \setminus \delta$ splits into two closed intervals (segments) symmetric to each other with respect to the point $\frac{1}{2}$: the left segment $\Delta_0 = [0, a]$ and the right segment $\Delta_1 = [1 - a, 1]$. They can be obtained by shrinking $[0, 1]$ by the factor a and are called segments of rank 1. Then the procedure is iterated: from Δ_0 and Δ_1 we remove the equal open middle intervals δ_0 and δ_1 of length $(1 - 2a)a$, so we are left with four segments $\Delta_{00}, \Delta_{01}, \Delta_{10}, \Delta_{11}$ of length a^2, the first subscript indicating the ambient segment.

The construction and labelling (by binary words in the alphabet $\{0, 1\}$) of further segments proceed by induction. Below, the number of letters in a binary word ε (the length of ε) is denoted by[1] $|\varepsilon|$.

Let Δ_ε and δ_ε be a segment and the corresponding open interval constructed at the kth step, so ε is a binary word of length k, i.e., $\varepsilon = \varepsilon_1 \ldots \varepsilon_k$ where each letter is either 0 or 1. The segments obtained at the kth step are called segments of rank k, and the open middle intervals of these segments are called intervals of rank k. Clearly, there are 2^k segments of rank k, each having length a^k, and the length of an interval of the same rank is equal to $(1 - 2a)a^k$.

At the $(k + 1)$th step, we consider the difference $\Delta_\varepsilon \setminus \delta_\varepsilon$, where $\varepsilon = \varepsilon_1 \ldots \varepsilon_k$. It consists of two equal segments of length a^{k+1}, which are called segments of rank $k + 1$. The left one is denoted by $\Delta_{\varepsilon 0}$, and the right one, by $\Delta_{\varepsilon 1}$ (or, in more detail, $\Delta_{\varepsilon_1 \ldots \varepsilon_k 0}$ and $\Delta_{\varepsilon_1 \ldots \varepsilon_k 1}$). Their open middle intervals of length $(1 - 2a)a^{k+1}$ are called intervals of rank $k + 1$ and labeled by the same words.

Note also that the intervals (of all ranks) to be removed are pairwise disjoint, and the union of all segments of a given rank is similar to the union of all segments of

---

[1] The reader should not confuse the order of a multi-index (the sum of its coordinates) and the length of a binary word, denoted by the same symbol.

the next rank contained in one of the halves of $[0, 1]$ (i.e., the union of segments Δ_ε with fixed ε_1).

To complete the definition of a generalized Cantor set, consider the sets $S_k = \bigcup\limits_{|\varepsilon|=k} \Delta_\varepsilon$ $(k = 1, 2, \ldots)$ and let

$$C_a = \bigcap_{k=1}^{\infty} S_k = \bigcap_{k=1}^{\infty} \bigcup_{|\varepsilon|=k} \Delta_\varepsilon.$$

Since the number of segments of rank k is 2^k, the length of each of them is a^k, and $a < \frac{1}{2}$, it is clear that the measure of the set S_k is equal to $(2a)^k \to 0$, so the Lebesgue measure of the set C_a vanishes.

Another representation of the set C_a follows from De Morgan's[2] laws:

$$C_a = [0, 1] \setminus \left(\delta \cup \bigcup_{k=1}^{\infty} \bigcup_{|\varepsilon|=k} \delta_\varepsilon \right).$$

In what follows, we will have to simultaneously consider two different sets C_a and C_b. The segments similar to Δ_ε arising in the construction of C_b will be denoted by $\widetilde{\Delta}_\varepsilon$.

5.3 Constructing an Auxiliary Function

Here we construct a t-smooth function on $\mathbb{R}$ with a given set of critical values.

Proposition *For $0 < a < \frac{1}{2}, t \geq 1$, and $0 < b < a^t$ there exists a strictly monotone function of class $C^t(\mathbb{R})$ such that its critical set coincides with C_a and its set of critical values coincides with C_b.*

Proof First, we construct not a required function itself, but its derivative, which must vanish on the set C_a.

Take a nonnegative function θ of class $C^\infty(\mathbb{R})$ that is positive only on the interval $(0, 1)$ and satisfies the condition

$$\int_0^1 \theta(u)\, du = 1 - 2b. \tag{1}$$

Denote by u_ε the left endpoint of the interval δ_ε, and by $d_k = a^k(1 - 2a)$ (where $k = |\varepsilon|$) its length; also, let $d_0 = 1 - 2a$ be the length of the interval δ (throughout the proof, we think of δ as an interval of rank 0). Define functions $\psi_0, \psi_1, \psi_2, \ldots$ on $\mathbb{R}$ as

---

[2] Augustus **De Morgan** (1806–1871) was a British mathematician and logician.

$$\psi_0(u) = \frac{1}{d_0}\theta\left(\frac{u-a}{d_0}\right), \quad \psi_k(u) = \frac{b^k}{d_k}\sum_{|\varepsilon|=k}\theta\left(\frac{u-u_\varepsilon}{d_k}\right). \tag{2}$$

It is easy to see that $\theta\left(\frac{u-a}{d_0}\right) > 0$ only if $u \in \delta$, and $\theta\left(\frac{u-u_\varepsilon}{d_k}\right) > 0$ only if $u \in \delta_\varepsilon$. Hence, for every u the sum defining ψ_k has at most one nonzero term. It is also important for what follows that the supports of the functions $\psi_0, \psi_1, \psi_2, \ldots$ are pairwise disjoint.

Set $\psi = \sum_{k=0}^{\infty} \psi_k$. The function ψ is positive on δ and on each of the intervals δ_ε, and its zeros belonging to $[0, 1]$ exhaust the set C_a.

We claim that $\psi \in C^{t-1}(\mathbb{R})$. It is clear that for all $u \in \mathbb{R}$ and $k, j \in \{0, 1, 2, \ldots\}$,

$$\left|\psi_k^{(j)}(u)\right| \leq \frac{b^k}{d_k^{j+1}}\max_{[0,1]}\left|\theta^{(j)}\right| = \frac{1}{(1-2a)^{j+1}}\left(\frac{b}{a^{j+1}}\right)^k \max_{[0,1]}\left|\theta^{(j)}\right|. \tag{3}$$

For $j < r = [t]$, obviously, $j + 1 \leq t$, and we obtain

$$\left|\psi_k^{(j)}(u)\right| \leq C\left(\frac{b}{a^t}\right)^k,$$

where the coefficient C does not depend on k and u. Hence, the sum of the series $\psi = \psi_0 + \psi_1 + \psi_2 + \ldots$ is continuous, and for $t \geq 2$ it is $r - 1$ times continuously differentiable.

To verify that $\psi \in C^{t-1}(\mathbb{R})$, set $t = r + \alpha$ where $0 < \alpha < 1$ (for $\alpha = 0$, this is already proved). Let $\Lambda = \max_{[0,1]}\left|\theta^{(r)}\right|$. Then inequality (3) for $j = r$ takes the form

$$\left|\psi_k^{(r)}(u)\right| \leq \Lambda\frac{b^k}{d_k^{r+1}}.$$

We apply it to estimate the difference $\psi_k^{(r-1)}(u) - \psi_k^{(r-1)}(v)$, first assuming that u and v belong to the same interval δ_ε of rank k, so $|u - v| < d_k$. Then

$$\left|\psi_k^{(r-1)}(u) - \psi_k^{(r-1)}(v)\right| \leq \Lambda\frac{b^k}{d_k^{r+1}}|u - v| = \Lambda\frac{b^k}{d_k^r}\frac{|u-v|}{d_k} \leq \Lambda\frac{b^k}{d_k^r}\left(\frac{|u-v|}{d_k}\right)^\alpha$$

$$= \Lambda\frac{b^k}{d_k^t}|u - v|^\alpha = \frac{\Lambda}{(1-2a)^t}\left(\frac{b}{a^t}\right)^k |u - v|^\alpha = L_k|u - v|^\alpha, \tag{3'}$$

where $L_k = \frac{\Lambda}{(1-2a)^t}\left(\frac{b}{a^t}\right)^k$.

Now let u and v belong to different intervals of rank k. For definiteness, we assume that $u < v$. Let $\bar{u}$ be the right endpoint of the interval containing u, and $\bar{v}$ be the left endpoint of the interval containing v. Then $u < \bar{u} < \bar{v} < v$. By the construction of ψ_k, all its partial derivatives vanish at $\bar{u}$ and $\bar{v}$. Hence, (3') implies the inequalities

$$|\psi_k^{(r-1)}(u)| = |\psi_k^{(r-1)}(u) - \psi_k^{(r-1)}(\bar{u})| \le L_k(\bar{u} - u)^\alpha \le L_k(v - u)^\alpha$$

and

$$|\psi_k^{(r-1)}(v)| = |\psi_k^{(r-1)}(v) - \psi_k^{(r-1)}(\bar{v})| \le L_k(v - \bar{v})^\alpha \le L_k(v - u)^\alpha,$$

which give the estimate $|\psi_k^{(r-1)}(u) - \psi_k^{(r-1)}(v)| \le 2L_k|u - v|^\alpha$ wherever the points u and v may be. It follows that

$$|\psi^{(r-1)}(u) - \psi^{(r-1)}(v)| \le \sum_{k=0}^{\infty} |\psi_k^{(r-1)}(u) - \psi_k^{(r-1)}(v)| \le \sum_{k=0}^{\infty} 2L_k|u-v|^\alpha = L|u-v|^\alpha,$$

where

$$L = 2\sum_{k=0}^{\infty} L_k = \frac{2\Lambda}{(1-2a)^t} \sum_{k=0}^{\infty} \left(\frac{b}{a^t}\right)^k < +\infty,$$

since $b < a^t$. Thus, $\psi \in C^{t-1}(\mathbb{R})$.

Now we verify that for every binary word ω with $|\omega| = j$,

$$\int_{\Delta_\omega} \psi(u)\, du = b^j. \tag{4}$$

Indeed, by the definition of ψ, we have

$$\int_{\Delta_\omega} \psi(u)\, du = \frac{1}{d_0} \int_{\Delta_\omega} \theta\left(\frac{u-a}{d_0}\right) du + \sum_{k=1}^{\infty} \frac{b^k}{d_k} \sum_{|\varepsilon|=k} \int_{\Delta_\omega} \theta\left(\frac{u-u_\varepsilon}{d_k}\right) du.$$

The first integral in the right-hand side vanishes, because $\Delta_\omega \cap \delta = \varnothing$ and the integrand vanishes outside δ. Besides, the terms with $k < j$ in the outer sum also vanish. Indeed, the integrands in the inner sum are nonzero only on pairwise disjoint intervals δ_ε. Since $\Delta_\omega \cap \delta_\varepsilon = \varnothing$ for $|\varepsilon| < |\omega|$, the integrals in the inner sum vanish if $k < |\omega|$. Therefore,

$$\int_{\Delta_\omega} \psi(u)\, du = \sum_{k=j}^{\infty} b^k \sum_{\substack{|\varepsilon|=k, \\ \Delta_\varepsilon \subset \Delta_\omega}} \frac{1}{d_k} \int_{\delta_\varepsilon} \theta\left(\frac{u-u_\varepsilon}{d_k}\right) du.$$

Due to (1), the terms in the inner sum are easy to calculate:

$$\frac{1}{d_k} \int_{\delta_\varepsilon} \theta\left(\frac{u-u_\varepsilon}{d_k}\right) du = \int_0^1 \theta(v)\, dv = 1 - 2b.$$

Since the segment Δ_ω contains 2^{k-j} segments Δ_ε of rank k, this inner sum is equal to $2^{k-j}(1-2b)$. This implies (4):

$$\int_{\Delta_\omega} \psi(u)\, du = (1-2b) \sum_{k=j}^{\infty} b^k 2^{k-j} = b^j(1-2b) \sum_{i=0}^{\infty} (2b)^i = b^j.$$

Besides, it follows from (1) that

$$\int_\delta^{1-a} \psi(u)\, du = \int_\delta^{1-a} \psi_0(u)\, du = \frac{1}{d_0} \int_a^{1-a} \theta\!\left(\frac{u-a}{d_0}\right) du = \int_0^1 \theta(v)\, dv = 1-2b,$$

and hence $\int_0^1 \psi(u)\, du = \int_{\Delta_0} \ldots + \int_\delta \ldots + \int_{\Delta_1} \ldots = b + (1-2b) + b = 1$.

Now consider the antiderivative φ of ψ satisfying the condition $\varphi(0) = 0$. It is clear that $\varphi \in C^t(\mathbb{R})$, the function φ is constant on $(-\infty, 0]$ and $[1, +\infty)$ and strictly increasing on $[0, 1]$, $\varphi(1) = 1$, and the set of critical points of φ belonging to $[0, 1]$ coincides with C_a.

Let us check that $\varphi(C_a) = C_b$. For this, it suffices to show that $\varphi(\Delta_\varepsilon) = \widetilde{\Delta}_\varepsilon$ for every ε, since the sets of endpoints of these segments are dense in C_a and C_b, respectively.

First of all, note that (4) implies that the segments $\varphi(\Delta_\varepsilon)$ and $\widetilde{\Delta}_\varepsilon$ have the same length. Since $\varphi(0) = 0$ and $\varphi(1) = 1$, it follows, in particular, that $\varphi(\Delta_0) = \widetilde{\Delta}_0$ and $\varphi(\Delta_1) = \widetilde{\Delta}_1$. Using these relations as the base case and bearing in mind that each segment of rank k contains two segments of rank $k+1$ adjacent to its endpoints, one can easily obtain the desired equality for segments of arbitrary rank. The details are left to the reader.

To complete the proof, it remains to get rid of redundant (not lying in $[0, 1]$) critical points of φ. For this, it suffices to redefine φ outside $[0, 1]$ so that it is still C^t-smooth on $\mathbb{R}$ and has a positive derivative on $(-\infty, 0) \cup (1, +\infty)$. □

Remark Multiplying the function φ, which is an antiderivative for ψ, by a C^∞-function that is equal to 1 on $(-\infty, 1]$, equal to 0 on $[2, +\infty)$, and has a negative derivative on $(1, 2)$, we obtain a compactly supported function (coinciding with φ on $[0, 1]$) whose set of critical values coincides with that of φ (i.e, is equal to C_b) and whose set of critical points contains, in addition to C_a, the half-lines $(-\infty, 0)$ and $[2, +\infty)$.

5.4 The Auxiliary Function (Continued)

It is clear from Proposition 5.3 that increasing the parameter a allows us to increase t as well and construct a smoother function with the same set of critical values C_b. Such an improvement is possible only for $a < \frac{1}{2}$, but we can modify it as to implement the inaccessible case $a = \frac{1}{2}$ "in the limit." As a result, the sets C_a get replaced by a more massive set C_*. It is constructed in a similar way, but instead of a common shrinking factor a applied at all steps of the construction, now we use a sequence of such factors increasing to $\frac{1}{2}$.

Let us briefly describe the construction of the set C_*, meaning that the segments and intervals of ranks $1, 2, \ldots$ are labelled in the same way as in the construction of the sets C_a. First (this is step 0), we remove from $[0, 1]$ the open interval $\delta^* = \left(\frac{1}{4}, \frac{3}{4}\right)$. Then, from the remaining two segments Δ_0^*, Δ_1^* of rank 1 we remove the open middle intervals δ_0^*, δ_1^* of length $\frac{1}{12}$ (intervals of rank 1); after that, in $[0, 1]$ there remain four segments of rank 2, etc. At the kth step ($k = 0, 1, 2 \ldots$), from the segments Δ_ε^* of rank k we remove the open middle intervals δ_ε^* of length $d_k = \frac{2^{-k}}{(k+1)(k+2)}$ (intervals of rank k). Since the lengths of these intervals sum to

$$\frac{1}{(k+1)(k+2)} = \frac{1}{k+1} - \frac{1}{k+2},$$

the total length of the intervals removed from $[0, 1]$ after the kth step is

$$\frac{1}{2} + \left(\frac{1}{2} - \frac{1}{3}\right) + \ldots + \left(\frac{1}{k} - \frac{1}{k+1}\right) = 1 - \frac{1}{k+1}.$$

Hence, the set S_k^* formed by the segments of rank k has measure $\frac{1}{k+1}$, and the length of each such segment is $\frac{2^{-k}}{k+1}$. By definition, the set C_* is the intersection of the sets S_k^*. Since the measures of these sets tend to 0, we conclude that C_* has zero Lebesgue measure.

With this construction, we have $\frac{d_k}{d_{k-1}} = \frac{k}{k+2} \cdot \frac{1}{2}$, and the similar ratio for the lengths of segments is $\frac{k}{k+1} \cdot \frac{1}{2}$. Thus, in contrast to the construction of the set C_a, where at each step the lengths of intervals and segments are multiplied by the same shrinking factor a, in the construction of C_* the shrinking factors vary with k and increase to $\frac{1}{2}$.

Although we will not need this fact, the set C_* has the greatest possible Hausdorff dimension. Indeed, the reader can easily check that the natural map from C_* onto C_a that sends the point in the intersection of segments

$$\Delta_{\varepsilon_1}^* \supset \Delta_{\varepsilon_1 \varepsilon_2}^* \supset \ldots \supset \Delta_{\varepsilon_1 \ldots \varepsilon_n}^* \supset \ldots$$

to the point in the intersection of the segments

$$\Delta_{\varepsilon_1} \supset \Delta_{\varepsilon_1 \varepsilon_2} \supset \ldots \supset \Delta_{\varepsilon_1 \ldots \varepsilon_n} \supset \ldots$$

from the construction of C_a satisfies the Lipschitz condition. Since the Hausdorff dimension does not increase under Lipschitz maps (see Property 3 at p. 261), we see that

$$1 \geq \dim_H(C_*) \geq \dim_H(C_a) = \log_{1/a} 2$$

(the last equality is established in Corollary V.3.6) for all $a \in \left(0, \frac{1}{2}\right)$. Since a is arbitrary, we conclude that $\dim_H(C_*) = 1$.

Below, by C^{t-0} we denote the set of functions belonging to every class C^s with $s < t$, i.e., $C^{t-0} = \bigcap_{1 < s < t} C^s$.

Proposition *For* $t > 1$ *there exists a strictly increasing function of class* $C^{t-0}(\mathbb{R})$ *such that its critical set coincides with* C_* *and its set of critical values coincides with* C_b *where* $b = 2^{-t}$.

Proof The proof can be obtained by slightly modifying the proof of Proposition 5.3. Using the function θ introduced at the beginning of that proof, by analogy with (2) we define functions $\psi_0, \psi_1, \psi_2, \ldots$ on $\mathbb{R}$ as follows (below, u^* and u^*_ε are the left endpoints of the intervals δ^* and δ^*_ε, while d_0 and d_k are their lengths, with $k = |\varepsilon|$):

$$\psi_0(u) = \frac{1}{d_0}\theta\left(\frac{u - u^*}{d_0}\right), \quad \psi_k(u) = \frac{b^k}{d_k}\sum_{|\varepsilon|=k}\theta\left(\frac{u - u^*_\varepsilon}{d_k}\right). \tag{2'}$$

It is easy to see that $\theta\left(\frac{u - u^*_\varepsilon}{d_k}\right) > 0$ only for $u \in \delta^*_\varepsilon$. Hence, for every u the sum defining ψ_k has at most one nonzero term. It is important for what follows that the supports of the functions $\psi_0, \psi_1, \psi_2, \ldots$ are pairwise disjoint. Set

$$\psi = \sum_{k=0}^{\infty} \psi_k.$$

The function ψ is positive on δ^* and on all intervals δ^*_ε, and its zeros from $[0, 1]$ exhaust the set C_*.

To verify that the function ψ is sufficiently smooth, assume that $t = r + \sigma$, where r is an integer and $0 < \sigma \leq 1$, and write inequality (3') for $\alpha < \sigma$. This yields inequalities showing that $\psi \in C^s$ for every $s < t - 1$.

Let φ be the antiderivative of the function ψ that vanishes at the origin. Obviously, $N_\varphi \cap [0, 1] = C_*$. The equality $\varphi(C_*) = C_b$ can be verified as the equality $\varphi(C_a) = C_b$ in the proof of Proposition 5.3 (see the argument starting from (4) up to the end of the proof). To obtain a function with the required properties, we redefine φ outside $[0, 1]$ so that it is still C^{t-0}-smooth on $\mathbb{R}$ and has a positive derivative on $(-\infty, 0) \cup (1, +\infty)$. $\qquad\qquad\square$

Remark Multiplying the function φ used in the proof by the function described at the end of the previous subsection, we obtain a compactly supported function φ_* of class C^{t-0} (coinciding with φ on $[0, 1]$) such that

$$N_{\varphi_*} = (-\infty, 0) \cup C_* \cup [2, +\infty), \quad \mathcal{K}_{\varphi_*} = \mathcal{K}_\varphi = C_{2-t}.$$

5.5 Counterexamples: Functions of Several Variables

According to the original Morse's result, which initiated further research, the set of critical values of a function of class $C^m(\mathbb{R}^m)$ has zero length, while for an $(m - 1)$-smooth function it can coincide with an interval. Now we know from the main theorem that the set of critical values of a function of class $C^t(\mathbb{R}^m)$ has zero

μ_p-measure for $p = \frac{m}{t}$. The proposition below shows that for functions the result of the main theorem is sharp and the smoothness conditions cannot be weakened.

We need a technical result related to arithmetic properties of points that constitute a Cantor-like set.

Lemma *Let* $0 < s < \frac{1}{2}$, $b = s^m$. *Then*

$$\frac{1-b}{1-s}C_s = C_b + sC_b + \ldots + s^{m-1}C_b.$$

Proof Using the definition of a Cantor-like set, one can easily check by induction that the left endpoint of the segment $\Delta_{\varepsilon_1 \ldots \varepsilon_k}$ arising in the construction of C_s is

$$(1-s)(\varepsilon_1 + \varepsilon_2 s + \ldots + \varepsilon_k s^{k-1}).$$

Hence, every point u of this set, being the limit of endpoints of such segments, can be represented in the form

$$u = (1-s)\sum_{i=0}^{\infty} \varepsilon_i s^i \quad \text{where } \varepsilon_i = 0 \text{ or } 1 \text{ for all } i. \tag{5}$$

Splitting this series into the sum of blocks consisting of m terms each, we can (using the equality $s^m = b$) write the result in the form

$$u = (1-s)\sum_{k=0}^{\infty}\sum_{j=0}^{m-1} \varepsilon_{j,k} s^{j+mk} = (1-s)\sum_{j=0}^{m-1} s^j \sum_{k=0}^{\infty} \varepsilon_{j,k} b^k,$$

where $\varepsilon_{j,k} = 0$ or 1 for all j, k. Since (by analogy with (5)) the sums

$$u_j = (1-b)\sum_{k=0}^{\infty} \varepsilon_{j,k} b^k$$

belong to C_b, for every point $u \in C_s$ we obtain

$$u = \frac{1-s}{1-b}\sum_{j=0}^{m-1} s^j u_j \in \frac{1-s}{1-b}(C_b + sC_b + \ldots + s^{m-1}C_b),$$

i.e.,

$$C_s \subset \frac{1-s}{1-b}(C_b + sC_b + \ldots + s^{m-1}C_b).$$

Starting from a point belonging not to the left-hand side, but to the right-hand side of the equality to be proved, one can easily check the inverse inclusion too, which completes the proof.   □

So far, speaking of the set C_s, we assumed that $0 < s < \frac{1}{2}$. Below, for consistency we allow the value $s = \frac{1}{2}$, assuming that $C_{1/2} = [0, 1]$. One can easily see that the lemma remains valid in this case too.

Proposition *For $t > 1$, $t \geq m$ there exists a compactly supported function of class $C^{t-0}(\mathbb{R}^m)$ such that its critical set contains $C_*^m = C_* \times \ldots \times C_*$ (m factors) and its set of critical values coincides with C_s where $s = 2^{-t/m}$ (so, by Proposition V.3.6, the $\mu_{m/t}$-measure of this set is equal to 1, and its Hausdorff dimension is equal to $\frac{m}{t}$).*

In particular, for $t = m > 1$ we obtain the following complement to Morse's result stated at the beginning of this subsection:

there exists a function of class $C^{m-0}(\mathbb{R}^m)$ whose set of critical values coincides with $[0, 1]$.

Proof Let $b = s^m = 2^{-t}$ and φ be a compactly supported function whose existence is proved in Proposition 5.4 and the corresponding remark. We define a function f of class $C^{t-0}(\mathbb{R}^m)$ as

$$f(x_1, \ldots, x_m) = \varphi(x_1) + s\varphi(x_2) + \ldots + s^{m-1}\varphi(x_m).$$

The critical set of f contains C_*^m, while the set of its critical values coincides with $f(C_*^m)$ and, consequently, with the sum $C_b + sC_b + \ldots + s^{m-1}C_b$, which, by the lemma, is nothing else than the set $\frac{1-b}{1-s}C_s$. Hence, the function $\frac{1-s}{1-b}f$ is as required.□

We note the following corollary of this proposition.

Corollary *For $t \geq m$ and $p = \frac{m}{t}$ there exists a function g of class $C^t(\mathbb{R}^m)$ such that $\dim_H(\mathcal{K}_g) = p$.*

Observe that here we allow t to be not only strictly greater than, but also equal to m.

Proof Set $t_j = t + \frac{1}{j}$, $s_j = 2^{-(1/m)t_j}$ ($j \in \mathbb{N}$) and use the above proposition with these parameters instead of t and s. By the proposition, there exist compactly supported functions f_j of class $C^{t_j-0}(\mathbb{R}^m)$ (so that they are t-smooth) such that $\dim_H(\mathcal{K}_{f_j}) = \frac{m}{t_j}$.

Let $\tilde{f}_1, \tilde{f}_2, \ldots$ be shifts of $f_1, f_2, \ldots$ such that the distances between their supports are bounded away from zero. Then for g we can take the sum of the series $\tilde{f}_1 + \tilde{f}_2 + \ldots$. In this case, $g \in C^t(\mathbb{R}^m)$, $\mathcal{K}_g = \bigcup_j \mathcal{K}_{f_j}$, and thus

$$\dim_H(\mathcal{K}_g) = \sup_j \dim_H(\mathcal{K}_{f_j}) = \sup_j \frac{m}{t_j} = p. \qquad \square$$

This corollary can be equivalently stated as follows: for $t \geq m$ there exists a function $g \in C^t(\mathbb{R}^m)$ such that $\mu_q(\mathcal{K}_g) = +\infty$ for every $q \in \left(0, \frac{m}{t}\right)$, i.e., for $qt < m$.

5.6 Counterexamples: Maps in the Case $n \geq m$

Using the counterexamples constructed in the previous subsections, which show that the conditions of the main theorem are sharp for functions, i.e., for maps from $\mathbb{R}^m$ to $\mathbb{R}^n$ where $n = 1$, we can verify that they are sharp for arbitrary n for every "fiber" of the critical set at which the rank of the Jacobian matrix is constant. In Sections 1 and 4, we denoted such a fiber where the rank is equal to i by $\mathcal{N}_i$ (as in the main theorem), and its image, by $\mathcal{K}_i$. To avoid confusion, when we have to deal with critical sets of different maps, the sets $\mathcal{N}_i$ corresponding to a map F will be denoted by $\mathcal{N}_{F,i}$, and their F-images, by $\mathcal{K}_{F,i}$.

Note that the existence of counterexamples illustrating the sharpness of the main theorem in the case $i = 0$ implies the existence of similar counterexamples for arbitrary $i < \min\{m, n\}$. More exactly, the following result holds.

Lemma *If $\widetilde{F}$ is a map of class $C^t(\mathbb{R}^{m-i}; \mathbb{R}^{n-i})$ such that $\dim_H(\mathcal{K}_{\widetilde{F},0}) = \frac{m-i}{t}$, then for the t-smooth map $F \colon \mathbb{R}^m \to \mathbb{R}^n$ defined by the formula*

$$F(x_1, \ldots, x_m) = \left(x_1, \ldots, x_i, \widetilde{F}(x_{i+1}, \ldots, x_m)\right),$$

we have $\dim_H(\mathcal{K}_{F,i}) = i + \frac{m-i}{t}$.

Proof The lemma follows from the obvious equalities

$$\mathcal{N}_{F,i} = \mathbb{R}^i \times \mathcal{N}_{\widetilde{F},0} \quad \text{and} \quad \mathcal{K}_{F,i} = \mathbb{R}^i \times \mathcal{K}_{\widetilde{F},0},$$

since (see Proposition V.4.5)

$$\dim_H(\mathbb{R}^i \times \mathcal{K}_{\widetilde{F},0}) = i + \dim_H(\mathcal{K}_{\widetilde{F},0}) = i + \frac{m-i}{t}. \qquad \square$$

For this reason, in what follows we restrict ourselves, without further explanations, to the case $i = 0$. Constructing maps from $\mathbb{R}^m$ to $\mathbb{R}^n$, we use the function constructed in Proposition 5.4.

A product of the form $A = [0,1]^i \times C_{a_1} \times \ldots \times C_{a_k}$, where $C_{a_1}, \ldots, C_{a_k}$ are Cantor-like sets (constructed as described in Section 5.2), will be called a *standard set* (it is assumed to lie in some space $\mathbb{R}^n$ with $n \geq i + k$). By Proposition V.4.5, the Hausdorff dimension p of such a product A is equal to the sum of the dimensions of the factors and $\mu_p(A) > 0$.

It turns out that it is useful to distinguish between the cases $m \leq n$ and $m > n$. In this subsection, we consider the simpler case $m \leq n$.

Proposition *For $t > 1$ and $m \leq n$, in $C^{t-0}(\mathbb{R}^m; \mathbb{R}^n)$ there is a compactly supported map F such that the set $\mathcal{K}_{F,0}$ contains a standard set of dimension $p = \frac{m}{t}$ and, consequently, $\mu_p(\mathcal{K}_{F,0}) > 0$.*

Note that, by the main theorem, under the assumptions of the proposition, $\mu_q(\mathcal{K}_{F,0}) = 0$ for arbitrary $q > p$ (since $q = \frac{m}{s}$ for $s < t$ and $F \in C^s$).

Proof Clearly, it suffices to construct a map with the required properties acting from $\mathbb{R}^m$ to $\mathbb{R}^m$. Hence, in what follows we assume without loss of generality that $n = m$.

According to Proposition 5.4, fix a function φ of class $C^{t-0}(\mathbb{R})$ such that its critical set $\mathcal{N}_\varphi$ coincides with C_* and its set of critical values coincides with C_b where $b = 2^{-t}$. Also, let θ be a compactly supported function that is infinitely smooth in $\mathbb{R}^m$ and equal to 1 on the cube $[0, 1]^m$.

Now consider the map $F \colon \mathbb{R}^m \to \mathbb{R}^m$ defined by the formula

$$F(x_1, \ldots, x_m) = \theta(x)\bigl(\varphi(x_1), \ldots, \varphi(x_m)\bigr).$$

Clearly, F has compact support, belongs to $C^{t-0}(\mathbb{R}^m; \mathbb{R}^m)$, and the set $\mathcal{N}_{F,0}$ contains the Cartesian power $\mathcal{N}_\varphi^m = C_*^m$. Hence,

$$\mathcal{K}_{F,0} = F(\mathcal{N}_{F,0}) \supset F(C_*^m) = (\varphi(C_*))^m = C_b^m.$$

As proved in Section V.3.6, $\dim_H(C_b) = \frac{1}{t}$ and $\mu_{1/t}(C_b) > 0$. It remains to observe that the dimension of the standard set C_b^m is equal to the sum of the dimensions of its factors, i.e., to $\frac{m}{t} = p$, and $\mu_p(C_b^m) > 0$ (see Proposition V.4.4). $\quad\square$

Using the construction described in the lemma, we can obtain counterexamples for $i = 0, 1, \ldots$. They are not compactly supported maps, but we can made them compactly supported by multiplying by an infinitely smooth compactly supported function equal to 1 in a neighborhood of the unit cube. Applying appropriate shifts of the argument, we obtain maps with disjoint supports, and the sum of such maps (over all i) gives an example of a map Φ for which all sets $\mathcal{K}_{\Phi,i}$ simultaneously have the maximum possible dimension.

5.7 Counterexamples: Maps in the Case $m > n$

Now we discuss the sharpness of the conditions of the main theorem in the remaining more complicated case. When constructing the corresponding counterexamples, we must bear in mind the following fact.

By the main theorem, for a t-smooth map $F \colon \mathbb{R}^m \to \mathbb{R}^n$ (with arbitrary m, n), we have $\mu_{p_i}(\mathcal{K}_{F,i}) = 0$, where $p_i = i + \frac{m-i}{t}$, for every i such that $0 \leq i < \min\{m, n\}$. However, if $m > n$, then these assertions are sometimes trivial. Namely, they are trivial for indices i such that $p_i > n$, i.e., for $i > \frac{nt-m}{t-1}$, or, which is the same, $t < \frac{m-i}{n-i}$. Such a case occurs, for example, if $m = 4$, $n = t = 2$, $i = 1$ or $m = 3$, $n = 2$, $t = \frac{3}{2}$, $i = 1$. The situation is "stable" with respect to t: it is preserved under small perturbations of this smoothness parameter. When constructing counterexamples, we disregard such inessential values of i and take into consideration only values of i for which $p_i \leq n$; such values will be said to be essential (for given m, n, t). So, below we always assume not only that $n < m$, but also that $i \leq \frac{nt-m}{t-1}$. In particular, this means that the maps under consideration are sufficiently smooth: $t \geq \frac{m}{n}$.

The result that will be obtained in this subsection is not so exhaustive as that for $m \le n$ and does not cover the case of fractional smoothness (although the assertion is valid for all $t > 1$). However, we will obtain examples showing that the conditions of the theorem are sharp in the classes of maps of integer smoothness.

Proposition *Let* $1 < n < m$, $t > 1$, *and assume that at least one of the numbers* $\frac{m}{n}$ *or* t *is integer. Then for* $p = \frac{m}{t} \le n$ *there exists a compactly supported map* F *of class* $C^{t-0}(\mathbb{R}^m; \mathbb{R}^n)$ *such that the set* $\mathcal{K}_{F,0}$ *contains a standard set of Hausdorff dimension* p *and, consequently,* $\mu_p(\mathcal{K}_{F,0}) > 0$.

Proof It suffices to construct a map F that has all the required properties except having compact support. Then it remains to multiply F by an appropriate compactly supported function, as in the proof of Proposition 5.6.

I. First, consider the case where $\frac{m}{n} = k \in \mathbb{N}$. Since $t \ge \frac{m}{n} = k$, the fraction $\frac{m}{t}$ can be written as the sum of n fractions $\frac{k}{t}$, which are less than 1. By Proposition 5.5, there exists a function f of class $C^{t-0}(\mathbb{R}^k)$ such that its critical set contains C_*^k and its set of critical values $f(C_*^k)$ coincides with C_b, where $b = 2^{-t/k}$. Recall that $\dim_H(C_b) = \frac{k}{t} = \frac{m}{nt}$.

Now we split the coordinates of the space $\mathbb{R}^m = \mathbb{R}^{nk}$ into n groups of k coordinates each and define a map F as

$$F(x_1, \ldots, x_{nk}) = \big(f(x_1, \ldots, x_k), f(x_{k+1}, \ldots, x_{2k}), \ldots, f(x_{(n-1)k+1}, \ldots, x_{nk})\big).$$

Clearly, the set $\mathcal{N}_{F,0}$ contains $(C_*^k)^n = C_*^m$, and $F(C_*^m) = C_b^n$. But the dimension of the latter set is equal to $n \frac{k}{t} = \frac{m}{t} = p$, which completes the proof in the case under consideration.

II. Turning to the case where $t \in \mathbb{N}$, we assume that m is not divisible by n: $m = kn + d$, where $k, d \in \mathbb{N}$ and $0 < d < n$. Then

$$t \ge \frac{m}{n} = \frac{nk + d}{n} > k.$$

Since t is an integer, we have $t \ge k + 1$ and, therefore,

$$\frac{k+1}{t} \le 1. \tag{6}$$

Now write $\frac{m}{t}$ as a sum of fractions as follows:

$$\frac{m}{t} = \frac{nk+d}{t} = \frac{(n-d)k + d(k+1)}{t} = \underbrace{\frac{k}{t} + \ldots + \frac{k}{t}}_{n-d \text{ terms}} + \underbrace{\frac{k+1}{t} + \ldots + \frac{k+1}{t}}_{d \text{ terms}}. \tag{7}$$

Each of the fractions in this sum does not exceed 1. This allows us to construct F by still taking the function f constructed at the first step of the proof for the first $n - d$ coordinate functions, and a similar function of $k + 1$ variables for the last d coordinate functions. Here each of the variables $x_1, \ldots, x_m$ occurs as an argument in only one coordinate function, i.e., all m variables are divided into n blocks: each of the first $n - d$ blocks consists of k variables, and each of the remaining d blocks consists of $k + 1$ variables. We leave it to the reader to fill in all the details. □

For noninteger t and essential values of i, the proposition remains valid if at least one of the following conditions is satisfied: either $m - i$ is a multiple of $n - i$, or t satisfies not only the inequality $t \geq \frac{m-i}{n-i}$, but also the stronger condition

$$t \geq \left\lceil \frac{m - i}{n - i} \right\rceil + 1.$$

It replaces inequality (6) and guarantees that all terms in the expansion (7) do not exceed 1. Then the proof can be completed in the same way as for positive integer t.

We note the following corollary of the proposition.

Corollary *Under the assumptions of the proposition, there exists a compactly supported map F of class $C^t(\mathbb{R}^m; \mathbb{R}^n)$ such that $\dim_H(\mathcal{K}_{F,0}) = \frac{m}{t}$.*

To see this, take a sequence $\{q_j\}$ that strictly increases to p and a map of the form $F = c_1 F_1 + c_2 F_2 + \ldots$ where c_j are positive constants and $F_1, F_2, \ldots$ are compactly supported t-smooth maps with disjoint supports such that $\mu_{q_j}(\mathcal{K}_{F_j,0}) > 0$ for every j. If the supports of all these maps are contained in a ball, then the sum of the series can be made sufficiently smooth by using sufficiently fast decaying coefficients c_j.

Lemma 5.6 implies that for essential values of i, results similar to this corollary are valid also for the sets $\mathcal{K}_{F,i}$.

6 *Whitney's Example

6.1 Statement of the Problem and the Main Result

It is well known that a smooth function in a domain is constant if its gradient vanishes everywhere. In this section we consider a finer question, assuming that the gradient vanishes only on a simple arc. Is it true that the function is constant on this arc? The affirmative answer is easy to prove if the arc is smooth (it suffices to replace it with an interval by "parametrizing" the function). But can we discard the smoothness condition? At first sight, it may seem that this assumption stems only from the method of proof but not from the essence of the matter. Yet this is not true. Our purpose in this section is to consider the famous Whitney's example (see [18]) which shows that the seemingly natural affirmative answer to the above question is in general false. The following theorem is due to Whitney.

Theorem *For every $t \in (1, 2)$ there exists a function of class $C^t(\mathbb{R}^2)$ that is not constant on a simple arc on which its gradient vanishes.*

For a C^2-function, this is no longer possible, since, by the main theorem (see Section 1.2), the length (i.e., the Hausdorff measure μ_1) of the set of its critical values vanishes. As follows from Lagrange's inequality, this is impossible also on a rectifiable arc (even if the function is not smooth, but only differentiable on this arc, see the corollary at the end of Section I.5.4).

The rest of this section is devoted to the proof of Whitney's theorem.

6.2 The Sets C_a^2

Assuming that a number $a \in \left(0, \frac{1}{2}\right)$ is fixed, we use the Cartesian squares of the sets C_a constructed in Section 5.2. We will need two-dimensional analogs of the notions and notation introduced there.

Analogs of segments Δ_ξ of rank j ($|\xi| = j$) used in the construction of C_a are all their Cartesian products. The resulting squares (there are 2^{2j} of them) will be called squares of rank j (see Fig. 1) and denoted by Q_ε (where ε is a binary word of length $2j$). By the initial (respectively, final) point of Q_ε we will mean the vertex of Q_ε whose both coordinates have the minimum (maximum) value.

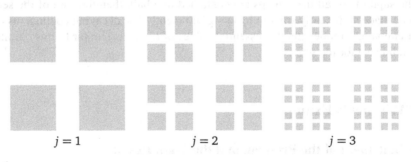

$$j = 1 \qquad\qquad\qquad j = 2 \qquad\qquad\qquad j = 3$$

Fig. 1

Now we describe a correspondence between such squares and binary words. The four squares of rank 1 are the products of two segments $\Delta_0 = [0, a]$ and $\Delta_1 = [1 - a, 1]$ of rank 1. For $\xi, \sigma = 0, 1$, set $Q_{\xi\sigma} = \Delta_\xi \times \Delta_\sigma$. The squares of higher ranks are labelled as follows. The square $\Delta_\xi \times \Delta_\sigma$, which is the product of squares $\Delta_\xi, \Delta_\sigma$ of rank j, where $\xi = \xi_1 \ldots \xi_j$, $\sigma = \sigma_1 \ldots \sigma_j$, will be denoted by $Q_{\overline{\xi\sigma}}$, assuming that $\overline{\xi\sigma}$ is the binary word obtained by alternating the letters of the words ξ and σ: $\overline{\xi\sigma} = \xi_1\sigma_1\xi_2\sigma_2 \ldots \xi_j\sigma_j$. The result is a bijection from the set of all binary words of even length onto the set of squares of all ranks.

Clearly, an arbitrary square Q_ε of rank j can be obtained from the unit square $[0, 1]^2$ by a homothety with factor a^j and the translation that maps the origin to the initial point of Q_ε. The resulting map sends each square of rank 1 to one of the four squares of rank $j + 1$ contained in Q_ε. We leave it to the reader to check by induction that the image of a square $\Delta_\xi \times \Delta_\sigma$ of rank 1 under this map coincides with $Q_{\varepsilon\xi\sigma}$, so the index of a square Q_ω of rank $j + 1$ consists of the index of the ambient square of the previous rank and two last letters showing the position of Q_ω in the ambient square.

Since $C_a = \bigcap\limits_{j=1}^{\infty} \bigcup\limits_{|\xi|=j} \Delta_\xi$ (see the end of Section 5.2), for the Cartesian square of this set we have

$$C_a^2 = \left(\bigcap_{j=1}^{\infty} \bigcup_{|\xi|=j} \Delta_\xi\right) \times \left(\bigcap_{j=1}^{\infty} \bigcup_{|\sigma|=j} \Delta_\sigma\right) = \bigcap_{j=1}^{\infty} \bigcup_{|\xi|=|\sigma|=j} \Delta_\xi \times \Delta_\sigma = \bigcap_{j=1}^{\infty} \bigcup_{|\varepsilon|=2j} Q_\varepsilon.$$

6.3 Constructing the Arc

The set C_a^2 is, of course, disconnected. To obtain a connected set, we will add to C_a^2 some line segments.

Given two squares Q_ε and $Q_{\varepsilon'}$ of the same rank, we say that the former is an immediate predecessor of the latter, and call these squares adjacent, if their labels satisfy this condition (with respect to the lexicographic ordering), i.e., if the binary word ε is a direct predecessor of ε'.

The final point of a square Q_ε and the initial point of its immediate successor of the same rank will be called adjacent points. Now we "link together" the four squares contained in each square of the previous rank as follows. Starting with the squares of rank 1, connect the final point of Q_{00} with the initial point of Q_{01} by a line segment L_0, the final point of Q_{01} with the initial point of Q_{10} by a segment L', and the final point of Q_{01} with the initial point of Q_{11} by a segment L_1. The segments L_0, L', L_1 without endpoints are said to be links of rank 1. In the general case, we use similarity. Let $j \geq 2$. Every square Q_ε of rank $j - 1$ can be obtained from $[0, 1]^2$ by a homothety with coefficient a^{j-1} and an appropriate translation. This map sends L_0, L', L_1 to segments connecting the adjacent points of the four squares of rank j contained in Q_ε. These segments without endpoints are said to be links of rank j and denoted by $L_{\varepsilon 0}, L'_\varepsilon, L_{\varepsilon 1}$. Adding all the constructed links to the set C_a^2, we obtain the set

$$\mathcal{L}_a = C_a^2 \cup L_0 \cup L' \cup L_1 \cup \bigcup_{j=1}^{\infty} \bigcup_{|\varepsilon|=2j} (L_{\varepsilon 0} \cup L'_\varepsilon \cup L_{\varepsilon 1}).$$

Lemma *The set $\mathcal{L}_a$ is a simple arc.*

Proof To obtain a homeomorphism between $[0, 1]$ and $\mathcal{L}_a$, we construct a sequence of paths $\gamma_1, \gamma_2, \ldots$ whose supports are polygonal lines approximating the set $\mathcal{L}_a$ with increasing accuracy. All these paths are simple, defined on the interval $[0, 1]$, and connect the points $(0, 0)$ and $(1, 1)$.

First, we describe the path γ_1. As u increases on the interval δ_0 (or δ_1), the point $\gamma_1(u)$ runs over the link L_0 (respectively, L_1) upwards, and as u increases on δ, it runs over the link L' downwards. As u increases on any of the remaining segments Δ_ε of rank 2, the point $\gamma(u)$ runs over the diagonal of Q_ε from the initial point to the final one. Thus, as u passes from the left interval to the right one in the representation

$$[0, 1] = \Delta_{00} \cup \delta_0 \cup \Delta_{01} \cup \delta \cup \Delta_{10} \cup \delta_1 \cup \Delta_{11}, \tag{1}$$

its image $\gamma(u)$ moves successively in the intervals

$$\mathcal{D}_{00}, L_0, \mathcal{D}_{01}, L', \mathcal{D}_{10}, L_1, \mathcal{D}_{11}, \tag{2}$$

where $\mathcal{D}_\varepsilon$ is the diagonal of the square Q_ε (which connects its initial and final points). Note that the constructed map is continuous. At the interior points of the intervals (1), this follows immediately from the definition, while at the endpoints, from the consistent way of labelling the intervals (1) and (2). Then the construction (see Fig. 2) proceeds by induction.

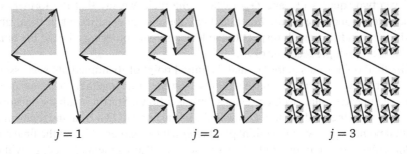

$j = 1$ $j = 2$ $j = 3$

Fig. 2

Assume that we have already constructed a simple path γ_{j-1}, and its support is a polygonal line consisting of the diagonals of all squares of rank $j - 1$ (these are the γ_{j-1}-images of segments of rank $2j - 2$) and all links of rank less than j (these are the γ_{j-1}-images of intervals of rank less than $2j - 2$). Then γ_j is obtained by correcting γ_{j-1} on segments of rank $2j - 2$ (between them, $\gamma_j = \gamma_{j-1}$) so as to traverse the links of rank j and the diagonals of the squares of the same rank. As u increases on the segment Δ_ε, $|\varepsilon| = 2j - 2$, the point $\gamma_j(u)$, like $\gamma_{j-1}(u)$, moves in the square Q_ε, but, in contrast to $\gamma_{j-1}(u)$, not along the diagonal, but along a seven-link polygonal line. More exactly, the restriction of γ_j to Δ_ε is the composition $V \circ \gamma_1 \circ U$ where U is a smooth strictly increasing function that sends Δ_ε onto $[0, 1]$ and V is the map (the composition of a translation and a homothety with coefficient a^{j-1}) that sends $[0, 1]^2$ to Q_ε. The result is a path γ_j defined on $[0, 1]$.

Obviously, each path γ_j is simple and $\gamma_j = \gamma_{j+1} = \gamma_{j+2} = \ldots$ on all intervals of rank $2j$ and $2j + 1$. Hence, links of rank j are contained not only in the polygonal line $\gamma_j([0, 1])$, but also in all the subsequent ones. Note also that the image of every segment Δ_ε, $|\varepsilon| = 2j$, is contained in the square Q_ε not only for the map γ_j, but for all the subsequent maps.

It is clear from construction that for all $u \in [0, 1]$ and all $j \in \mathbb{N}$, the value $\|\gamma_{j+1}(u) - \gamma_j(u)\|$ does not exceed the length of the diagonal of a square of rank j, i.e., does not exceed $\sqrt{2}\, a^j$. Hence, the sequence $\{\gamma_j\}$ converges uniformly to a continuous map $\gamma \colon [0, 1] \to [0, 1]^2$. The support of γ is a "polygonal line with infinitely many links" containing the links of all ranks. Obviously, the γ-image of the union of all intervals of all ranks consists of such links, i.e.,

$$\gamma([0, 1] \setminus C_a) = \mathcal{L}_a \setminus C_a^2.$$

The endpoints of links belong to the set C_a^2 and are dense in it. On the other hand, they are the values of γ at the endpoints of intervals (of all ranks), which belong to the set C_a and are dense in it. Hence, $\gamma(C_a) = C_a^2$; since $\gamma([0, 1] \setminus C_a) = \mathcal{L}_a \setminus C_a^2$, we conclude that $\gamma([0, 1]) = \mathcal{L}_a$.

As we have already mentioned, the values of all maps $\gamma_j, \gamma_{j+1}, \dots$ on a segment Δ_ε, $|\varepsilon| = 2j$, do not leave the square Q_ε, hence the limiting map also has this property, i.e., $\gamma(\Delta_\varepsilon) \subset Q_\varepsilon$ for every index ε.

It remains to prove that the map γ is one-to-one. Since the sets $\gamma([0, 1] \setminus C_a)$ and $\gamma(C_a)$ are disjoint, for $\gamma(u) = \gamma(v)$ and $u \neq v$ we have either $u, v \notin C_a$ or $u, v \in C_a$. Let us verify that neither case can happen.

By construction, at every point $u \notin C_a$, the values of all maps γ_j for sufficiently large j coincide, and, consequently, they coincide with $\gamma(u)$. Hence, for $u, v \notin C_a$, for sufficiently large j we have $\gamma_j(u) = \gamma(v)$ and $\gamma_j(v) = \gamma(v)$, so $\gamma_j(u) = \gamma_j(v)$, because $\gamma(u) = \gamma(v)$. But for $u \neq v$ the equality $\gamma_j(u) = \gamma_j(v)$ is impossible, since the path γ_j is simple.

Consider the remaining case where $\gamma(u) = \gamma(v)$, $u \neq v$, and $u, v \in C_a$. Then the points u, v belong to different segments of a sufficiently large even rank: $u \in \Delta_\varepsilon$, $v \in \Delta_\omega$, and $\Delta_\varepsilon \cap \Delta_\omega = \varnothing$, where $|\varepsilon| = |\omega|$ is a large even number. Hence, $\gamma(u) \in \gamma(\Delta_\varepsilon) \subset Q_\varepsilon$ and $\gamma(v) \in \gamma(\Delta_\omega) \subset Q_\omega$, so the point $\gamma(u) = \gamma(v)$ belongs to both squares Q_ε and Q_ω, which is impossible, because they are disjoint.

So, γ is a continuous one-to-one map from the interval $[0, 1]$ onto the set $\mathcal{L}_a$. By Theorem 0.10, it is a homeomorphism, i.e., $\mathcal{L}_a$ is a simple arc. $\square$

Note that, by construction, every link contained in $\mathcal{L}_a \setminus C_a^2$ is parallel either to L', or to L_0, L_1. Below, the following simple fact will be useful: for $a = \frac{1}{4}$, these three links are parallel to each other. More exactly, they lie on the lines $2x_1 + x_2 = \frac{3}{4}i$, $i = 1, 2, 3$. Hence, all links of the arc $\mathcal{L}_{1/4}$ lie on lines of the form $2x_1 + x_2 = \text{const}$.

6.4 Constructing the Function

Let $t \in (1, 2)$, $a \in \left(2^{-2/t}, \frac{1}{2}\right)$, so $\frac{1}{4} < a^t$, and let φ be the function of class $C^t(\mathbb{R})$, constructed in Proposition 5.3, such that it strictly increases from 0 to 1, its critical set coincides with C_a, and its set of critical values coincides with $C_{1/4}$, i.e., $\varphi(C_a) = C_{1/4}$.

Define a map $\Phi \colon \mathbb{R}^2 \to \mathbb{R}^2$ by the formula

$$\Phi(x_1, x_2) = (\varphi(x_1), \varphi(x_2)).$$

Clearly, $\Phi \in C^t(\mathbb{R}^2; \mathbb{R}^2)$, and since the coordinate functions of Φ are strictly monotone, it is a homeomorphism of the square $[0, 1]^2$ onto itself. Besides, obviously,
$\Phi(C_a^2) = C_{1/4}^2$. It follows from the properties of φ that Φ maps the squares Q_ε corresponding to C_a onto the squares $\widetilde{Q}_\varepsilon$ corresponding to $C_{1/4}$, preserving not only the
rank, but also the label and, sending adjacent points to adjacent points. As noted at
the end of the previous subsection, all links arising in the construction of the arc $\mathcal{L}_{1/4}$
lie on lines of the form $2y_1 + y_2 = \text{const}$. Therefore, the function $F(y_1, y_2) = 2y_1 + y_2$
is constant on every such link.

Now set $f = F \circ \Phi$, in other words,

$$f(x) = 2\varphi(x_1) + \varphi(x_2), \qquad x = (x_1, x_2) \in \mathbb{R}^2.$$

Clearly, $f \in C^t(\mathbb{R}^2)$, $N_f = C_a^2$, and $f(0, 0) = 0$, $f(1, 1) = 3$, so f is not constant on
the set of critical points. Since the first-order Taylor polynomials at these points are
constants, by Corollary 1 at p. 78 (see inequality (4′) for $m = 2, n = 1$, and $\alpha = t - 1$)
we have

$$|f(x) - f(y)| \le A\|x - y\|^t \quad \text{for every } y \in N_f = C_a^2, x \in \mathbb{R}^2, \tag{3}$$

where A is a constant depending only on φ.

The values of f at adjacent points coincide, because this is so for the function F,
which (as we have already observed) is constant on every link connecting adjacent
squares. We claim that there exists a function of the same smoothness as f that
coincides with f on C_a^2 and whose critical set contains not only C_a^2, but the entire
arc $\mathcal{L}_a$. Obviously, this function must be constant on every link of $\mathcal{L}_a$, because such
a link consists of its critical points, and the first-order Taylor polynomials of this
function corresponding to points of $\mathcal{L}_a$ must be constants.

Thus, we seek a smooth function whose first-order Taylor polynomials at the points
of $\mathcal{L}_a$ have degree 0, i.e., are constants. Besides, the function itself coincides with f
on C_a^2 and is constant on every link of the arc $\mathcal{L}_a$. This data allows us to describe
a family of polynomials of degree 0 (i.e., a family of constants) needed to apply
Whitney's theorem I.11.2. It should be the family of first-order Taylor polynomials
of the extended function. Note that, as follows from Whitney's theorem, the points to
which there correspond zero-degree polynomials of this family, i.e., the points of the
arc $\mathcal{L}_a$, are critical points of the extended function. We define this family $\{P_y\}_{y \in \mathcal{L}_a}$
as follows: for every $x \in \mathbb{R}^2$,

$$P_y(x) = \begin{cases} f(y) & \text{if } y \in C_a^2, \\ f(v_\ell) = f(u_\ell) & \text{if } y \in \ell \subset (\mathcal{L}_a \setminus C_a^2), \end{cases}$$

where u_ℓ, v_ℓ are the endpoints of a link ℓ contained in $\mathcal{L}_a$.

One can easily see that the map $(x, y) \mapsto P_y(x)$ is continuous on $\mathcal{L}_a \times \mathcal{L}_a$. We
will apply Whitney's theorem to the family $\{P_y\}_{y \in \mathcal{L}_a}$, thus constructing a smooth

extension of the function $y \mapsto P_y(y)$ ($y \in \mathcal{L}_a$), which coincides with f on C_a^2 and is constant on the links of $\mathcal{L}_a$. Let us check that the constructed family satisfies the assumption of Whitney's extension theorem (see Theorem I.11.2) for $m = 2$, $r = 1$, $K = E = \mathcal{L}_a$, and $\omega_K(u) = \widetilde{A} u^{t-1}$, where $\widetilde{A}$ is a constant. In our case, this assumption means that

$$|P_x(y) - P_y(y)| = |P_x(x) - P_y(y)| \le \widetilde{A}\|x - y\|^t \quad \text{for } x, y \in \mathcal{L}_a \qquad (4)$$

(the inequalities for derivatives are trivially satisfied, since all derivatives of P_y are identically zero). This inequality (it will be proved in the next subsection) allows us to apply Whitney's extension theorem, which says (see also Corollary I.11.4 for $\theta = t - 1$) that there exists a t-smooth function g such that its first-order Taylor polynomials at points $y \in \mathcal{L}_a$ coincide with P_y and hence are constants. Thus, $\operatorname{grad} g = 0$ on $\mathcal{L}_a$. In other words, the arc $\mathcal{L}_a$ is contained in the critical set of g. Since $g(0, 0) = f(0, 0) = 0$ and $g(1, 1) = f(1, 1) = 3$, the function g is not constant on this arc. Therefore, g is as required. To complete the construction of Whitney's example, it remains to prove inequality (4).

6.5 Proving Inequality (4)

Note that if both points x and y belong to the set C_a^2, then inequality (4) takes the form

$$|f(x) - f(y)| \le \widetilde{A}\|x - y\|^t$$

and, therefore, follows from (3) with $\widetilde{A} = A$.

Since the set $\mathcal{L}_a \setminus C_a^2$ of all links is dense in $\mathcal{L}_a$, it suffices to prove (4) (bearing in mind that both sides are continuous) only in the case where $x, y \in \mathcal{L}_a \setminus C_a^2$, i.e., when both these points belong to links of the arc $\mathcal{L}_a$: $x \in \ell$, $y \in \widetilde{\ell}$. We will assume that $\ell \ne \widetilde{\ell}$, since otherwise the inequality is trivial: its left-hand side vanishes.

Let j be the rank of ℓ and i be the rank of $\widetilde{\ell}$. We assume without loss of generality that $j \le i$. Also, let z be an endpoint of ℓ and w be an endpoint of $\widetilde{\ell}$ (the choice of z, w will be made more precise later). Then, by (3) we have

$$|P_x(y) - P_y(y)| = |f(z) - f(w)| \le A\|z - w\|^t$$
$$\le A(\|x - y\| + \|z - x\| + \|w - y\|)^t. \qquad (5)$$

To prove (4), we will estimate the norms $\|z - x\|$ and $\|w - y\|$ in terms of $\|x - y\|$. Below, $A_1, A_2, \ldots$ are positive coefficients depending only on a and f.

If y belongs to one of the squares (denote it by Q) connected by the link ℓ, then let z be the endpoint of this link belonging to Q. It is easy to see that

$$\|z - x\| \le A_1 \operatorname{dist}(x, Q) \le A_1\|x - y\|.$$

Now we specify the choice of w and estimate $\|w - y\|$. Since $y \in \tilde{\ell} \subset Q$, it is clear from elementary geometric considerations that at least one endpoint of $\tilde{\ell}$ is at distance at most $\sqrt{2}\|z - y\|$ from z. Take it for w. Then we have

$$\|w - y\| \le \|w - z\| + \|z - y\| \le (1 + \sqrt{2})\|z - y\| \le 3(\|x - y\| + \|z - x\|)$$
$$\le 3(1 + A_1)\|x - y\|.$$

If y does not belong to the squares connected by ℓ, then $\|x - y\| \ge A_2 a^j$. On the other hand, $\|z - x\| \le \sqrt{2}a^{j-1}$, since the points z and x, as well as the whole link ℓ, lie in a square of rank $j - 1$. Therefore,

$$\|z - x\| \le \sqrt{2}a^{j-1} \le A_3\|x - y\|.$$

For the same reason, $\|w - y\| \le \sqrt{2}a^{i-1}$ and, therefore,

$$\|w - y\| \le \sqrt{2}a^{j-1} \le A_3\|x - y\|.$$

Thus, in both cases the norms $\|z - x\|$ and $\|w - y\|$ are dominated by $\|x - y\|$ (whatever the position of y). Together with (5), this gives the inequality

$$|P_x(y) - P_y(y)| \le A_4\|x - y\|^t \quad \text{if } x, y \in \mathcal{L}_a \setminus C_a^2.$$

As we have already mentioned, this suffices for inequality (4) to hold with a constant $\widetilde{A} \ge A_4$. Now we can apply Whitney's extension theorem and thus complete the proof of Theorem 6.1.   □

6.6 Generalization to the Case of Several Variables

The constructed examples can be generalized to the case of functions of any number of variables. More exactly, the following result holds.

Theorem *For every* $t \in (1, m)$ *there exists a function of class* $C^t(\mathbb{R}^m)$ *that is not constant on a simple arc on which its gradient vanishes.*

The proof of this theorem is similar to that of Theorem 6.1. Now, the parameter $a < \frac{1}{2}$ should be chosen so that $b = 2^{-m} < a^t$, and the squares Q_ε should be replaced by m-dimensional cubes. The cubes of rank j are all the products of segments of the same rank arising in the construction of the set C_a. By the initial (final) point of a cube we mean its vertex with minimal (maximal) coordinates. Cubes whose labels immediately succeed each other (in the lexicographic ordering) are said to be adjacent. For adjacent cubes, the final point of the cube with the smaller label and the initial point of the cube with the larger label are said to be adjacent. A simple arc contained in an m-dimensional cube and passing through the set C_a^m can be obtained by analogy with the plane arc $\mathcal{L}_a$ constructed in Section 6.3 (with a modification of labels). This spatial arc will still be denoted by $\mathcal{L}_a$. It turns out that links connecting adjacent cubes corresponding to $b = 2^{-m}$ lie in parallel planes

$$2^{m-1}x_1 + 2^{m-2}x_2 + \ldots + x_m = \text{const},$$

so the function

$$x = (x_1, \ldots, x_m) \mapsto F(x) = 2^{m-1}x_1 + 2^{m-2}x_2 + \ldots + x_m,$$

which is a counterpart of the function F from Section 6.4, is constant on these planes and, in particular, takes the same values at adjacent points of the cubes arising in the construction of the set C_b^m. By analogy with what we did in the proof of Theorem 6.1, consider the homeomorphism

$$\Phi \colon [0,1]^m \to [0,1]^m \quad \text{where} \quad \Phi(x) = (\varphi(x_1), \ldots, \varphi(x_m)),$$

and for $x \in [0,1]^m$ set

$$f(x) = (F \circ \Phi)(x) = 2^{m-1}\varphi(x_1) + 2^{m-2}\varphi(x_2) + \ldots + \varphi(x_m).$$

One can easily check that $f(0, \ldots, 0) = 0$, $f(1, \ldots, 1) = 2^m - 1$ and the critical set of f coincides with C_a^m.

Note that at a critical point of f, not only the first-order derivatives of this function vanish, but all derivatives which exist according to the smoothness; this easily follows from the fact that the function φ has this property. Hence, as in the case of two variables, the Taylor polynomials of f of degree $[t]$ at critical points have zero degree and thus are constants.

A counterexample can be obtained, as in Theorem 6.1, by "correcting" f on the part of $\mathcal{L}_a$ outside C_a^m so as to make the set of critical values of the corrected function contain not only C_a^m, but the (connected) arc $\mathcal{L}_a$. As in the case of two variables, the existence of such a correction is guaranteed by Whitney's extension theorem. Here, at the points of the set C_a^m one should use polynomials coinciding with the Taylor polynomials of degree $[t]$ of the original functon (i.e., f), which, as noted above, have zero degree.

Chapter V. Addenda

1 *Smooth Partitions of Unity

Here we establish a useful result which often makes it possible to simplify a problem by making it local. Everywhere below, O is a nonempty open subset in $\mathbb{R}^m$, $O \neq \mathbb{R}^m$, and

$$\rho_x = \operatorname{dist}(x, \partial O) \quad \text{for } x \in O.$$

We introduce notation for balls used in what follows:

$$B_x = B\left(x, \frac{1}{15}\rho_x\right), \quad B_x^* = B\left(x, \frac{1}{3}\rho_x\right), \quad B_x^\circ = B\left(x, \frac{1}{2}\rho_x\right).$$

1.1 Auxiliary Inequalities

We need two simple inequalities, the first of which follows from elementary geometric considerations and the second one makes it possible to estimate higher-order partial derivatives of the product and quotient of two smooth functions.

To begin with, recall (see Section 0.9) that the function $x \mapsto \rho_x$ satisfies the Lipschitz condition:

$$|\rho_x - \rho_y| = |\operatorname{dist}(x, \partial O) - \operatorname{dist}(y, \partial O)| \leq \|x - y\|.$$

Besides, since the distance between the centers of intersecting balls does not exceed the sum of their radii, in the case where $B_x^\circ \cap B_y^\circ \neq \varnothing$ we have $\|x - y\| \leq \frac{1}{2}(\rho_x + \rho_y)$, and hence $|\rho_x - \rho_y| \leq \frac{1}{2}(\rho_x + \rho_y)$, which is equivalent to the double inequality

$$\frac{1}{3}\rho_x \leq \rho_y \leq 3\rho_x \quad \text{if } B_x^\circ \cap B_y^\circ \neq \varnothing. \tag{1}$$

© The Author(s), under exclusive license to Springer Nature Switzerland AG 2021
B. M. Makarov, A. N. Podkorytov, *Smooth Functions and Maps*, Moscow Lectures 7,
https://doi.org/10.1007/978-3-030-79438-5_5

Lemma 1 *Let* $x \in O$ *and* $Y \subset O$. *If the balls* $\{B_y\}_{y \in Y}$ *are pairwise disjoint, then at most* 10^{2m} *balls among* B_y° *intersect* B_x°.

Proof We assume without loss of generality that all balls B_y° intersect B_x°. Then $\|x - y\| \le \frac{1}{2}(\rho_x + \rho_y)$ and, by the right inequality in (1), we have $\|x - y\| \le 2\rho_x$ for every y. Hence, all balls $B(y, r)$ of radius $r = \frac{1}{45}\rho_x$ are contained in the ball $B(x, R)$ with $R = 2\rho_x + r$. Besides, these balls are pairwise disjoint, because $B(y, r) \subset B_y$ due to the left inequality in (1) and the balls B_y are disjoint by assumption. Since the volume of $B(x, R)$ is greater than the total volume of the equal balls $B(y, r)$ contained in it, the number of these balls does not exceed $\left(\frac{R}{r}\right)^m = 91^m$.   □

Below, as usual, $\alpha \in \mathbb{Z}_+^m$ is a multi-index and $|\alpha|$ is its order, i.e., the sum of the coordinates.

Lemma 2 *If* f *and* F *are functions of class* $C^\infty(O)$ *that for all multi-indices* α *satisfy the inequalities*

$$\left| f^{(\alpha)}(x) \right| \le \frac{C_\alpha'}{\rho_x^{|\alpha|}}, \quad \left| F^{(\alpha)}(x) \right| \le \frac{C_\alpha''}{\rho_x^{|\alpha|}} \quad \text{for } x \in O, \tag{2}$$

where C_α', C_α'' *are constants depending only on the order of the derivative, then the product* fF *and, if* $F \ge 1$, *the quotient* $\frac{f}{F}$ *also satisfy inequalities of this type.*

Proof The first assertion follows from Leibniz's rule for differentiating a product (see Section I.6.5).

To prove the second assertion, we set $\varphi = \frac{f}{F}$ and use induction on $i = |\alpha|$ taking $i = 0$ as the base case. In the induction step, we assume that all partial derivatives of φ of order up to i satisfy the required inequality. Let α be an arbitrary multi-index of order $i + 1$. Differentiating (via Leibniz's rule) the identity $f = F\varphi$, we write the result in the form

$$F\varphi^{(\alpha)} = f^{(\alpha)} - \sum_{\beta + \gamma = \alpha, \beta \ne 0} \frac{\alpha!}{\beta! \gamma!} F^{(\beta)} \varphi^{(\gamma)}.$$

All derivatives of φ in the right-hand side are of order at most i. Applying to them the induction hypothesis, we obtain the required inequality.   □

1.2 The Partition of Unity Theorem

We will prove that the characteristic function of every open set $O \subset \mathbb{R}^m$ can be represented as a sum of a series of nonnegative infinitely smooth functions each of which is nonzero only in a ball with closure in O. Since the characteristic function itself is discontinuous at boundary points of O, near these points the series must have terms with large partial derivatives. The construction we use allows us to control their values. More exactly, the following result holds.

Theorem *There exist a sequence of nonnegative functions φ_j of class $C^\infty(\mathbb{R}^m)$ and a sequence of open balls B_j ($j \in \mathbb{N}$) satisfying the following conditions:*

a) *the sum $\varphi_1 + \varphi_2 + \varphi_3 + \ldots$ is equal to 1 in O and 0 outside O;*

b) *if $x \in O$ and $\rho_x = \text{dist}(x, \partial O)$, then $\left|\varphi_j^{(\alpha)}(x)\right| \leq \frac{C_\alpha}{\rho_x^{|\alpha|}}$ ($j = 1, 2, \ldots$) for every multi-index α;*

c) *$\varphi_j = 0$ outside B_j, $\overline{B}_j \subset O$, and at most N (where $N = 10^{2m}$) balls among $B_1, B_2, B_3, \ldots$ intersect B_j.*

Here C_α are constants depending on the order of the derivative (but not on j).

Remark As we will see in the proof, the balls B_j are chosen so that their radii R_j are comparable with the distance from the ball to the boundary of O:

$$R_j \leq \text{dist}(B_j, \partial O) \leq 2R_j. \tag{3}$$

Since $O = \bigcup_j B_j$, it follows from c) that on every compact subset in O, only finitely many of the functions φ_j do not vanish.

Proof We keep the notation introduced at the beginning of this section. Since $O = \bigcup_{y \in O} B_y$, by the preliminary covering theorem (see Section 2.1 below), in O there exists a sequence $\{y_j\}$ such that

$$\text{the balls } B_{y_j} \text{ are pairwise disjoint and } O = \bigcup_{j=1}^\infty B_{y_j}^*.$$

Using the balls $B_j = B_{y_j}^\circ = B\left(y_j, \frac{1}{2}\rho_{y_j}\right)$ (which are obviously contained in O), we expand the characteristic function of O into a sum of nonnegative infinitely smooth functions each of which is nonzero only in its own ball B_j.

As the reader can easily verify, the radii of these balls satisfy inequality (3). Note also that, as we have established in Lemma 1, each ball B_k intersects at most $N = 10^{2m}$ other balls, and, consequently, at most N functions can be nonzero in the same ball.

Now we turn to the construction of a required partition of unity. Consider a function ω of class $C^\infty(\mathbb{R})$ such that

$$\omega(t) = 1 \qquad \text{for } t \leq 1,$$

$$0 \leq \omega(t) \leq 1 \quad \text{for } t \in \left[1, \frac{3}{2}\right], \quad \text{and}$$

$$\omega(t) = 0 \qquad \text{for } t \geq \frac{3}{2}.$$

First, we use it to define in $\mathbb{R}^m$ the auxiliary functions

$$f_j(x) = \omega\left(3\frac{\|x - y_j\|}{\rho_{y_j}}\right) \quad (j = 1, 2, \ldots).$$

Clearly, $f_j \in C^\infty(\mathbb{R}^m)$, $0 \le f_j \le 1$ everywhere, $f_j = 1$ on $B^*_{y_j}$, and $f_j = 0$ outside $B^\circ_{y_j}$. Since $O = \bigcup_j B^*_{y_j}$, at every point of O at least one of the functions $f_1, f_2, f_3, \ldots$ takes the value 1.

Note also that for every multi-index α, all values of the partial derivative $f_j^{(\alpha)}$ in the ball B_j are dominated by a value of order $\frac{1}{\rho_{y_j}^{|\alpha|}}$ (the reader is encouraged to verify this). Besides, $\frac{1}{3}\rho_x \le \rho_{y_j}$ for $x \in B_j$ by the double inequality (1). Hence, in B_j we have

$$\left| f_j^{(\alpha)}(x) \right| \le \frac{A_\alpha}{\rho_x^{|\alpha|}}, \tag{4}$$

where the coefficients A_α depend only on α and ω, but not on j and x. Since outside B_j the function f_j and all its derivatives vanish, they satisfy (2).

Now consider the sum

$$F(x) = \sum_{j=1}^\infty f_j(x) \quad (x \in O). \tag{5}$$

All its terms are nonnegative, and at least one of them is not less than 1, so $F(x) \ge 1$. As we have already noted, in each ball B_k at most N functions φ_j are nonzero, and hence in B_k the series (5) coincides with a sum of at most N terms, so $F \in C^\infty(B_k)$. Besides, it follows from (4) that

$$\left| F^{(\alpha)}(x) \right| \le N \frac{A_\alpha}{\rho_x^{|\alpha|}} \quad \text{for } x \in B_k. \tag{6}$$

Since k and α are arbitrary, it is clear that $F \in C^\infty(O)$ and (6) holds everywhere in O.

Finally, we define functions $\varphi_1, \varphi_2, \ldots$ as follows:

$$\varphi_j = \frac{f_j}{F} \quad \text{in } O \quad \text{and} \quad \varphi_j = 0 \quad \text{outside } O.$$

Then, obviously, $\varphi_j \in C^\infty(\mathbb{R}^m)$, $\varphi_j \ge 0$ everywhere, $\varphi_j = 0$ outside B_j, and $\sum_j \varphi_j = 1$ in O, which proves a) and c). Due to (4) and (6), we can apply Lemma 2 to the functions φ_j and $F \ge 1$, which implies b).   □

The theorem remains valid for $O = \mathbb{R}^m$. In this case, one can repeat the proof (with some simplifications) initially taking the cover of $\mathbb{R}^m$ by all balls of the same radius. This gives functions φ_j whose all derivatives of a given order are uniformly bounded.

Preserving the notation from the theorem, we now prove the following result.

Corollary (smoothing) *Let E be an arbitrary closed subset in O and*

$$\Phi = \sum_{\{j \,:\, B_j \cap E \ne \varnothing\}} \varphi_j.$$

Then the function Φ is equal to 1 on E, equal to 0 outside O, and infinitely differentiable in the whole space.

Thus, the characteristic function of every closed set E can be "smoothened" by modifying it within an arbitrary neighborhood of E so as to obtain a function that is infinitely differentiable in $\mathbb{R}^m$, is equal to 1 on E, and vanishes outside a neighborhood of E.

Proof The equality $\Phi(x) = 0$ for $x \notin O$ is obvious, and the equality $\Phi(x) = 1$ for $x \in E$ follows from the definition of Φ, because $\varphi_j = 0$ outside B_j.

The infinite differentiability of Φ in O can be established in the same way as the infinite differentiability of F in the proof of the theorem. To prove the infinite differentiability of Φ in the whole space $\mathbb{R}^m$, we use the double inequality (3).

Clearly, it suffices to check that $\Phi = 0$ in some neighborhood of ∂O. We will prove that $\Phi = 0$ in every ball $B\left(c, \frac{d}{4}\right)$ where $c \in \partial O$, $d = \text{dist}(c, E)$. In other words (see the definition of Φ), neither of the balls B_j intersects both the set E and the ball $B\left(c, \frac{d}{4}\right)$.

Assume to the contrary that $z_0 \in B_j \cap E$, $z \in B_j \cap B\left(c, \frac{d}{4}\right)$. Then, by (3) we have

$$R_j \le \text{dist}(B_j, \partial O) \le \|z - c\| < \frac{d}{4}.$$

Besides, since $z_0, z \in B_j$, we obtain

$$\|z_0 - z\| < 2R_j < \frac{d}{2}.$$

Therefore,

$$d = \text{dist}(c, E) \le \|z_0 - c\| \le \|z_0 - z\| + \|z - c\| < \frac{d}{2} + \frac{d}{4},$$

a contradiction. Hence, the function Φ vanishes in ∂O, which implies that it is infinitely differentiable in $\mathbb{R}^m$. $\square$

2 *Covering Theorems

Here we establish two results about coverings of subsets of a Euclidean space by balls. Although the first of them involves only simple geometric considerations, this "elementary" result often proves to be a convenient tool in function theory. Emphasizing the importance of this result, we quote the monograph [15], where it is stated as a lemma: "Not only the simplicity of its statement, or the application we use it for, but also the many variants of it that can be found in the mathematical literature attest to the fundamental character of this lemma." The second one, the famous Vitali covering theorem, is a beautiful and deep result. In particular, it allows one to justify the intuitively plausible conjecture that a subset in a Euclidean space cannot have many "rarefaction" points (they form a set of zero volume).

As we have already mentioned, the prerequisites for Chapters IV and V are a knowledge of the basic properties of the Lebesgue measure in $\mathbb{R}^m$. We will denote this measure by λ_m and call it also the m-dimensional volume. A detailed exposition of the facts we use can be found in [16, 17, 9, 14].

Below, given a ball B, by $r(B)$ and B^* we denote the radius of B and the ball of radius $5r(B)$ with the same center.

2.1 A Preliminary Theorem

The proof of the following result uses an intuitively clear property of the volume: the sum of the volumes of pairwise disjoint sets does not exceed the volume of an ambient set.

Theorem *Let $\mathcal{B}$ be a collection of balls covering a bounded set E, i.e., $E \subset \bigcup_{B \in \mathcal{B}} B$. It contains a sequence $B_1, B_2, \ldots$ (finite or otherwise) of pairwise disjoint balls such that*

$$E \subset \bigcup_{k \geq 1} B_k^*.$$

Proof We assume without loss of generality that $E \cap B \neq \emptyset$ for all balls $B \in \mathcal{B}$. Besides, the assertion is trivial if $\mathcal{B}$ contains arbitrarily large balls, since for $r(B) \geq \operatorname{diam}(E)$ and $E \cap B \neq \emptyset$ we obviously have $E \subset B^*$. So, below we assume that the bounded set E is covered by balls whose radii are bounded from above. Clearly, the set $\widetilde{E} = \bigcup_{B \in \mathcal{B}} B$ is then bounded.

A sequence with the required properties will be constructed by induction. Let $R_1 = \sup\{r(B) \colon B \in \mathcal{B}\}$. Pick a ball $B_1 \in \mathcal{B}$ of radius greater than $\frac{1}{2}R_1$. Assume that we have already chosen pairwise disjoint sets $B_1, \ldots, B_{n-1}$. If all the other balls from $\mathcal{B}$ intersect one of them, the construction is completed. Otherwise, in $\mathcal{B}$ there are balls that do not intersect $B_1, \ldots, B_{n-1}$. Let R_n be the supremum of their radii and B_n be one of these balls such that $r(B_n) > \frac{1}{2}R_n$.

Thus, either our construction eventually stops, or it consists of infinitely many steps described above. The result is a sequence (finite or otherwise) of pairwise disjoint balls $B_1, B_2, \ldots$ satisfying the condition $R_n < 2r(B_n)$.

We claim that an arbitrary point $x \in E$ belongs to the union $B_1^* \cup B_2^* \cup \ldots$. Let B be a ball from $\mathcal{B}$ such that $x \in B$. First, we verify that it has a nonempty intersection with one of the balls chosen. This is obvious if there are finitely many of them, since otherwise the construction could be continued. If the number of balls is infinite and none of them intersects B, then, by the definition of R_n, we have

$$0 < r(B) \leq R_n < 2r(B_n) \quad \text{for every } n,$$

so $r(B_n) \nrightarrow 0$. But, on the other hand, $B_1, B_2, \ldots$ are pairwise disjoint balls lying in the bounded set $\widetilde{E}$. Hence,

$$\lambda_m(B_1) + \lambda_m(B_2) + \ldots \le \lambda_m(\widetilde{E}) < +\infty, \tag{1}$$

and, consequently, $r(B_n) \to 0$. This contradiction shows that the intersection $B \cap B_n$ is not empty for some n.

Let k be the smallest among the numbers $n > 1$ such that $B \cap B_n \ne \emptyset$, so

$$B \cap (B_1 \cup \ldots \cup B_{k-1}) = \emptyset.$$

Then $r(B) \le R_k$ (for $k = 1$ this inequality is trivial, and for $k > 1$ it follows from the fact that the ball B does not intersect the union $B_1 \cup \ldots \cup B_{k-1}$). Since $R_k < 2r(B_k)$, it is clear that

$$\text{diam}(B) = 2r(B) \le 2R_k < 4r(B_k).$$

We claim that $x \in B_k^*$. Indeed, let x_k be the center of B_k. Since $B \cap B_k \ne \emptyset$, for $y \in B \cap B_k$ we obtain

$$\|x - x_k\| \le \|x - y\| + \|y - x_k\| \le \text{diam}(B) + r(B_k) < 5r(B_k).$$

Thus, every point $x \in E$ belongs to some ball B_k^*. $\qquad\qquad\qquad\square$

It is seen from the proof that in the case where the radii of the balls are bounded, the conclusion of the theorem holds for every infinite sequence of pairwise disjoint balls $B_n \in \mathcal{B}$ satisfying the inequalities

$$r(B_1) > \frac{1}{2} \sup\{r(B) : B \in \mathcal{B}\}$$

and

$$r(B_n) > \frac{1}{2} \sup\{r(B) : B \in \mathcal{B}, \, B \cap (B_1 \cup \ldots \cup B_{n-1}) = \emptyset\} \quad (n > 1). \tag{2}$$

2.2 The Vitali Covering Theorem

The result proved in the previous subsection can be refined substantially if the cover satisfies an additional condition.

Definition A collection $\mathcal{B}$ of open balls is called a *Vitali cover*[1] of a set E if for every point $x \in E$ in $\mathcal{B}$ there is an arbitrarily small ball containing x.

Theorem *Every Vitali cover $\mathcal{B}$ of a bounded set $E \subset \mathbb{R}^m$ contains a sequence (finite or otherwise) of balls B_k with the following properties:*
    1) $B_1, B_2, \ldots$ are pairwise disjoint;

---

[1] Giuseppe **Vitali** (1875–1932) was an Italian mathematician.

2) $E \subset \bigcup_{k \geq 1} B_k^*$;

3) $\lambda_m \left(E \setminus \bigcup_{k \geq 1} B_k \right) = 0$.

Note that E is not assumed to be measurable.

Proof Discarding if necessary balls with too large radii, we assume that $r(B) < 1$ for all $B \in \mathcal{B}$.

Let $B_1, B_2, \ldots$ be the balls from the proof of Theorem 2.1. The sequence $\{B_k\}_{k \geq 1}$ has properties 1 and 2. Let us verify that it has property 3 too.

If this sequence is finite and consists of n balls, then

$$B \cap (B_1 \cup \ldots \cup B_n) \neq \emptyset \quad \text{for every } B \in \mathcal{B}.$$

Since every point in E belongs to balls with arbitrarily small radii, this means that it belongs to the closure of the set $B_1 \cup \ldots \cup B_n$ and, consequently, $E \subset \bar{B}_1 \cup \ldots \cup \bar{B}_n$. Therefore,

$$E \setminus (B_1 \cup \ldots \cup B_n) \subset \partial B_1 \cup \ldots \cup \partial B_n,$$

and hence condition 3 is satisfied, because the boundaries of balls have zero volume.

Now consider the case where the sequence $B_1, B_2, \ldots$ is infinite. It is easy to verify that for every n, the subcollection in $\mathcal{B}$ (we will denote it by $\mathcal{B}_n$) consisting of the balls that do not intersect $B_1, \ldots, B_n$ is a Vitali cover of the difference

$$E_n = E \setminus (\bar{B}_1 \cup \ldots \cup \bar{B}_n)$$

and, besides, the balls $B_{n+1}, B_{n+2}, \ldots$ satisfy inequality (2). By Theorem 2.1 with E and $\mathcal{B}$ replaced by E_n and $\mathcal{B}_n$, we have $E_n \subset B_{n+1}^* \cup B_{n+2}^* \cup \ldots = C_n$. Besides,

$$\lambda_m(C_n) \leq \sum_{k > n} \lambda_m(B_k^*) = 5^m \sum_{k > n} \lambda_m(B_k) \xrightarrow[n \to \infty]{} 0,$$

because the series $\sum_{k \geq 1} \lambda_m(B_k)$ converges (see inequality (1)).

Since

$$E \setminus \bigcup_{k \geq 1} \bar{B}_k \subset E_n \subset C_n$$

for every n, we obtain

$$E \setminus \bigcup_{k \geq 1} \bar{B}_k \subset C = \bigcap_{n \geq 1} C_n,$$

with $\lambda_m(C) = 0$. Therefore,

$$E \setminus \bigcup_{k \geq 1} B_k \subset \left(E \setminus \bigcup_{k \geq 1} \bar{B}_k \right) \cup \left(\bigcup_{k \geq 1} \partial B_k \right) \subset C \cup \bigcup_{k \geq 1} \partial B_k.$$

This completes the proof of property 3, since all sets in the right-hand side have zero volume. $\square$

Due to the Vitali covering theorem, it is easy to see that an open set can be almost entirely exhausted by arbitrarily small balls (another, more involved, proof of this result, which does not use the Vitali theorem, can be found in [9, Section II.6.5]).

Corollary *Let $\varepsilon > 0$ and G be a nonempty open subset in $\mathbb{R}^m$. There exist a (possibly finite) sequence of pairwise disjoint balls B_j with radii at most ε and a set e of zero volume such that*

$$G = e \cup \bigcup_{j \geq 1} B_j.$$

Proof Consider the system $\mathcal{B}$ of "small" open balls contained in G:

$$\mathcal{B} = \{B : B \subset G, \ r(B) \leq \varepsilon\}.$$

It is obviously a Vitali cover of G. If this set is bounded, then, by the Vitali theorem, in $\mathcal{B}$ there is a sequence $\{B_j\}_{j \geq 1}$ such that the difference $e = G \setminus \bigcup_j B_j$ has zero volume.

If the set G is not bounded, one should divide it (up to a set of zero volume) into bounded open subsets and apply the result already proved to each of them. □

2.3 Density Points

An important corollary of the Vitali covering theorem is related to density points.

Definition Let $E \subset \mathbb{R}^m$ and λ_m^* be the Lebesgue outer measure in $\mathbb{R}^m$. A point x is called a *density point* of E if

$$\frac{\lambda_m^*(E \cap B(x,r))}{\lambda_m(B(x,r))} \xrightarrow[r \to 0]{} 1.$$

Here (as well as in the theorem below), E is not assumed to be measurable. Obviously, a density point of E may either belong to E or not.

Theorem *Almost every point of a set $E \subset \mathbb{R}^m$ is a density point of this set: if E' is the set of density points of E, then $\lambda_m(E \setminus E') = 0$.*

Proof Assume without loss of generality that E is bounded. Since the difference $E \setminus E'$ is exhausted by the sets

$$E_\theta = \left\{ x \in E : \lim_{r \to 0} \frac{\lambda_m^*(E \cap B(x,r))}{\lambda_m(B(x,r))} < \theta \right\}, \quad \theta \in (0,1),$$

which increase with θ, it suffices to show that each of them has zero volume. Fix θ up to the end of the proof. Note that the set E_θ is bounded, since $E_\theta \subset E$.

It easily follows from the definition of the Lebesgue outer measure that for every positive ε there is a neighborhood G of the set E_θ such that

$$\lambda_m(G) < \lambda_m^*(E_\theta) + \varepsilon.$$

Besides, for every point $x \in E_\theta$ there exist arbitrarily small balls $B(x, r)$ such that

$$\lambda_m^*(E_\theta \cap B(x, r)) < \theta \lambda_m(B(x, r)).$$

Below, we take into consideration only those among them which are contained in G. They form a Vitali cover of the set E_θ and hence contain a sequence $B_1, B_2, \ldots$ of pairwise disjoint balls such that the difference $e = E_\theta \setminus (B_1 \cup B_2 \cup \ldots)$ has zero volume. Since

$$E_\theta \subset e \cup (E_\theta \cap B_1) \cup (E_\theta \cap B_2) \cup \ldots,$$

we obtain

$$\lambda_m^*(E_\theta) \le \lambda_m^*(e) + \sum_{k \ge 1} \lambda_m^*(E_\theta \cap B_k) \le \sum_{k \ge 1} \theta \lambda_m(B_k)$$

$$= \theta \lambda_m(B_1 \cup B_2 \cup \ldots) \le \theta \lambda_m(G) < \theta(\lambda_m^*(E_\theta) + \varepsilon).$$

Thus, $\lambda_m^*(E_\theta) < \frac{\theta}{1-\theta} \varepsilon$. Since ε is arbitrary, this implies that $\lambda_m^*(E_\theta) = 0$, which means that E_θ is a set of zero volume. $\qquad\square$

The theorem we have proved implies that not only $E \setminus E'$, but also $E' \setminus E$ has zero volume (otherwise, $E' \setminus E$ would contain density points of the difference $\mathbb{R}^m \setminus E$, which is impossible).

3 *Hausdorff Measures and Hausdorff Dimension

In this section, we introduce a family of Hausdorff measures μ_p ($p > 0$) which are defined on subsets of Euclidean spaces and generalize the Lebesgue measure. For $p = m$ the measure μ_p in $\mathbb{R}^m$ is proportional to λ_m, and for $p = 1, 2, \ldots, m - 1$ these measures are generalizations of Lebesgue measures which can be defined in a natural way on p-dimensional affine subspaces in $\mathbb{R}^m$ due to their being isometric to $\mathbb{R}^p$.

3.1 Outer Measures

The construction of Hausdorff measures will be based on the notion of outer measure, so we recall its definition.

An *outer measure* (on a set X) is a nonnegative function μ (which is allowed to take the value $+\infty$) defined on all subsets in X that vanishes on the empty set ($\mu(\varnothing) = 0$) and is countably subadditive, i.e., has the following property:

if $E, E_1, E_2, \ldots$ are arbitrary subsets such that $E \subset \bigcup_{k \ge 1} E_k$, then

$$\mu(E) \le \sum_{k \ge 1} \mu(E_k).$$

This property (for $E_2 = E_3 = \ldots = \varnothing$) implies that an outer measure is monotone: if $E \subset E_1$, then $\mu(E) \le \mu(E_1)$.

An outer measure is not necessarily an additive function: for some disjoint sets E, E' in X, the equality $\mu(E \cup E') = \mu(E) + \mu(E')$ may fail to hold. A subset A in X is said to be *measurable* (more exactly, μ-measurable) if it splits any set additively:

$$\mu(E) = \mu(E \cap A) + \mu(E \setminus A) \quad \text{for every set } E \subset X.$$

One can prove (see, e.g., [16, 17, 9, 14]) that the collection of all measurable sets is an algebra on which the outer measure μ is additive, and the restriction of μ to this algebra is a measure (i.e., is countably additive).

3.2 The Construction of the Hausdorff Measures

The construction of the measures μ_p is based on a clear geometric characteristic of a set, its diameter. Recall that the diameter of a set E is the value

$$\operatorname{diam}(E) = \sup\{\|x - y\| : x, y \in E\}.$$

The diameter of the empty set is assumed to be zero. The diameter of an interval coincides with its length.

Let $\varepsilon > 0$, $E \subset \mathbb{R}^m$. A family of sets $\{e_\alpha\}_{\alpha \in A}$ is called an ε-*cover* of E if

$$E \subset \bigcup_{\alpha \in A} e_\alpha \quad \text{and} \quad \operatorname{diam}(e_\alpha) \le \varepsilon \quad \text{for every } \alpha \in A.$$

Since in what follows we need only at most countable covers, below we assume, without explicitly mentioning this, that the set A is countable. It may be assumed without loss of generality that $A = \mathbb{N}$, as we usually do, yet sometimes it is convenient to consider also other sets of indices. Clearly, for every $\varepsilon > 0$ the space $\mathbb{R}^m$ and, consequently, every subset in $\mathbb{R}^m$ has a countable ε-cover.

For arbitrary $p > 0$, $\varepsilon > 0$, and $E \subset \mathbb{R}^m$, set

$$\mu_p(E; \varepsilon) = \inf\Big\{\sum_{j \ge 1} \operatorname{diam}^p(e_j) : \{e_j\}_{j \ge 1} \text{ is an } \varepsilon\text{-cover of } E\Big\}. \tag{1}$$

Since the diameter of a set remains unchanged when we pass to its closure or convex hull, it may be assumed without loss of generality that the sets e_j are compact, or convex, or both.

Obviously, the function $\varepsilon \mapsto \mu_p(E; \varepsilon)$ (which is allowed to take infinite values) increases as ε decreases, hence there exists a limit

$$\lim_{\varepsilon \to +0} \mu_p(E; \varepsilon) = \sup_{\varepsilon > 0} \mu_p(E; \varepsilon),$$

which will be denoted by $\mu_p(E)$ and called the *Hausdorff measure*, or, in more detail, *p-dimensional Hausdorff measure* of the set E.

One can easily see that using only open ε-covers in the formula for $\mu_p(E; \varepsilon)$ yields the same limit $\lim_{\varepsilon \to +0} \mu_p(E; \varepsilon)$.

In a similar way, for a fixed $\varepsilon > 0$, the function $E \mapsto \mu_p(E; \varepsilon)$ (which is, like μ_p, defined on all subsets in $\mathbb{R}^m$) will be called the *approximating p-dimensional Hausdorff measure*.

The term "Hausdorff measure" is not quite accurate. Actually, the functions μ_p and $\mu_p(\,\cdot\,; \varepsilon)$ are not measures, but outer measures. The countable subadditivity of μ_p will be established in the next subsection, and then the reader will easily prove the same property for $\mu_p(\,\cdot\,; \varepsilon)$ by analogy. To obtain the Hausdorff measure in the strict sense of the word, one should consider μ_p only on measurable subsets, which include, in particular, all null sets, i.e., sets of zero measure, as well as all Borel sets, i.e., sets from the minimal σ-algebra containing all open sets (see [9, § I.1, I.6]).

Note also that, interpreting the space $\mathbb{R}^m$ as a subspace in $\mathbb{R}^n$ (for $n > m$), we can regard every set E in $\mathbb{R}^m$ as a subset in $\mathbb{R}^n$. Calculating the diameter of E in $\mathbb{R}^m$ and $\mathbb{R}^n$, obviously, gives the same result. Thus, the value $\mu_p(E)$ does not depend on the ambient space. That is why, we need not indicate the space in which E is assumed to lie. So, we usually omit mentioning this space, specifying it only if necessary. In this connection note that for subsets in $\mathbb{R}^m$, the (outer) measures μ_p are of interest only for $p \le m$, since otherwise μ_p is the zero measure in $\mathbb{R}^m$ (see Theorem 3.4 and Lemma 3.5).

Usually, one defines the Hausdorff measures not in terms of the sums from (1), but in terms of the values

$$\frac{\pi^{p/2}}{2^p \Gamma\left(\dfrac{p}{2} + 1\right)} \sum_{j \ge 1} \operatorname{diam}^p(e_j),$$

which are proportional to them. Due to the factor in front of the sum, the resulting Hausdorff (outer) measure for $p = m$ coincides with the Lebesgue outer measure λ_m^*. However, we are interested mainly in sets of zero measure, and the collection of such sets does not depend on the proportionality factor. That is why, we chose to use the somewhat unconventional, but simpler formula (1).

3.3 The Basic Properties of the Hausdorff Measures

First of all, note that

$$0 \le \mu_p(E) \le +\infty, \quad \mu_p(\varnothing) = 0,$$

and the function μ_p is monotone:

$$\text{if } E \subset F, \quad \text{then } \mu_p(E) \leq \mu_p(F).$$

These properties are obvious. The following result justifies the term "measure" for the function μ_p.

Theorem *The function μ_p is countably subadditive. In more detail*:

$$\text{if } E \subset \bigcup_{k \geq 1} E_k, \quad \text{then } \mu_p(E) \leq \sum_{k \geq 1} \mu_p(E_k)$$

(*and, consequently, μ_p is an outer measure*).

Proof We will assume that $\sum\limits_{k \geq 1} \mu_p(E_k) < +\infty$, since otherwise the inequality to be proved is trivial. Fix $\varepsilon > 0$ and for every k consider an ε-cover $\{e_j^{(k)}\}_{j \geq 1}$ of E_k such that

$$\sum_{j \geq 1} \text{diam}^P(e_j^{(k)}) < \mu_p(E_k; \varepsilon) + \frac{\varepsilon}{2^k} \quad (k = 1, 2, \ldots).$$

Obviously, the union of all such covers, i.e., the family $\{e_j^{(k)}\}_{k,j \geq 1}$, is an ε-cover of E and, therefore,

$$\mu_p(E; \varepsilon) \leq \sum_{k,j \geq 1} \text{diam}^P(e_j^{(k)}) \leq \sum_{k \geq 1}\left(\mu_p(E_k; \varepsilon) + \frac{\varepsilon}{2^k}\right) \leq \sum_{k \geq 1} \mu_p(E_k) + \varepsilon.$$

Taking the limit as $\varepsilon \to 0$ in the leftmost and rightmost sides of this inequality completes the proof. $\square$

We mention several other simple yet important properties of the Hausdorff measures.

1. *On sets that are "far apart" from each other, the function μ_p is additive. More exactly, we say that sets E are F are separated if*

$$\inf\{\|x - y\| : x \in E, \ y \in F\} > 0.$$

For separated sets, $\mu_p(E \cup F) = \mu_p(E) + \mu_p(F)$.

Proof Since $\mu_p(E \cup F) \leq \mu_p(E) + \mu_p(F)$ by the subadditivity of μ_p, we must prove only the opposite inequality.

Let $0 < \varepsilon < \inf\{\|x - y\| : x \in E, \ y \in F\}$. Consider an arbitrary ε-cover $\{e_j\}_{j \geq 1}$ of the set $E \cup F$. By the choice of ε, for every j at least one of the intersections $e_j \cap E, e_j \cap F$ is empty, hence

$$\sum_{j \geq 1} \text{diam}^P(e_j) \geq \sum_{e_j \cap E \neq \varnothing} \text{diam}^P(e_j) + \sum_{e_j \cap F \neq \varnothing} \text{diam}^P(e_j).$$

Since $\{e_j\}_{e_j \cap E \neq \varnothing}$ and $\{e_j\}_{e_j \cap F \neq \varnothing}$ are ε-covers of E and F, we have

$$\sum_{j \geq 1} \text{diam}^P(e_j) \geq \mu_p(E; \varepsilon) + \mu_p(F; \varepsilon).$$

Taking the infimum of the left-hand side over all ε-covers, we conclude that $\mu_p(E \cup F; \varepsilon) \geq \mu_p(E; \varepsilon) + \mu_p(F; \varepsilon)$. To complete the proof, it suffices to take the limit as $\varepsilon \to 0$. $\square$

2. *Let $E \subset \mathbb{R}^m$ and assume that a map $\Phi \colon E \to \mathbb{R}^n$ satisfies the Lipschitz condition*:
$$\|\Phi(x) - \Phi(y)\| \leq L\|x - y\| \quad for\ x, y \in E,$$
where L is a constant. Then
$$\mu_p(\Phi(E)) \leq L^p \mu_p(E).$$

In particular, Lipschitz maps preserve null sets for every $p > 0$: if $\mu_p(E) = 0$, then $\mu_p(\Phi(E)) = 0$.

Proof Let $\mu_p(E) < +\infty$ and $\{e_j\}_{j \geq 1}$ be an ε-cover of E such that
$$\sum_{j \geq 1} \operatorname{diam}^p(e_j) < \mu_p(E; \varepsilon) + \varepsilon.$$

We will assume that $e_j \subset E$ for all j (otherwise, replace e_j with $e_j \cap E$). Since $\operatorname{diam}(\Phi(e_j)) \leq L \operatorname{diam}(e_j) \leq L\varepsilon$, the sets $\Phi(e_j)$ form an $L\varepsilon$-cover of $\Phi(E)$, hence
$$\mu_p(\Phi(E); L\varepsilon) \leq \sum_{j \geq 1} \operatorname{diam}^p(\Phi(e_j)) \leq L^p \sum_{j \geq 1} \operatorname{diam}^p(e_j) \leq L^p(\mu_p(E; \varepsilon) + \varepsilon).$$

Taking the limit as $\varepsilon \to 0$ yields the required inequality. $\square$

Remark Null sets are preserved not only by Lipschitz maps, but also by locally Lipschitz maps (for example, maps that are smooth in a neighborhood of a set). To see this, divide the set into countably many parts on each of which the map satisfies the Lipschitz condition (with different constants for different parts), apply the obtained result to each of these parts, and then use the countable subadditivity of μ_p.

Property 2 applied both to a map and its inverse implies two useful properties of the Hausdorff measures.

3. *If a map Φ preserves distances between points of E, then $\mu_p(\Phi(E)) = \mu_p(E)$. In particular, the Hausdorff outer measure is invariant under translations and orthogonal transformations.*

4. *The Hausdorff measures of similar sets are proportional. More exactly,*
$$\mu_p(aE) = |a|^p \, \mu_p(E), \quad where\ aE = \{ax \colon x \in E\}\ (a \in \mathbb{R}).$$

3.4 Relationship With the Lebesgue Measure

Here we show that for $p = m$ the Hausdorff measure is proportional to the m-dimensional Lebesgue measure λ_m^*. In the proof we rely on the isodiametric inequality, which says that among all measurable sets with the same diameter the ball has the largest volume. For the proof of this result, see, e.g, [9, Section II.8.3].

Theorem *For every subset E in $\mathbb{R}^m$,*

$$\mu_m(E) = \frac{2^m}{\alpha_m} \lambda_m^*(E),$$

where α_m stands for the volume of the unit ball in $\mathbb{R}^m$.

In particular, $\mu_m([0, 1]^m) = \frac{2^m}{\alpha_m}$.

Proof For $\varepsilon > 0$, consider an arbitrary ε-cover of E by compact sets e_j. The isodiametric inequality implies that

$$\lambda_m^*(E) \leq \sum_{j \geq 1} \lambda_m(e_j) \leq \sum_{j \geq 1} \frac{\alpha_m}{2^m} \operatorname{diam}^m(e_j).$$

Taking the infimum over all such covers yields an upper bound on $\lambda_m^*(E)$:

$$\lambda_m^*(E) \leq \frac{\alpha_m}{2^m} \mu_m(E; \varepsilon) \leq \frac{\alpha_m}{2^m} \mu_m(E).$$

Now we prove the opposite inequality:

$$\lambda_m^*(E) \geq \frac{\alpha_m}{2^m} \mu_m(E).$$

It follows from the definition of λ_m^* that

$$\lambda_m^*(E) = \inf\{\lambda_m(G) : G \text{ is a neighborhood of } E\}.$$

Hence, it suffices to prove that

$$\lambda_m(G) \geq \frac{\alpha_m}{2^m} \mu_m(E; \varepsilon)$$

for every $\varepsilon > 0$ and every neighborhood G of E. By the corollary of the Vitali covering theorem (see Section 2.2), this neighborhood can be represented in the form $G = e \cup B_1 \cup B_2 \cup \dots$, where $B_1, B_2, \dots$ are pairwise disjoint balls with diameter at most ε and e is a set of zero volume. The reader can easily check that it can be covered by cubes with arbitrarily small total volume. Obviously, the diameters of these cubes are also arbitrarily small, so $\mu_m(e; \varepsilon) = 0$. Since $\{B_j\}_{j \geq 1}$ is an ε-cover of the difference $G \setminus e$ and, consequently, of $E \setminus e$, it is clear that

$$\mu_m(E;\varepsilon) \le \mu_m(e;\varepsilon) + \mu_m(E \setminus e;\varepsilon) \le \sum_{j \ge 1} \text{diam}^m(B_j)$$

$$= \frac{2^m}{\alpha_m} \sum_{j \ge 1} \lambda_m(B_j) = \frac{2^m}{\alpha_m} \lambda_m(G \setminus e) = \frac{2^m}{\alpha_m} \lambda_m(G).$$

Taking the limit as $\varepsilon \to 0$, we conclude that $\mu_m(E) \le \frac{2^m}{\alpha_m} \lambda_m(G)$ for every neighborhood G of E. Therefore, $\mu_m(E) \le \frac{2^m}{\alpha_m} \lambda_m^*(E)$. Together with the opposite inequality already proved above this completes the proof. $\square$

3.5 The Hausdorff Dimension

Now we discuss the dependence of $\mu_p(E)$ on p. Obviously, the measure $\mu_p(E)$ decreases as p grows. This statement can be considerably refined.

Lemma I. *If $\mu_p(E) < +\infty$ and $q > p$, then $\mu_q(E) = 0$.*
   II. *If $\mu_p(E) > 0$ and $q < p$, then $\mu_q(E) = +\infty$.*

Thus, the measure $\mu_p(E)$ can take a finite nonzero value for at most one value of p.

Proof One can easily see that properties I and II imply each other. Let us prove property I.

Since $\mu_p(E) < +\infty$, for every $\varepsilon > 0$ there is an ε-cover $\{e_j\}_{j \ge 1}$ of E such that

$$\sum_{j \ge 1} \text{diam}^p(e_j) < C = \mu_p(E) + 1.$$

Therefore, for $\sigma = q - p > 0$ we have

$$\mu_q(E) \le \sum_{j \ge 1} \text{diam}^q(e_j) \le \varepsilon^\sigma \sum_{j \ge 1} \text{diam}^p(e_j) \le \varepsilon^\sigma C.$$

It remains to take the limit as $\varepsilon \to 0$. $\square$

The lemma implies that for every set E (if $\mu_q(E) > 0$ for at least one q) we have

$$\inf\{q > 0: \mu_q(E) = 0\} = \sup\{q > 0: \mu_q(E) = +\infty\}.$$

This critical value which characterizes the set E is of special interest. It is called the *Hausdorff dimension* of E and denoted by $\dim_H(E)$. If $\mu_q(E) = 0$ for all $q > 0$, then, by definition, $\dim_H(E) = 0$.

Here are the basic properties of Hausdorff dimension.

1. *The Hausdorff dimension is monotone: if $E \subset F$, then $\dim_H(E) \le \dim_H(F)$.*

2. *If $0 < \mu_p(E) < +\infty$, then $\dim_H(E) = p$.*

These properties immediately follow from the lemma and the definition of $\dim_H$.

3. *The Hausdorff dimension does not increase under Lipschitz (and locally Lipschitz) maps (since the image of a set of zero measure also has zero measure).*

4. *The Hausdorff dimension of a union is easy to compute*:

$$\dim_H \left(\bigcup_{j \geq 1} E_j \right) = \sup_{j \geq 1} \dim_H (E_j). \tag{2}$$

Indeed, let $E = \bigcup_{j \geq 1} E_j$ and $p = \sup_{j \geq 1} \dim_H (E_j)$. Since the Hausdorff dimension is monotone, $\dim_H (E) \geq p$. To prove the opposite inequality, assume that $p < +\infty$. In this case, if $q > p$, then $\mu_q(E_j) = 0$ for all j and, therefore, $\mu_q(E) = 0$, so, by the definition of Hausdorff dimension, $\dim_H (E) \leq q$. Since q is arbitrary, it follows that $\dim_H (E) \leq p$, and together with the opposite inequality this proves (2).

5. *We have* $\dim_H ([0,1]^m) = \dim_H (\mathbb{R}^m) = m$ (this follows from Theorem 3.4 and formula (2)). Thus, the Hausdorff dimension of every subset in $\mathbb{R}^m$ is at most m.

3.6 The Hausdorff Measure and Hausdorff Dimension of Cantor-Like Sets

Let us find the Hausdorff dimension of the sets C_a constructed in Section IV.5.2.

Proposition *Let* $0 < a < \frac{1}{2}$, $a^p = \frac{1}{2}$. *Then* $\mu_p(C_a) = 1$.

Due to property 2 from the previous subsection, this immediately implies the following result.

Corollary *We have* $\dim_H (C_a) = p$.

Proof First, we estimate $\mu_p(C_a)$ from above. Since C_a is contained in the union of 2^k segments of rank k, which have length a^k, if follows from the definition of the approximating measure $\mu_p(\cdot; \varepsilon)$ (see Section 3.2) that for every $k \in \mathbb{N}$ we have

$$\mu_p(C_a; a^k) \leq 2^k (a^k)^p = (2a^p)^k = 1.$$

This implies the upper bound $\mu_p(C_a) \leq 1$.

Now, we prove the more complicated opposite inequality. It suffices to check that for every cover $\mathcal{F}$ of C_a by open intervals, the sum of the pth powers of their lengths is not less than 1. Obviously, we may assume that the number of these intervals is finite and that all of them intersect C_a. Besides, we may assume that they are pairwise disjoint. Indeed, otherwise $\mathcal{F}$ contains a pair of intervals e_1, e_2 with a nonempty intersection. Disregarding the trivial case where one of them is contained in the other, we pick a point in $e_1 \cap e_2$ that does not belong to C_a. It divides $e_1 \cup e_2$ into two new shorter intervals, one contained in e_1 and the other one in e_2. Obviously, replacing the original intervals in $\mathcal{F}$ by these new intervals, we not only obtain a cover of C_a again, but decrease the sum of the pth powers of the lengths of the intervals.

So, we will assume that the cover $\mathcal{F}$ consists of finitely many pairwise disjoint intervals intersecting C_a and e is the shortest of them. Let Δ be a segment of minimum rank entirely contained in e and Δ' be the segment of the previous rank containing it. Then $\Delta' \not\subset e$. For definiteness, assume that the segments Δ and Δ' have the same right endpoint. Let k be the rank of Δ'.

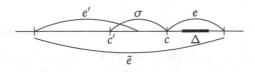

Fig. 1

Denote the left endpoint of e by c, so $\min \Delta' < c$ (see Fig. 1). Also, let c' be the point from $C_a \cap [0, c]$ closest to e and e' be the interval from $\mathcal{F}$ containing c'. Then $c' < c$ (since $e \cap e' = \varnothing$) and the interval $\sigma = (c', c)$ contains no point of C_a. Clearly, $\sigma \subset \Delta'$ and, consequently, σ is a part of one of the additional intervals, contained in Δ', arising in the construction of C_a. The ranks of these intervals are not less than the rank of Δ', i.e., than k. Thus, the length of σ is at most $(1 - 2a)a^k$.

Let $\widetilde{e}$ be the smallest interval containing the union $e' \cup e$. Clearly,

$$e' \cup \sigma \cup \{c\} \cup e = \widetilde{e}.$$

Denote the lengths of the intervals e, e', σ, and $\widetilde{e}$ by u, v, ℓ, and w. Since the length of $\widetilde{e}$ is equal to the sum of the lengths of e, e' and the gap between them (lying in σ), it is clear that

$$w \le u + v + \ell.$$

Besides, $u \ge a^{k+1}$ (since e contains the segment Δ of rank $k + 1$) and, as we have already noted, $\ell \le (1 - 2a)a^k$. Hence,

$$w \le u + v + (1 - 2a)a^k \le v + u + \frac{1 - 2a}{a}u = v + \frac{1 - a}{a}u$$

$$= v\left(1 + \frac{1 - a}{a}\frac{u}{v}\right) \le v\left(1 + \left(\frac{u}{v}\right)^p\right)^{1/p} = (v^p + u^p)^{1/p}.$$

The last inequality follows from the concavity of the function $t \mapsto (1 + t^p)^{1/p}$ on $[0, +\infty)$ (its derivative, equal to $(1 + t^{-p})^{(1/p)-1}$, decreases for $t \ge 0$). Therefore, the line segment connecting the points $(0, 1)$ and $(1, 2^{1/p})$ of the graph lies below the graph:

$$1 + (2^{1/p} - 1)t \le (1 + t^p)^{1/p} \quad \text{for } t \in [0, 1].$$

Since $2^{1/p} = 1/a$, this inequality with $t = u/v$ implies the required one (here $u/v \le 1$, because u is the length of the shortest interval in $\mathcal{F}$).

So, $w \le (u^p + v^p)^{1/p}$, i.e., $w^p \le v^p + u^p$. Hence, replacing the intervals e and e' in the original cover by the interval $\widetilde{e}$, we obtain a new cover consisting of a smaller number of intervals and having a smaller sum of the pth powers of their lengths.

Repeating this procedure sufficiently many times, we arrive at a cover of $[0, 1]$ consisting of a single interval of length greater than 1. Since at each step the sum of the pth powers of the lengths of the intervals did not increase, we obtain a lower bound on the Hausdorff measure: $\mu_p(C_a) \geq 1$. Together with the upper bound already proved, this completes the proof.   □

3.7 The Cantor Function Corresponding to the Set C_a

In this subsection, with every Cantor-like set C_a we associate a function φ_a similar to the classical Cantor function. It is of interest for us, since the corresponding Lebesgue–Stieltjes measure can be used to calculate the Hausdorff measure of subsets in C_a.

The construction of the function φ_a follows the same pattern as that of the set C_a, and we keep the notation from that construction. By definition, $\varphi_a(0) = 0, \varphi_a(1) = 1$. On the closure of the interval δ, the function φ_a is equal to the half sum of these values: $\varphi_a(x) = \frac{1}{2}$ if $x \in \overline{\delta} = [a, 1 - a]$. On the segments Δ_0 and Δ_1 complementary to δ, the same procedure is repeated: on $\overline{\delta}_0$, the value of φ_a is equal to the half sum of its values at the endpoints of Δ_0 (i.e., to $\frac{1}{4}$), and on $\overline{\delta}_1$, it is equal to the half sum of its values at the endpoints of Δ_1 (i.e., to $\frac{3}{4}$). And so on: at the kth step, we define φ_a on the closure of each interval of rank k by setting the value of φ_a on $\overline{\delta}_\varepsilon$, $|\varepsilon| = k$, to be the half sum of its values at the endpoints of the segment Δ_ε (which are defined at the previous step of the construction).

The result is a continuous nondecreasing function φ_a defined on a dense subset in $[0, 1]$. It can be uniquely extended to $[0, 1]$ as a monotone or continuous function. By construction, the increment of φ_a on every segment Δ_ε, $|\varepsilon| = k$, is 2^{-k}. Recall that the length of this segment is a^k. Let $\Delta_\varepsilon = [s_\varepsilon, t_\varepsilon]$, so $s_\varepsilon - t_\varepsilon = a^k$. Then the increment of φ_a on Δ_ε can be written in the form

$$\varphi_a(t_\varepsilon) - \varphi_a(s_\varepsilon) = 2^{-k} = a^{kp} = (t_\varepsilon - s_\varepsilon)^p, \quad \text{where } p = \log_{1/a} 2.$$

Lemma *For* $0 \leq u < v \leq 1$, *we have*

$$0 \leq \varphi_a(v) - \varphi_a(u) \leq 4(v - u)^p. \tag{3}$$

Proof Take an integer $k \geq 0$ such that $a^k \geq v - u > a^{k+1}$. Then

$$(v - u)^p > (a^p)^{k+1} = \frac{1}{2^{k+1}}.$$

On the other hand, since the length of the segment $[u, v]$ does not exceed a^k, i.e., does not exceed the length of segments of rank k, it has a nonempty intersection with at most two such segments. Assume that there are two of them, Δ' and Δ''. Since the segment $[0, 1]$ is formed by all segments of rank k and all intervals of smaller ranks, on which the function φ_a is constant, we see that the increment of φ_a on $[u, v]$ does not exceed the sum of its increments on Δ' and Δ''. As we have observed before the statement of the lemma, these increments are equal to 2^{-k}. Thus,

$$0 \le \varphi_a(v) - \varphi_a(u) \le 2 \cdot 2^{-k} = \frac{4}{2^{k+1}} \le 4(v - u)^p. \qquad \square$$

A more involved argument shows that the coefficient 4 in (3) can be replaced by 1.

Let v_p be the Lebesgue–Stieltjes measure associated with the function φ_a (recall that $p = \log_{1/a} 2$). We will prove the following result.

Proposition *On Borel subsets in C_a, the measures μ_p and v_p coincide.*

Proof We will use the equality $\mu_p(C_a) = 1$ established in Proposition 3.6. Applying similarity considerations and the translational invariance of the Hausdorff measure, one can easily deduce from it that

$$\mu_p(C_a \cap \Delta_\varepsilon) = 2^{-k} = v_p(C_a \cap \Delta_\varepsilon) \qquad (4)$$

for every segment Δ_ε of rank k.

Thus, the measures μ_p and v_p coincide on all intersections $C_a \cap \Delta_\varepsilon$. One can easily check that these intersections form a semiring of subsets in C_a generating the Borel σ-algebra. By the uniqueness theorem (see [9, Section I.5.1]), if finite measures coincide on a semiring, then they coincide on the σ-algebra generated by this semiring, i.e., on the Borel σ-algebra in C_a. $\qquad \square$

Corollary *Let E be a Borel subset in C_a. Then*

$$\mu_p(E) \le 4 \operatorname{diam}^p(E).$$

Proof Set $u = \inf E$, $v = \sup E$, so $E \subset [u, v]$ and $v - u = \operatorname{diam}(E)$. Since $\mu_p(E) = v_p(E)$ and v_p is the Lebesgue–Stieltjes measure associated with the function φ_a, we have

$$\mu_p(E) = v_p(E) \le v_p([u, v]) = \varphi_a(v) - \varphi_a(u).$$

To complete the proof, it suffices to apply inequality (3). $\qquad \square$

4 *Comparing the Measures $\mu_p \times \mu_q$ and μ_{p+q}

In this section, we seek estimates between $\mu_p \times \mu_q$, the product of the Hausdorff measures μ_p and μ_q, and the measure μ_{p+q}. Below, all Hausdorff (outer) measures are regarded only on the Borel σ-algebra, so these restrictions are measures.

4.1 The Upper Bound

Now we state a clear geometric condition under which the product of Hausdorff measures satisfies a simple inequality.

Definition We say that on a subset A of a Euclidean space the measure μ_p satisfies the *upper bound with coefficient M* if there exists a constant $M = M_{A,p}$ such that

$$\mu_p(e) \le M \operatorname{diam}^p(e)$$

for every Borel set e in A.

Clearly, a measure satisfying the upper bound on a bounded set A is finite, and a measure satisfying the upper bound on an arbitrary set A is σ-finite (i.e., A can be represented as the union of a sequence of sets of finite measure). For $p \in \mathbb{N}$, the Hausdorff measure μ_p (proportional to the Lebesgue measure) satisfies the upper bound in $\mathbb{R}^p$ with coefficient 1. Besides, as we established in the previous addendum (see Corollary 3.7), the measure μ_p with $0 < p < 1$ satisfies the upper bound on the Cantor-like set C_a, $a = 2^{-\frac{1}{p}}$, with coefficient 4.

Remark For $p < m$, in $\mathbb{R}^m$ there exist compact sets of finite μ_p-measure on which it does not satisfy the upper bound. To see this, we need the following result.

Theorem *Every Borel set that has infinite μ_p-measure contains a compact subset of arbitrarily large finite measure.*

The proof of this result can be found in [6, Chapter II, Theorem 3].

Now we can construct the above-mentioned counterexample. Since m-dimensional balls have infinite μ_p-measure, it follows from the theorem above that every ball contains compact subsets of arbitrarily large (finite) μ_p-measure. We will assume that

$$0 \in Q_i, \quad \operatorname{diam}(Q_i) \le 1, \quad \text{and} \quad V_i = \mu_p(Q_i) \ge 2^i \quad \text{for all } i \in \mathbb{N}.$$

Set $K_i = V_i^{-\frac{2}{p}} Q_i$, so

$$\operatorname{diam}(K_i) \le V_i^{-\frac{2}{p}}, \quad \mu_p(K_i) = V_i^{-2} \mu_p(Q_i) = \frac{1}{V_i} \le 2^{-i}.$$

We claim that the union of these sets (denote it by K) is as required. Indeed, this set is compact, and its μ_p-measure is finite:

$$\mu_p(K) \le \sum_{i=1}^{\infty} \mu_p(K_i) = \sum_{i=1}^{\infty} \frac{1}{V_i} < +\infty.$$

Meanwhile, the measure μ_p does not satisfy the upper bound on K, since for every i we have

$$\mu_p(K_i) = \frac{1}{V_i} = V_i \left(V_i^{-\frac{2}{p}} \right)^p \ge V_i \operatorname{diam}^p(K_i) \ge 2^i \operatorname{diam}^p(K_i).$$

4.2 Estimating $\mu_p \times \mu_q$ From Above

We need the following simple lemma.

Lemma *Let μ and ν be σ-finite measures defined on the minimal σ-algebra $\mathfrak{A}$ containing a semiring $\mathcal{P}$. If $\mu \leq \nu$ on $\mathcal{P}$, i.e.,*

$$\mu(P) \leq \nu(P) \quad \text{for all } P \in \mathcal{P},$$

then $\mu \leq \nu$ on $\mathfrak{A}$.

Proof Let A be an arbitrary set from $\mathfrak{A}$. By the uniqueness extension theorem (see [9, Section I.5.1]),

$$\mu(A) = \inf_{\{P_j\}} \sum_{j=1}^{\infty} \mu(P_j) \quad \text{and} \quad \nu(A) = \inf_{\{P_j\}} \sum_{j=1}^{\infty} \nu(P_j)$$

(the infima are taken over all countable covers $\{P_j\}$ of A by elements of the semiring $\mathcal{P}$). Since, by assumption, $\mu(P_j) \leq \nu(P_j)$ for all P_j, this implies the required inequality $\mu(A) \leq \nu(A)$.   □

Below, $\mu \times \nu$ stands for the product of measures μ and ν. We also stick to the convention $0 \cdot \infty = \infty \cdot 0 = 0$.

Proposition *Let $A \subset \mathbb{R}^m$, $B \subset \mathbb{R}^n$ be Borel sets, $\dim_H(A) = p > 0$, $\mu_q(B) < \infty$. If μ_p satisfies on A the upper bound with coefficient M, then on Borel subsets of $A \times B$ we have*

$$\mu_p \times \mu_q \leq M \mu_{p+q}. \tag{1}$$

Proof First, we estimate the measure $\mu_{p+q}(A \times B)$ from below. To do this, for arbitrary $\varepsilon > 0$ we consider a cover of $A \times B$ by sets $e_1, e_2, \ldots$ whose diameters $d_1, d_2, \ldots$ are at most ε, and estimate the corresponding sum

$$S = \sum_{k \geq 1} d_k^{p+q}$$

from below. Denote the projections of the sets e_k to the first and second factors in the product $\mathbb{R}^m \times \mathbb{R}^n$ by e_k' and e_k'', respectively. The upper bound on μ_p on the set A implies that

$$d_k^p = \mathrm{diam}^p(e_k) \geq \mathrm{diam}^p(e_k') \geq \mathrm{diam}^p(e_k' \cap A) \geq \frac{1}{M}\mu_p(e_k' \cap A).$$

This allows us to estimate the sum S from below (as usual, χ_E stands for the characteristic function of a set E: $\chi_E = 1$ on E and $\chi_E = 0$ outside E):

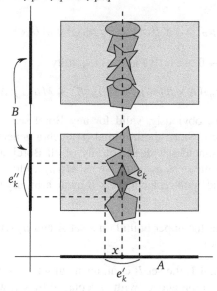

Fig. 2

$$S = \sum_{k \geq 1} d_k^q d_k^p \geq \frac{1}{M} \sum_{k \geq 1} d_k^q \mu_p(e_k' \cap A) = \frac{1}{M} \sum_{k \geq 1} d_k^q \int_A \chi_{e_k'}(x)\, d\mu_p \geq$$

$$\geq \frac{1}{M} \int_A \left(\sum_{k \geq 1} \operatorname{diam}^q(e_k'') \chi_{e_k'}(x) \right) d\mu_p. \quad (2)$$

Let us estimate the sum in the integrand from below. Note that for every $x \in A$, the "cross section" $\{x\} \times B$ of the product $A \times B$ is covered by those of the sets e_k whose projections e_k' contain x (see Fig. 2). Therefore, for every x the values of k corresponding to nonzero terms of the sum are such that the sets e_k with these indices cover the set $\{x\} \times B$, and hence their projections e_k'' form a cover of the set B. It follows by the definition of the approximating measure (see Section 3.2) that for every $x \in A$ the sum in the integrand satisfies the inequality

$$\sum_{k \geq 1} \operatorname{diam}^q(e_k'') \chi_{e_k'}(x) \geq \mu_q(B; \varepsilon).$$

Substituting it into (2) yields

$$S \geq \frac{1}{M} \int_A \left(\sum_{k \geq 1} \operatorname{diam}^q(e_k'') \chi_{e_k'}(x) \right) d\mu_p \geq \frac{1}{M} \int_A \mu_q(B; \varepsilon)\, d\mu_p$$

$$= \frac{1}{M} \mu_p(A) \mu_q(B; \varepsilon).$$

Since $\{e_k\}_{k \geq 1}$ is an arbitrary ε-cover, it follows that

$$\mu_{p+q}(A \times B; \varepsilon) \geq \frac{1}{M} \mu_p(A)\mu_q(B; \varepsilon).$$

Taking the limit as $\varepsilon \to 0$, we arrive at the inequality

$$(\mu_p \times \mu_q)(A \times B) = \mu_p(A)\,\mu_q(B) \leq M\mu_{p+q}(A \times B).$$

A similar inequality is, obviously, valid for any Borel sets A' and B' contained in A and B, respectively. Since the minimal σ-algebra generated by the semiring of products $A' \times B'$ coincides with the family of all Borel subsets of $A \times B$, the preliminary lemma (with $\mu = \mu_p \times \mu_q$, $\nu = M\mu_{p+q}$, etc.) implies that the last inequality remains valid upon replacing $A \times B$ by an arbitrary (Borel) subset of this set.   □

Remark If μ_p satisfies the upper bound on a set A and $\mu_p(A) > 0$, $\mu_q(B) = +\infty$, then $\mu_{p+q}(A \times B) = +\infty$.

Indeed, by Theorem 4.1, the set B contains a subset E of arbitrarily large finite measure. By the proposition above (with B replaced by E), we have (below, M is still the coefficient of the upper bound of μ_p on A)

$$\mu_p(A) \cdot \mu_q(E) = (\mu_p \times \mu_q)(A \times E) \leq M\mu_{p+q}(A \times E) \leq M\mu_{p+q}(A \times B).$$

Since the left-hand side can be made arbitrarily large (by choosing E), this inequality holds only if

$$\mu_{p+q}(A \times B) = +\infty.$$

4.3 Estimating $\mu_p \times \mu_q$ From Below

Here we consider products $A \times B$ of special form, where B is still an arbitrary Borel subset in $\mathbb{R}^n$ and A is the closed unit interval or a Cantor-like set C_a, $0 < a < \frac{1}{2}$ (these sets are constructed in Section IV.5.2). As in Section IV.5.5, we do not exclude the case $a = \frac{1}{2}$, assuming that $C_{1/2} = [0, 1]$. Recall (see Section 3.6) that

$$\dim_H(C_a) = \log_{1/a} 2 \quad \text{and} \quad \mu_p(C_a) = 1 \quad \text{for } p = \dim_H(C_a).$$

Below, Δ is a segment of some rank arising in the construction of the set C_a, $0 < a < \frac{1}{2}$. For $a = \frac{1}{2}$, by Δ we mean a dyadic segment, i.e., a segment of the form $[k2^{-i}, (k+1)2^{-i}]$ ($i \in \mathbb{N}$, $k = 0, 1, \ldots, 2^i - 1$). It is natural to call i the rank of this segment. Thus, for every $a \in (0, \frac{1}{2}]$ the length of a segment of rank i is a^i.

First, keeping the notation introduced above, we establish a preliminary inequality.

Lemma Let $C = C_a \cap \Delta$ and e be a set in $\mathbb{R}^n$ whose diameter δ is less than the length of Δ. Then for $s = p + q$ we have

$$\mu_s(C \times e; 2\delta) \leq 2^{s+1}\mu_p(C)\,\delta^q.$$

Proof First, we divide the set C into subsets of diameter close to δ. Consider an index i such that $a^{i+1} < \delta \le a^i$. Then the length of Δ is greater than a^{i+1}, and, consequently, its rank is at most i. Let $\Delta_1, \ldots, \Delta_J$ be the segments of rank $i + 1$ contained in Δ, so $C = C_a \cap \Delta = C_a \cap (\Delta_1 \cup \ldots \cup \Delta_J)$. Hence,

$$\mu_p(C) = J\mu_p(C_a \cap \Delta_1) = \frac{J}{2^{i+1}}$$

(the first equality holds thanks to the translational invariance of the Hausdorff measure, and the second one follows from formula (4) in Section 3.7). Since the lengths of the segments Δ_j, equal to a^{i+1}, are less than δ, we obtain

$$\text{diam}(\Delta_j \times e) \le \text{diam}(\Delta_j) + \text{diam}(e) < 2\delta.$$

Besides,

$$C \times e = (C_a \cap \Delta) \times e \subset \bigcup_{j=1}^{J} \Delta_j \times e.$$

Thus, the congruent sets $\Delta_j \times e$ form a (2δ)-cover of the product $C \times e$. Taking into account that

$$\delta^p \le (a^i)^p = (a^p)^i = 2^{-i} \quad \text{and} \quad J = 2^{i+1}\mu_p(C),$$

we see that $J\delta^p \le 2\mu_p(C)$. This gives the required inequality:

$$\mu_s(C \times e; 2\delta) \le \sum_{j=1}^{J} \text{diam}^s(\Delta_j \times e) \le J(2\delta)^s = 2^s J\delta^p \delta^q \le 2^{s+1}\mu_p(C)\delta^q. \qquad \square$$

Now we are ready to prove the main result of this subsection.

Proposition Let $A = C_a \left(0 < a \le \frac{1}{2}\right)$ be the closed unit interval or a Cantor-like set, $p = \dim_H(A)$, $B \subset \mathbb{R}^n$ be a Borel set, $\mu_q(B) < +\infty$, $s = p + q$. Then on Borel subsets in the product $A \times B$,

$$\mu_s \le 2^{s+1}\mu_p \times \mu_q. \tag{3}$$

Proof First, we prove that inequality (3) holds on sets of the form $C \times e$ where $C = A \cap \Delta$ is the set considered in the lemma and e is a Borel subset in B. Fixing a positive number ε less than the length of the segments Δ, we find an ε-cover $\{e_k\}$ of e such that

$$\sum_{k=1}^{\infty} \text{diam}^q(e_k) < \mu_q(e; \varepsilon) + \varepsilon.$$

Since $C \times e \subset \bigcup_k (C \times e_k)$, the lemma implies that

$$\mu_s(C \times e; 2\varepsilon) \leq \sum_{k=1}^{\infty} \mu_s(C \times e_k; 2\varepsilon) \leq 2^{s+1} \mu_p(C) \sum_{k=1}^{\infty} \text{diam}^q(e_k)$$

$$\leq 2^{s+1} \mu_p(C)(\mu_q(e; \varepsilon) + \varepsilon).$$

It remains to take the limit of the leftmost and rightmost sides of this inequality as $\varepsilon \to 0$.

Since the sets of the form under consideration constitute a semiring generating the σ-algebra of Borel subsets in $A \times B$, it follows from Lemma 4.2 that inequality (3) holds also on this σ-algebra.   □

Corollary *Under the assumptions of the proposition, if the measure μ_q satisfies the upper bound on B, then the measure μ_s satisfies the upper bound on $A \times B$.*

Proof Let E be a bounded Borel subset in $A \times B$ and $d = \text{diam}(E)$. Clearly, the projections of E to A and B (which are not necessarily Borel) are contained in closed balls A_1 and B_1 of radius d ($A_1 \subset \mathbb{R}$, $B_1 \subset \mathbb{R}^n$). Since these balls are not necessarily contained in the sets A, B, let $A' = A \cap A_1$, $B' = B \cap B_1$. Then $E \subset A' \times B'$ and, by the proposition above,

$$\mu_s(E) \leq 2^{s+1}(\mu_p \times \mu_q)(E) \leq 2^{s+1}(\mu_p \times \mu_q)(A' \times B')$$

$$= 2^{s+1}\mu_p(A')\mu_q(B') \leq 2^{s+1} M_p(2d)^p M_q(2d)^q = 2^{2s+1} M_p M_q d^s,$$

where M_p and M_q are the coefficients of the upper bounds of μ_p and μ_q, respectively. Since E is arbitrary, this proves the corollary.   □

4.4 Standard Sets

Recall (see Section IV.5.6) that by a standard set we mean a product of several closed unit intervals and Cantor-like sets, i.e. (taking into account the convention from Section IV.5.5 that $C_{1/2} = [0, 1]$), products of several sets C_a, where $0 < a \leq 1/2$. In the proposition below, such sets are denoted by $A_1, \ldots, A_N$.

Proposition *Let $E = A_1 \times \ldots \times A_N$ be a standard set. Then*

$$\dim_H(E) = \dim_H(A_1) + \ldots + \dim_H(A_N)$$

and there exist constants L and $\delta > 0$ such that

$$\delta \mu_{p_1} \times \ldots \times \mu_{p_N} \leq \mu_s \leq L \mu_{p_1} \times \ldots \times \mu_{p_N}, \tag{4}$$

where $p_k = \dim_H(A_k)$ ($k = 1, \ldots, N$), $s = p_1 + \ldots + p_N$. In particular, $0 < \mu_s(E) < +\infty$ and $s = \dim_H(E)$. Besides, the measure μ_s satisfies the upper bound on E.

Since the direct product of standard sets is again a standard set, the proposition remains valid if among $A_1, \ldots, A_N$ there are standard sets.

Proof First, consider the main case $N = 2$. Then $s = p_1 + p_2$. Obviously, the sets $A = A_1$, $B = A_2$ satisfy all assumptions of Propositions 4.2 and 4.3 and, besides, $\mu_{p_1}(A_1) = \mu_{p_2}(A_2) = 1$. Hence, it follows from (1) and (3) that

$$\frac{1}{4} \le \mu_s(A_1 \times A_2) \le 2^{s+1} < +\infty$$

(here we have used the result of Corollary 3.7, which shows that the Hausdorff measures satisfy the upper bound with coefficient 4 on Cantor-like sets). The double inequality $0 < \mu_s(A_1 \times A_2) < +\infty$ implies (see property 2 of Hausdorff dimension in Section 3.5) that $\dim_H(A_1 \times A_2) = s = p_1 + p_2$. The fact that the measure μ_s on $A_1 \times A_2$ satisfies the upper bound follows from Corollary 4.3. The proof is completed by induction. The details are left to the reader. □

Propositions 4.2 and 4.3 remain valid for every standard set A. More exactly, the following result holds.

Corollary *Let A be a standard set, $\dim_H(A) = p$, B be a Borel set, $\mu_q(B) < +\infty$, $s = p + q$. Then Proposition 4.2 remains valid for the sets A, B, and, besides, on Borel subsets in $A \times B$ we have*

$$\mu_s \le C\mu_p \times \mu_q \tag{3'}$$

for some constant C.

Proof First of all, note that, by the proposition above, the measure μ_p satisfies the upper bound and, consequently, the assumptions of Proposition 4.2 are fulfilled.

Let us verify that the part concerning Proposition 4.3 is true. Let A be a product of several factors, $A = A_1 \times A_2 \times \ldots \times A_N = A_1 \times A'$, and assume that the corollary holds for the standard sets formed by $N - 1$ factors: $A' = A_2 \times \ldots \times A_N$, $p' = \dim_H(A')$. By the induction hypothesis, the conclusion of the corollary holds for $B' = A' \times B$, and hence the measure $\mu_{s'}$ satisfies on this set (see (4)) the inequality $\mu_{s'} \le L\mu_{p'} \times \mu_q$, where $\mu_{p'}(A') < \infty$, $\mu_q(B) < \infty$, $s' = p' + q$, and the measure $\mu_{p'}$ satisfies the upper bound. To prove the induction step, we represent the set $A \times B$ in the form $A \times B = A_1 \times (A' \times B) = A_1 \times B'$ and apply Proposition 4.3 to the product $A_1 \times B'$; the details are left to the reader. Thus, we have established the validity of Proposition 4.3 and, in particular, the fact that the measure μ_s is finite in the case where A is an arbitrary standard set. Since $\mu_s(A \times B)$ is finite, Proposition 4.2 is also valid. □

4.5 The Hausdorff Dimension of a Cartesian Product

It is known that for arbitrary Borel sets,

$$\dim_H(A \times B) \ge \dim_H(A) + \dim_H(B).$$

There exist examples of compact sets A, B such that this inequality is strict (see [7]). We will show that if one of the factors is a standard set, then the dimension of the product is equal to the sum of the dimensions of the factors. For more general and complete results, we refer the reader to [13].

Due to Corollary 4.4 and Remark 4.2, one can easily describe cases where the measure $\mu_s(A \times B)$ takes extreme values. First, we prove the following simple result.

Lemma *Let A be a standard set, B be an arbitrary Borel set, p and q be their Hausdorff dimensions, $s = p + q$. Then*

$$\mu_s(A \times B) < +\infty \Leftrightarrow \mu_q(B) < +\infty; \quad \mu_s(A \times B) = 0 \Leftrightarrow \mu_q(B) = 0.$$

Proof Since $0 < \mu_p(A) < +\infty$, the implications $\Rightarrow$ follow from inequality (1), and the implications $\Leftarrow$ follow from (3). $\square$

The next result immediately follows from this lemma and the definition of Hausdorff dimension.

Proposition *If at least one of Borel sets A, B is standard, then*

$$\dim_H(A \times B) = \dim_H(A) + \dim_H(B).$$

In particular, if Q is an i-dimensional cube, then for every Borel set B we have

$$\dim_H(\mathbb{R}^i \times B) = \dim_H(Q \times B) = i + \dim_H(B).$$

The first equality follows from the formula for the dimension of a union (see formula (2) at p. 261), since $\mathbb{R}^i \times B$ is a countable union of sets congruent to $Q \times B$.

5 *Estimates for Smooth Maps Related to the Hausdorff and Lebesgue Measures

We keep the notation introduced in Section 3 (λ_m and λ_m^* are the Lebesgue measure and the Lebesgue outer measure in $\mathbb{R}^m$, μ_p is the p-dimensional (outer) Hausdorff measure, O is an open subset in $\mathbb{R}^m$).

For $i = 1, \ldots, m - 1$, the space $\mathbb{R}^m$ is identified with the Cartesian product $\mathbb{R}^i \times \mathbb{R}^{m-i}$. Accordingly, a point of $\mathbb{R}^m$ is written in the form (x, y), where $x \in \mathbb{R}^i$, $y \in \mathbb{R}^{m-i}$. The symbol $B^k(w, \rho)$ stands for the open k-dimensional ball of radius ρ centered at a point w (for $k = m$, the subscript indicating the dimension is omitted). The volume (Lebesgue measure) of the unit k-dimensional ball is denoted by α_k (for consistency of notation, we put $\alpha_0 = 1$).

5.1 Estimating the Hausdorff Measure of an Image

Our aim here is to estimate the "massiveness" of the image of a set under a map satisfying the following condition: the main contribution to its increments caused by small increments of the variable comes from changes in one group of coordinates, while the other coordinates have a substantially smaller effect.

Lemma *Let $E \subset O$ and $F: O \to \mathbb{R}^n$ be a map such that for some $L, \Lambda > 0$ and $t \geq 1$, the following inequality holds at every point $(a, b) \in E$ (here $x, a \in \mathbb{R}^i$, $y, b \in \mathbb{R}^{m-i}$):*

$$\|F(x, y) - F(a, b)\| \leq L\|x - a\| + \Lambda\|y - b\|^t$$
$$\text{for all } (x, y) \in O \text{ sufficiently close to } (a, b). \tag{1}$$

Then for $p = i + \frac{m-i}{t}$ we have

$$\mu_p(F(E)) \leq C_m L^i \Lambda^{p-i} \lambda_m^*(E). \tag{2}$$

This result remains valid also for $i = 0$ or $i = m$. One must remove from the right-hand side of (1) the first term in the case $i = 0$, and the second term in the case $i = m$.

As we will see from the proof, the constant C_m in (2) can be taken, for example, equal to $20^m/\alpha_m$. Note also that inequality (2) is meaningful only for $p \leq n$, because $\mu_p = 0$ in $\mathbb{R}^n$ for $p > n$.

Proof We assume without loss of generality that the set E is bounded. Fix an arbitrary number $\varepsilon > 0$ and choose an open set G_ε such that

$$E \subset G_\varepsilon \subset O \quad \text{and} \quad \lambda_m(G_\varepsilon) < \lambda_m^*(E) + \varepsilon.$$

For $\delta > 0$ and $(a, b) \in E$, set $D_\delta(a, b) = B^i(a, \Lambda\delta) \times B^{m-i}(b, (L\delta)^{1/t})$. Clearly,

$$\lambda_m(D_\delta(a, b)) = \alpha_i(\Lambda\delta)^i \, \alpha_{m-i} (L\delta)^{(m-i)/t} = \alpha_i \alpha_{m-i} \Lambda^i L^{p-i} \delta^p$$

and $D_\delta(a, b) \subset G_\varepsilon$ for small δ. Besides, by assumption, for sufficiently small δ, say $\delta < \delta_{a,b} < \varepsilon$, every point $(x, y) \in D_\delta(a, b)$ satisfies the inequality

$$\|F(x, y) - F(a, b)\| < L\Lambda\delta + \Lambda L\delta = 2L\Lambda\delta,$$

i.e., the F-image of the set $D_\delta(a, b)$ is contained in the ball of radius $2L\Lambda\delta$ centered at $F(a, b)$. These balls form a Vitali cover of the set $F(E)$. Hence (see Theorem 2.2), it contains a sequence $B_1, B_2, \ldots$ of pairwise disjoint balls such that $F(E) \subset \bigcup_j B_j^*$, where the ball B_j^* has the same center as B_j and five times the radius. Let $2L\Lambda\delta_j$ be the radius of B_j. Since $\delta_j < \varepsilon$ by construction, the diameter of B_j^* is at most $20L\Lambda\varepsilon$. Hence, for the approximating Hausdorff measure (see Section 3.2) we obtain the inequality

$$\mu_p(F(E); 20L\Lambda\varepsilon) \leq \sum_j (20L\Lambda\delta_j)^p = (20L\Lambda)^p \sum_j \delta_j^p.$$

Let $F(a_j, b_j)$ be the center of B_j. Since the balls B_j are pairwise disjoint, the same is true for the sets $D_{\delta_j}(a_j, b_j)$, which are contained in $\Phi^{-1}(B_j)$ (recall that, by construction, $D_{\delta_j}(a_j, b_j) \subset G_\varepsilon$ for all j). Therefore,

$$\lambda_m(G_\varepsilon) \geq \sum_j \lambda_m(D_{\delta_j}(a_j, b_j)) = \alpha_i \alpha_{m-i} L^{p-i} \Lambda^i \sum_j \delta_j^p.$$

Together with the previous inequality, this yields

$$\mu_p(F(E); 20L\Lambda\varepsilon) \leq \frac{(20L\Lambda)^p}{\alpha_i \alpha_{m-i}} L^{i-p} \Lambda^{-i} \lambda_m(G_\varepsilon) \leq \frac{20^p}{\alpha_i \alpha_{m-i}} L^i \Lambda^{p-i} (\lambda_m^*(E) + \varepsilon).$$

Taking the limit as $\varepsilon \to 0$, we obtain

$$\mu_p(F(E)) \leq \frac{20^p}{\alpha_i \alpha_{m-i}} L^i \Lambda^{p-i} \lambda_m^*(E).$$

Clearly, $\alpha_i \alpha_{m-i} = \lambda_m(B^i(0,1) \times B^{m-i}(0,1))$. This Cartesian product contains the m-dimensional unit ball, so $\alpha_i \alpha_{m-i} \geq \alpha_m$. Hence,

$$\mu_p(F(E)) \leq \frac{20^p}{\alpha_m} L^i \Lambda^{p-i} \lambda_m^*(E) \leq \frac{20^m}{\alpha_m} L^i \Lambda^{p-i} \lambda_m^*(E). \qquad \square$$

5.2 On Images of Zero Measure

Now we derive corollaries of inequality (2). We keep the notation introduced in the previous subsection. In particular, below $p = i + \frac{m-i}{t}$.

Corollary 1 If $\lambda_m(E) = 0$ and $F(x, y) - F(a, b) = O(\|x - a\| + \|y - b\|^t)$ near every point $(a, b) \in E$, then $\mu_p(F(E)) = 0$.

Proof To see this, it suffices to represent E as the union of a sequence of sets on each of which condition (1) holds with some coefficients L and Λ. $\qquad \square$

It is also useful to consider the following modification of the lemma, corresponding to the case where the coefficient Λ can be taken arbitrarily small.

Corollary 2 Let $1 \leq i < m, t \geq 1, E \subset O$, and $F: O \to \mathbb{R}^n$ be a map such that at every point $c = (a, b) \in E$ condition (1) holds for some $L > 0$ and arbitrarily small $\varepsilon > 0$, i.e.,

$$\exists L > 0 \,\forall \varepsilon > 0 \,\exists \delta > 0: B(c, \delta) \subset O \quad \text{and}$$
$$\|F(x, y) - F(a, b)\| \leq L\|x - a\| + \varepsilon \|y - b\|^t \quad \text{if } (x, y) \in B(c, \delta). \tag{3}$$

Then $\mu_p(F(E)) = 0$.

In what follows, with some abuse of notation, we write (3) in the form

$$\|F(x,y) - F(a,b)\| \underset{(x,y)\to(a,b)}{=} O(\|x-a\|) + o(\|y-b\|^t).$$

Proof We will assume without loss of generality that the set E is bounded. For $j, k \in \mathbb{N}$, consider the subset $E_{j,k}$ of E at every point (a, b) of which condition (1) holds with $L = j$ and $\Lambda = \frac{1}{k}$. It follows from (3) that $E = \bigcup_j \bigcap_k E_{j,k}$, and hence

$$F(E) = \bigcup_j F\left(\bigcap_k E_{j,k}\right) \subset \bigcup_j \bigcap_k F(E_{j,k}).$$

By condition (1) of the lemma, we have

$$\mu_p(F(E_{j,k})) \leq C_m j^i \left(\frac{1}{k}\right)^{p-i} \lambda_m^*(E_{j,k}) \leq C_m j^i \left(\frac{1}{k}\right)^{p-i} \lambda_m^*(E) \underset{k\to\infty}{\longrightarrow} 0.$$

Hence, the Hausdorff measure of the intersection $A_j = \bigcap_k F(E_{j,k})$ vanishes for every j. Thus,

$$\mu_p(F(E)) \leq \sum_j \mu_p(A_j) = 0. \qquad \square$$

Remark One can see from the proof of Corollary 2 that it holds also for $i = 0$. In this case, $p = \frac{m}{t}$, and the right-hand side of the inequality in (3) contains only the second term. Hence, $\mu_{m/t}(F(E)) = 0$ if

$$\|F(y) - F(b)\| \underset{y\to b}{=} o(\|y-b\|^t) \text{ at every point } b \in E.$$

5.3 A Refined Estimate of the Increment

Our aim here is to prove that the "smallness" of the gradient of a function f on a set E, more exactly, the relation $\|\operatorname{grad} f(x)\| = O(\|x-c\|^{t-1})$ (for $t > 1$) near every point $c \in E$, implies not only the standard estimate $f(x) - f(c) = O(\|x-c\|^t)$, but also a refined one: $f(x) - f(c) \underset{x\to c}{=} o(\|x-c\|^t)$ for almost all $c \in E$ (hereafter, "almost all" is understood with respect to the corresponding Lebesgue measure). This result, in turn, allows one to use Remark 5.2.

We begin with a simple observation concerning density points.

Lemma 1 *If c is a density point of a set $E \subset \mathbb{R}^m$, then*

$$\operatorname{dist}(x, E) \underset{x\to c}{=} o(\|x-c\|).$$

Proof Assume to the contrary that there exist a sequence of points x_j and a number $\theta \in (0, 1)$ such that $r_j = \|x_j - c\| \to 0$ and $\mathrm{dist}(x_j, E) \geq \theta r_j$ for all j. Then the set E does not intersect the ball $B(x_j, \theta r_j)$, which, obviously, lies in the ball $B_j = B(c, (1 + \theta)r_j)$. Therefore,

$$\frac{\lambda_m(E \cap B_j)}{\lambda_m(B_j)} \leq 1 - \frac{\lambda_m(B(x_j, \theta r_j))}{\lambda_m(B_j)} = 1 - \left(\frac{\theta}{1 + \theta}\right)^m,$$

which cannot happen since the balls B_j shrink to the point c, which is a density point of E (and hence the left-hand side of the inequality tends to 1). $\qquad\square$

Lemma 2 *Let $t > 1$, $g \in C^1(O)$, and let $E \subset O$ be the subset of points $c \in O$ such that*

$$\|\operatorname{grad} g(z)\| = O(\|z - c\|^{t-1}) \quad as \ z \to c.$$

Then for almost all $c \in E$,

$$g(z) - g(c) \underset{z \to c}{=} o(\|z - c\|^t). \tag{4}$$

Of course, for $t = 1$ the conclusion of the lemma is false, since then $E = O$. However, at points of the critical set $\{c \in O : \operatorname{grad} g(c) = 0\}$, relation (4) holds in this case too (just by the definition of differentiability).

Proof Clearly, E is exhausted by a sequence of sets E_k of the form (below, $k \in \mathbb{N}$)

$$E_k = \left\{c \in O : \|\operatorname{grad} g(z)\| \leq k \|z - c\|^{t-1} \text{ for } z \in B\left(c, \frac{1}{k}\right) \cap O\right\}.$$

Hence, it suffices to prove (4) for almost all points of each set E_k. Fixing k up to the end of the proof, we will show that (4) holds at every density point of E_k (recall that, by Theorem 2.3, almost all points of E_k are density points), which completes the proof.

Since $\|\operatorname{grad} g(z)\| \leq k \|z - c\|^{t-1}$ if $c \in E_k$ and $z \in B\left(c, \frac{1}{k}\right)$, for $\mathrm{dist}(z, E_k) < \frac{1}{k}$ we have

$$\|\operatorname{grad} g(z)\| \leq \inf\left\{k\|z - c\|^{t-1} : c \in E_k, \ \|z - c\| < \frac{1}{k}\right\} = k \ \mathrm{dist}^{t-1}(z, E_k).$$

If c is a density point of E_k, by Lemma 1 this inequality implies that

$$\|\operatorname{grad} g(z)\| \leq k \ \mathrm{dist}^{t-1}(z, E_k) \underset{z \to c}{=} o(\|z - c\|^{t-1})$$

(the last equality holds because $t > 1$). By Lagrange's inequality, this implies (4), so this relation holds at every density point of every set E_k. $\qquad\square$

In the next lemma we assume, as earlier, that the space $\mathbb{R}^m$ is identified with the Cartesian product $\mathbb{R}^i \times \mathbb{R}^{m-i}$. Below, as in Section 2.2, by $f_y'(x, y)$ we denote the projection of the gradient of a function f to the second factor of this product, i.e., the vector $\left(\frac{\partial f}{\partial y_1}(x, y), \ldots, \frac{\partial f}{\partial y_{m-i}}(x, y)\right)$.

Lemma 3 *Let $i < m, t > 1, f \in C^1(O)$, and let E be the subset of points $(a, b) \in O$ such that*

$$\|f_y'(a, y)\| = O(\|y - b\|^{t-1}) \quad as \ y \to b. \tag{5}$$

Then for almost all points $(a, b) \in E$,

$$f(x, y) - f(a, b) \underset{(x,y)\to(a,b)}{=} O(\|x - a\|) + o(\|y - b\|^t). \tag{6}$$

A formal interpretation of this relation is given in (3).

Note that if in (5) we replace O by o, then (6) holds at every point of E, not only almost everywhere.

Proof For an arbitrary point $a \in \mathbb{R}^i$, let $O_a, E_a, \dots$ be the corresponding cross sections of the sets $O, E, \dots$, so $O_a = \{y \in \mathbb{R}^{m-i} : (a, y) \in O\}$, etc.

Set $g(y) = f(a, y)$ for $y \in O_a$. Since grad $g(y) = f_y'(a, y)$, the function g satisfies the assumptions of Lemma 2 (with O replaced by O_a and E replaced by E_a), and hence for almost all (with respect to the measure λ_{m-i}) points $b \in E_a$ we have $g(y) - g(b) \underset{y\to b}{=} o(\|y - b\|^t)$, i.e.,

$$f(a, y) - f(a, b) \underset{y\to b}{=} o(\|y - b\|^t) \tag{7}$$

almost everywhere on E_a.

Consider the set H of all points $(a, b) \in O$ for which (6) holds. It follows from (7) that the difference $E_a \setminus H_a \subset \mathbb{R}^{m-i}$ has zero measure for every $a \in \mathbb{R}^i$.

Now we claim that the set $E \setminus H$ also has zero measure (but the m-dimensional one). Assume for the moment that the sets E and H are measurable. In this case, using the theorem on computation of the measure of a set via the measures of its cross sections (see [9, Theorem V.2.2]), we can claim that $\lambda_m(E \setminus H) = 0$, so (7) holds for almost all $(a, b) \in E$. Let us show that this implies (6).

Indeed, let B be an arbitrary ball in O centered at (a, b). Then for every point $(x, y) \in B$, the point (a, y) also belongs to B, which allows us to represent the increment of f in the form

$$f(x, y) - f(a, b) = f(x, y) - f(a, y) + f(a, y) - f(a, b).$$

Clearly, the first difference in the right-hand side for any $(a, b) \in E$ can be estimated as $f(x, y) - f(a, y) = O(\|x - a\|)$ by the smoothness of f, and the second difference satisfies (7) for almost all $(a, b) \in E$, which yields the required result.

To complete the proof, it suffices to show that the sets E and H are measurable. To prove the measurability of E, we introduce auxiliary sets

$$E_j = \left\{ c = (a, b) \in O : \|f_y'(a, y)\| \leq j\|y - b\| \quad \text{for } (a, y) \in B\left(c, \frac{1}{j}\right) \cap O \right\}.$$

As the reader can easily check, the sets E_j are relatively closed in O and $E = \bigcup\limits_{j=1}^{\infty} E_j$, which implies the measurability of E.

The measurability of H can be established in a similar (though somewhat more complicated) way. One should consider the sets

$$H_{L,k,j} = \left\{ c = (a, b) \in O : |f(x, y) - f(a, b)| \le L\|x - a\| + \frac{1}{k}\|y - b\|^t \right.$$

$$\left. \text{for } (x, y) \in B\left(c, \frac{1}{j}\right) \cap O \right\}.$$

Like E_j, they are relatively closed in O, and the set H can be represented in the form

$$H = \bigcup_{L=1}^{\infty} \bigcap_{k=1}^{\infty} \bigcup_{j=1}^{\infty} H_{L,k,j},$$

which implies that it is measurable.   □

References

1. Arnold, V. I.: A mathematical trivium. Russian Math. Surveys **46**(1), 271–278 (1991).
2. Arnold, V. I.: Arnold's Problems. Springer (2005).
3. Bates, S. M.: Toward a precise smoothness hypothesis in Sard theorem. Proc. Amer. Math. Soc. **117**(1), 279–283 (1993).
4. Bates, S. M., Moreira, C. G.: A generalisation of Sard's theorem. Preprint IMPA (1994).
5. Borisovich, Yu. G., Bliznyakov, N. M., Fomenko, T. N., Izrailevich, Ya. A. Introduction to Differential and Algebraic Topology. Kluwer, Dordrecht (1995).
6. Carleson, L.: Selected Problems on Exceptional Sets. Van Nostrand (1967).
7. Federer, H.: Geometric Measure Theory. Springer (1996).
8. Gordon, W. B.: On the diffeomorphisms of Euclidean space. Amer. Math. Monthly **79**(7), 755–759 (1972).
9. Makarov, B. M.:, Podkorytov, A. N.: Real Analysis: Measures, Integrals and Applications. Springer (2013).
10. Moreira, C. G.: Hausdorff measures and the Morse–Sard theorem. Publ. Mat. **45**, 149–162 (2001).
11. Morse, A. P.: The behavior of a function on its critical set. Ann. of Math. **40**(1), 62–70 (1939).
12. Norton, A.: A critical set with nonnull image has large Hausdorff dimension. Trans. Amer. Math. Soc. **296**(1), 367–376 (1986).
13. Pesin, Ya. B.: Dimension Theory in Dynamical Systems: Contemporary Views and Applications. University of Chicago Press (1997).
14. Reshetnyak, Yu. G.: A Course in Mathematical Analysis [in Russian]. Sobolev Institute of Mathematics, Novosibirsk (1999).
15. Stein, E. M.: Singular Integrals and Differentiability Properties of Functions. Princeton University Press (1971).
16. Vinogradov, O. L.: Mathematical Analysis [in Russian]. BHV-Petersburg, St. Petersburg (2017).
17. Vulikh, B. Z.: A Brief Course in the Theory of Functions of a Real Variable. Mir, Moscow (1976).
18. Whitney, H.: A function not constant on a connected set of critical points. Duke Math. J. **1**, 514–517 (1935).
19. Zorich, V. A.: Mathematical Analysis, vol. I, II. Springer (2015, 2016).

B. M. Makarov, A. N. Podkorytov, *Smooth Functions and Maps*, Moscow Lectures 7,
https://doi.org/10.1007/978-3-030-79438-5

Index

© The Author(s), under exclusive license to Springer Nature Switzerland AG 2021
B. M. Makarov, A. N. Podkorytov, *Smooth Functions and Maps*, Moscow Lectures 7,
https://doi.org/10.1007/978-3-030-79438-5